THE BIG BOOK OF FACTS

Terri Schlichenmeyer

THE BIG BOOK OF FACTS

Terri Schlichenmeyer

ABOUT THE AUTHOR

Terri Schlichenmeyer has been reading since she was three years old, and she took that even further when she started reviewing books as The Bookworm Sez. Terri's book reviews are read in dozens of newspapers and magazines around the world.

Terri collects nonfiction books (especially trivia books), unusual bookends, unique bookmarks, and dust. She lives on a Wisconsin prairie with one ever-patient man, two not spoiled little dogs, and 15,000 books.

DEDICATION

To Mark Moen. Every girl needs her own Great American Redheaded Goofball. I'm lucky I got mine, and now your name is in this book, too.

To Carol Munk, ever my Fearless Leader.

And to Marty. Thanks for the ideas. Ha!

ALSO FROM VISIBLE INK PRESS

The Handy American Government Answer Book: How Washington, Politics, and Elections Work
by Gina Misiroglu
ISBN: 978-1-57859-639-3

***The Handy Anatomy Answer Book,* 2nd edition**
by Patricia Barnes-Svarney and Thomas E. Svarney
ISBN: 978-1-57859-542-6

***The Handy Answer Book for Kids (and Parents),* 2nd edition**
by Gina Misiroglu
ISBN: 978-1-57859-219-7

***The Handy Astronomy Answer Book,* 3rd edition**
by Charles Liu, Ph.D.
ISBN: 978-1-57859-419-1

***The Handy Biology Answer Book,* 2nd edition**
by Patricia Barnes-Svarney and Thomas E. Svarney
ISBN: 978-1-57859-490-0

The Handy Chemistry Answer Book
by Ian C. Stewart and Justin P. Lamont
ISBN: 978-1-57859-374-3

The Handy Civil War Answer Book
by Samuel Willard Compton
ISBN: 978-1-57859-476-4

***The Handy Dinosaur Answer Book,* 2nd edition**
by Patricia Barnes-Svarney and Thomas E. Svarney
ISBN: 978-1-57859-218-0

***The Handy Geography Answer Book,* 3rd edition**
by Paul A. Tucci
ISBN: 978-1-57859-576-1

The Handy Geology Answer Book
by Patricia Barnes-Svarney and Thomas E. Svarney
ISBN: 978-1-57859-156-5

The Handy History Answer Book: From the Stone Age to the Digital Age
by Stephen A. Werner, Ph.D.
ISBN: 978-1-57859-680-5

***The Handy Math Answer Book,* 2nd edition**
by Patricia Barnes-Svarney and Thomas E. Svarney
ISBN: 978-1-57859-373-6

***The Handy Physics Answer Book,* 3rd edition**
by Charles Liu, Ph.D.
ISBN: 978-1-57859-695-9

***The Handy Presidents Answer Book,* 2nd edition**
by David L. Hudson Jr.
ISBN: 978-1-57859-317-0

***The Handy Science Answer Book,* 5th edition**
by The Carnegie Library of Pittsburgh, James Bobick, and Naomi Balaban
ISBN: 978-1-57859-691-1

The Handy Technology Answer Book
by Naomi Balaban and James Bobick
ISBN: 978-1-57859-563-1

***The Handy Weather Answer Book,* 2nd edition**
by Kevin Hile
ISBN: 978-1-57859-221-0

PLEASE VISIT US AT WWW.VISIBLEINKPRESS.COM.

THE BIG BOOK OF FACTS

Visible Ink Press®
43311 Joy Rd., #414
Canton, MI 48187-2075

Visible Ink Press is a registered trademark of Visible Ink Press LLC.

Most Visible Ink Press books are available at special quantity discounts when purchased in bulk by corporations, organizations, or groups. Customized printings, special imprints, messages, and excerpts can be produced to meet your needs. For more information, contact Special Markets Director, Visible Ink Press, www.visibleinkpress.com, or 734-667-3211.

Managing Editor: Kevin S. Hile
Cover Design: Graphikitchen
Page Design: Cinelli Design
Typesetting: Marco Divita
Proofreaders: Christa Gainor and Shoshana Hurwitz
Indexer: Larry Baker

ISBN: 978-1-57859-732-4

Cataloging-in-Publication data is on file at the Library of Congress.

Printed in the United States of America.

10 9 8 7 6 5 4 3 2 1

TABLE OF CONTENTS

HISTORY

AMERICAN HISTORY

ANIMALS

BRITISH AND AMERICAN HISTORY

CANADIAN HISTORY

CULTURE

SCIENCE

ANIMALS

BIOLOGY

BIOLOGY AND BOTANY

BOTANY

CHEMISTRY

THE ENVIRONMENT

PHOTO SOURCES

Aarkwilde (Wikicommons): p. 152.
Akinom (Wikicommons): p. 227.
Alarnsen (Wikicommons): p. 25 (top).
Avispa Marina: p. 222.
Bain News Service: p. 99.
Boltor (Wikicommons): p. 89.
Boyd's Cove Beothuk Interpretation Centre: p. 5.
George Chernilevsky: p. 192.
Collier's: p. 338.
Crocker Art Museum: p. 208.
Dominiklenne (Wikicommons): p. 195.
El Comandante (Wikicommons): p. 304.
Mikhail Evstafiev: p. 18.
Francis J. Petrie Photograph Collection: p. 41.
Frederick Brothers Agency: p. 175.
Gerald R. Ford Presidential Library: p. 148.
Bobak Ha'Eri: pp. 32, 133.
The Hawaiian Gazette: p. 115.
Hectonicus (Wikicommons): p. 252.
Henrietta Benedictis Health Sciences Library, Massachusetts College of Pharmacy and Health Sciences: p. 262.
D. Herdemerten: p. 162.
Michael Holley: p. 258.
Irv Nahan Philadelphia Management: p. 77.
David Jackson: p. 71.
Kanal Lyudi (Channel People): p. 146.
Kennedy Library, The White House: p. 58.
Kunsthistorisches Museum: p. 122.
Library and Archives Canada: p. 308.
Library of Congress: pp. 40, 53, 83,
Joe Mabel: p. 268.
Miksu (Wikicommons): p. 316.
Moorland-Spingarn Research Center: p. 9.
Museo Nacional del Prado: p. 48.
NASA: pp. 183, 210, 230.
National Archives at College Park: p. 112.
National Portrait Gallery, London: pp. 36, 69.
New York Public Library: p. 91.
Nobel Foundation: pp. 178, 200, 327.
Orange County (California) Archives: p. 173.
Christoph Päper: p. 255.
Pesotsky (Wikicommons): p. 108.
Patryk Reba: p. 326.
Royal Collection Trust: p. 29.
Sears Modern Homes: p. 117.
Shutterstock: pp. 2, 12, 21, 23, 25 (bottom), 31, 43, 45, 47, 50, 55, 59, 62, 64, 66, 74, 79, 92, 94, 96, 101, 128, 141, 158, 163, 166, 168, 170, 181, 185, 187, 198, 203, 207, 212, 217, 219, 221, 225, 234, 237, 240, 243, 245, 248, 264, 265, 271, 274, 276, 278, 284, 286, 287, 297, 302, 303, 311, 331, 333, 336, 340, 343.
SRI International: p. 259.
Harald Süpfle: p. 313.
Tretyakov Gallery: p. 20.
U.K. Parliament: p. 120.
University of California at Berkeley: p. 160.
University of North Florida: p. 130.

U.S. Camel Corps: p. 80.
U.S. Department of Energy: p. 293.
U.S. Government: p. 85.
U.S. Mint, National Numismatic Collection: p. 16.
U.S. Naval Historical Center: p. 111.
U.S. Navy: pp. 39, 347.
U.S. Senate: p. 145.
Tim Vickers: p. 319.
Ed Welter: p. 68.
White House Photographic Collection: p. 8.
Wikicommons: p. 34.
Wikicuda (Wikicommons): p. 13.
Public domain: pp. 3, 11, 24, 27, 46, 51, 73, 75, 81, 87, 104, 109, 116, 126, 135, 138, 142, 144, 155, 171, 189, 191, 194, 197, 214, 290, 307, 322, 324.

INTRODUCTION

Contrary to what people say, the internet is not a "rabbit hole."

When people call the web a "rabbit hole," they're referring to Alice's little journey down into a maze of amazement in *Alice's Adventures in Wonderland.* She falls into the hole, which leads to things that become more and more bizarre and perplexing. But remember: a rabbit hole—even a real one—has a bottom and an end.

Nope, the internet is no rabbit hole. It's really a black hole from which there is no escape (and I don't know about you, but that can be more fun than you can imagine sometimes). So, how about we use that for something constructive? Also contrary to what people say, DO try this at home: go to your favorite search engine and type in two words that interest you.

Let's try "deadly snakes" because, well, you might want to be aware of some facts, right?

At the time of this writing, those two words bring up nearly 13 million possible places to explore the subject of deadly snakes. And you know what'll happen: you'll be on one website and that'll lead you to another and another and—oh, my—look at the clock! Four hours will have passed, bedtime is long gone, and you've discovered ten more things to search for tomorrow.

Yep, the internet is like that.

Books are like that, too.

Take, for instance, the one you have in your hand.

The Big Book of Facts isn't exactly a black hole of information, but it might lead you down a long and pleasant path, and it might make you lose track of time. Here, you'll read about how a breed of dog was created and disappeared because of technology. You'll learn about the weird things you do without even knowing you're doing them. You'll find out about the colors you love and the ones you just can't seem to see. You'll be taken down Memory Lane a little bit in both history and science. Your eyes will be opened to possibilities that could be or that probably won't. You'll read about fierce women who changed their own little cubbyholes of history and brave men who solved some of the worlds' mysteries. And, of course, it could be the other way around.

You'll find things that will surprise you, make you laugh, make you wonder, make you gasp, and things that may make you think. You might find some mistakes (oops, okay, consider yourself superior, then) and a few argument-starters, as well as some things that will send you to your own computer or library to find out more. And while you're learning about those things that might spur you to go deeper, you'll also get some relevant, hard information courtesy of a number of *The Handy Answer Books* from Visible Ink Press.

Oh, and one more thing: don't be surprised if you find science-y things in the history section and historical facts in the science part because you know the world doesn't put a fence around such things. That's the best part of the kind of journey you're about to take in *The Big Book of Facts.*

This book is divided into two sections: History and Science. Here are some tips on how to approach the content in this book....

HISTORY

So, here's the very first thing you need to know about history: you're making it right now. It might not be the most exciting thing, and it might be of little importance at this moment, but you're making it. You started making history the minute you were born, in fact, and if you leave behind children, important work, or some sort of legacy, you'll make it long after you're gone.

So, think about that—and then imagine this: there are more than seven billion people in the world right now, and *each of them has a story to tell*. Granted, you'll only ever hear from about a fraction of those folks, but they lived interesting lives, guaranteed.

In the end, it really all boils down to this: history is a series of stories. Here, it's the tale of a fearless Wild West stagecoach driver who left everyone surprised. It's the story of a winning racehorse and a confused jockey. It's a tale of fashions, fads, and packing up your stuff every year to move. And it's stories of bravery, foolhardiness, small inventions that became big deals, and people just doing their jobs. History is all about the story and, like any good story, it will only make you want more.

DEFINITIONS AND EXPLANATIONS

With the above in mind, toss aside all the dates and events you were forced to learn in high school, and let's look at the little things that make history interesting:

There were lots of them in the relatively short time since America began, so we'll look at American history quite a bit in this section. We'll jump around from the Old West to things that happened practically yesterday, here, there, and everywhere, but I promise: no boring dates, charters, or edicts to memorize.

Because we're inextricably linked with other countries throughout time, we'll also look across the pond, with a spot of British history, and we'll look north at our Canadian neighbors. Again, it's all about the little things that make you say, "Wow!"

You can't have history without the historymakers. In those segments, you'll read about some interesting people and the remarkable reasons why you'll want to know about them. Likewise, you'll read about animals because they're a big part of history, too.

Within those categories, you'll read about culture, which is a broad way to describe the interesting things we've done, the fun we have, and that which entertains us. We'll also take peeks at the food we eat, the things we wear, and the places we visit.

SCIENCE

When it comes to science, I'm always surprised at the number of adults who wrinkle their noses. They say they "don't like" or "don't care about" science. Chances are that they really don't *understand* science.

Maybe they had a teacher who made the subject seem deadly dull, or maybe they had a class or two that skipped over the good parts of science, or made it seem inaccessible, or made it seem complicated, or there was too much to remember, it was just *so hard.* And there are too many "ologies," and how am I *ever* going to use this stuff in real life? And will there be a quiz on Friday? Argh!

No, wait! Come back here.

Dip into our science section of *The Big Book of Facts* and find out the basics to get you started. Then tiptoe through physics and see that, despite what you thought in the seventh grade, you really *will* use it somewhere in your life. Surf through the biology pages and learn a thing or two about how your body works, how your mind works, and how you're even cooler than you thought you were. See how truly fascinating chemistry really is. Learn a thing or two about the environment that you didn't know before. Dive in and learn about technologies from yesterday and tomorrow. Flip back and find out more about botany. Read cover to cover or jump around with wild abandon. That's the fun of science!

DEFINITIONS AND EXPLANATIONS

To make *The Big Book of Facts* as browsable as possible for you, here are some of the fields we'll cross together:

If you're a pet lover, you probably study animals all the time, but here we're going to look at the furry and finned a little differently. Learn about creatures that you really wouldn't want lounging on your sofa. Find out about animals you'll never see outside of books and pictures and discover creatures that swarm.

Biology is the science of living things, so you'll learn what makes you tick in those entries, why you favor your mother or your father, why some people are truly blue, and why we do the dumb things we do. Pick up your feet, stick out your tongue, and have a laugh at this section.

Botany is a fancy word for information about plants. That's all. Why would you be worried about flowers and vegetables, anyhow? Nothing scary here.

When someone says "chemistry," you mght immediately think of Bunsen burners and beakers, but chemistry can save your life. Knowing more about your potato chip craving is chemistry. Understanding why you should be glad you were never a royal taster is also chemistry.

The environment is one of those topics that we all discuss, mostly without thinking about it. What's the weather going to be like? Can you expect rain? Look at those clouds! And what about climate change? In these chapters, we'll take a peek at Earth science and weather topics, some of which will make your day easier, while others are just interesting to know.

It's pretty hard for the average person to go through the day without some sort of math. Here, you'll learn about timekeeping, about the difference between a numeral and a number and, along with a chapter in physics, you'll never complain about a lack of time again.

Physics seems like an intimidating word to describe something you think you'll never understand, but nothing could be further from the truth. It's easy, actually: physics explains how the world works. If you know why time flies, if you understand how glaciers are important to our very existence, and you know about the colors of the rainbow, then congratulations! You know physics.

Space is pretty self-explanatory, but we're not just going to look at the Moon here. Instead, we'll look at the little-known things that got us beyond our atmosphere; a few stories about what's out there, once we left this orbit; what would happen if you fell into a black hole; and more.

Technology might seem to be a subject that mostly raises your hackles. Yes, there are times when technology is overwhelming, but give it a chance. You know that you already use it in your everyday life, undoubtedly without even thinking about it. If you turn on the lights at night, you're using technology. Even this book, yep, was brought to you through many, many levels of technology.

Are you ready? Let's go!

HISTORY

AMERICAN HISTORY: CAN I BORROW A STAMP?

When the Pony Express started delivering mail in April of 1860, everyone immediately hoped that this new way of delivering messages and letters was here to stay. It's true that there'd been a Constitutionally mandated postmaster general and a way to send mail since the Constitution was created nearly a hundred years prior, but sending and getting mail was a complicated process.

In a time when messaging is instant, it's sometimes hard to remember that in the mid- to later part of the nineteenth century, mail was carried and passed from person to person or delivered by a human via horse or coach to a post office. Sending a letter wasn't easy: until 1858, when letter boxes were invented, you had to find the post office building and drop your envelope off to be mailed. Receiving mail could take days, if not weeks, and that, too, had to be retrieved directly from a post office clerk until 1863, when home delivery began in cities and towns.

So, yes, the Pony Express was faster, but alas, though it looms large in our collective imagination, the Express only lasted a scant not-quite-seventeen months. Once the transcontinental telegraph wire was finished being strung from coast to coast in October of 1861, it didn't seem reasonable to send urgent messages any other way. On the other

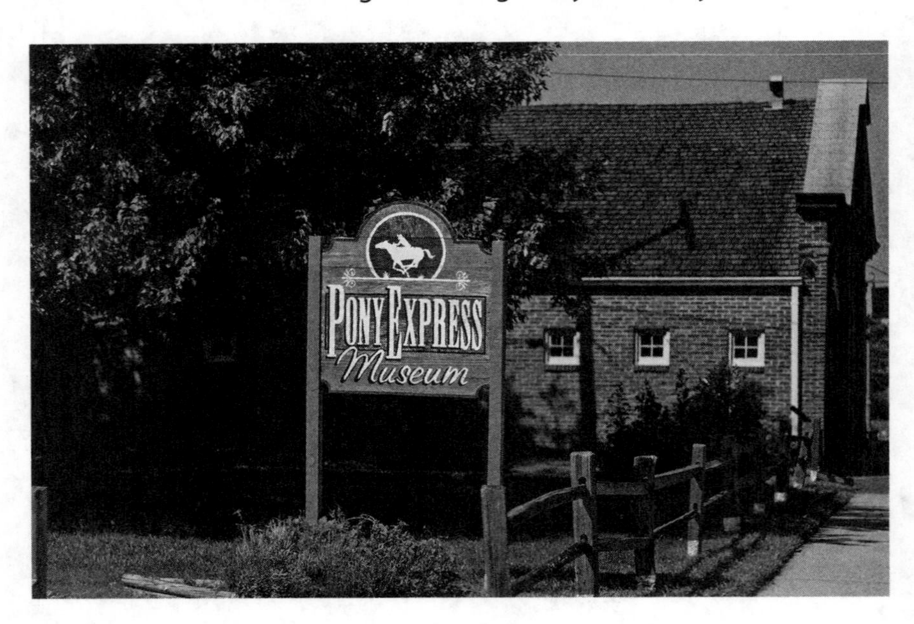

The Pony Express Museum is located in St. Joseph, Missouri.

hand, the telegraph was great for speed, but it would cost you, per word, which meant that long, rambling missives from home, love letters, news, and gossip were almost always still sent by horse and buggy or, increasingly, by boat, train, and the Postal Service.

As Americans spread out over the country and the post office might've been a half day's ride away, rural free delivery was a boon to the country's farmers and ranchers. Starting then, in 1896, the U.S. Postal Service (USPS) relied more and more on individuals who formally or informally carried the mail along the chain to their rural customers. After all, someone had to get those letters and packages from boat or train to local post offices to the far-flung households in their area, right? News from friends and family was always welcome; catalogs were a great diversion for long evenings on the farm; and everyone looked forward to newspapers and magazines in the mailbox.

On January 1, 1913, the USPS started delivering packages that were over four pounds. For rural residents and those in far-flung villages, this was a godsend: it meant that they could get goods and send things through a safe, stable government program, and it wouldn't cost a lot. It didn't take long for a lot of odd-shaped things to pass through a carrier's hands: a casket, bicycles, parasols, and other items that defied easy boxing. But like most good things with rules, somebody saw a loophole: later that month, a farmer and his wife in rural Ohio paid fifteen cents postage and a few more pennies for fifty bucks insurance and mailed their baby son to his grandparents a short distance away. Nobody, as it turned out, said they couldn't, and besides—a stamp or five was vastly less expensive than a train ticket.

Although it wasn't an everyday occurrence, sending one's children through the mail was common enough for the USPS to notice. A six-year-old girl was mailed a con-

Born a slave in about 1815, Henry Brown was distraught: at some point prior to 1849, his wife and children were sold away from him. This was the final straw in a series of indignities and trauma, and so he devised a daring escape. First, Brown hired a carpenter to make a box that was big enough to hold Brown's body (the box was roughly the size of a small coffee table; no record of Brown's size). Then he enlisted the help of another black man from his church and a white cobbler, who helped Brown into the box and sealed it, taking it to the Adams Express Company, where Brown mailed himself to a Quaker abolitionist in Philadelphia and to freedom.

Henry Brown

In 1913, it cost two cents an ounce to send a regular-sized letter or note through the USPS. Back then, the residents of many cities and towns saw twice-a-day delivery. That particular service officially ended in April of 1950, although it continued for several years in smaller towns and villages.

siderable distance: over 700 miles from Florida to Virginia. A four-year-old girl named May was mailed from her Idaho home to her grandparents' house, 73 miles away, accompanied by her postal-worker cousin—which brings up a fact that modern readers should note: back then, these kids weren't exactly put in the hands of strangers to be passed down the road. People weren't reckless then: most of them knew their mail carriers because he (almost always he) was a neighbor, too.

All in all, it's been estimated that a half dozen or so children were actually "mailed" to someone else in the years 1913–1915, despite the Postal Service banning the mailing of human beings in 1914; at least two families tried to mail Junior between 1915 and 1920, but they were turned down each time. Countless other kids were set up for publicity photos and never got mailed, which means that some of the photos are genuine fake news.

Interestingly enough, even today, you can send live baby chicks, some live fish and reptiles, some adult birds, and honeybees through the mail. Just not your kid.

HANDY FACT

One of the things rural readers relied on in the late 1800s and early 1900s was the newspaper. Local papers offered the news, farm reports for crops and livestock, recipe columns for farm wives, cartoons for the kids, and reading material to those for whom electricity hadn't yet reached their farm.

According to *The Handy History Answer Book*, 3rd edition, "The first so-called penny newspaper (or one-cent paper) was published by Benjamin H. Day (1810–1889): The debut issue of the daily *New York Sun* appeared in 1833. The American newspaper industry was off and running. Now reaching a mass audience, publishers worked feverishly to outdo each other in order to keep their readers.

"Population growth (spurred by increasing immigration at the end of the nineteenth century and early in the twentieth century) meant there were plenty of readers for the now thousands of newspapers. During the first decade of the 1900s, before the proliferation of radio (invented 1895), the number of American newspapers peaked at about 2,600 dailies and 14,000 weeklies."

HISTORYMAKERS: THE BEOTHUK

Lots of press is given to endangered animals, and rightly so: when the very last of a nearly extinct creature dies, that's it, at least for now. The pure genetic material of *that species* is no longer alive, but there's hope. Every year, scientists learn a little more about genetics and about animal husbandry. But what about extinct people?

Here's the story of one of them.

Maybe....

Long before Europeans discovered Newfoundland, the Beothuk people lived and thrived there along the very upper east coast of Canada. They are considered today to be descendants of the Little Passage Complex, having come through Newfoundland's Little Passage.

The word "complex" is an archaeological term used to describe a pattern of toolmaking by people in a given area when archaeologists don't know much else about those people. In the case of the Beothuk, their arrowheads are definitive.

There were never many of the Beothuk; historians say that no more than 2,000 (but probably fewer) populated the area after having

An exhibit at the Boyd's Cove Beothuk Interpretation Centre in Newfoundland gives visitors an impression of what this extinct culture was like.

migrated from Labrador. Their lifestyle was entirely that of mostly coastal hunter-gatherers with a main diet that consisted of caribou, seal, salmon, and plants they could forage in-season or store for the winter months. There were likely no villages, per se; instead, the Beothuk appeared to have lived in tents in the summertime and partially subterranean in the winter in small family groups of thirty to fifty individuals. Artifacts such as carved bone and antler have been identified as having come from the Beothuk; one of the hallmarks of their culture was the use of red ochre, which was liberally used on nearly everything, including bodies. It's also possible that they traded with other Native groups.

By most accounts, the Beothuk were not a warring people, although it's been theorized that they had violent skirmishes with the Vikings in 800–1000 C.E., which might have made them suspicious of Europeans.

And yet, they appeared to be semiaccepting toward the Europeans who arrived in the sixteenth century in Newfoundland. There was evidence of trade with the newcomers, but it was often done by "silent trade" (leaving goods at a neutral site, to be replaced by the traded item) due, perhaps, to a lack of complete trust. In exchange for the furs that the Beothuk had collected in their hunts, they received iron implements, which surely made their lives easier.

But then, even the shakiest relations fell apart.

In the late eighteenth century, there was trouble, possibly due to lack of land as European settlers moved further into Beothuk territory, as did other First Nation people. Naturally, there were competitions for resources, and it didn't take long for the Beothuk to begin avoiding contact with the Europeans altogether (though they were apparently will-

HANDY FACT

"The Vikings, also called Norsemen, were fierce, seafaring warriors who originated in Scandinavia (today the countries Norway, Sweden, and Denmark). Beginning in the late 700s, they raided England, France, Germany, Ireland, Scotland, Italy, Russia, and Spain. They also reached Greenland, Iceland, and even North America long before the Europeans. (Ruins of a Norse settlement were found on the northeastern coast of Newfoundland, Canada.) The Viking raiders were greatly feared."

—*The Handy History Answer Book*, 4th edition

ing to retrieve the discards that European hunters and trappers left behind in order to fashion better tools for their hunting and fishing expeditions). The Europeans are said to have sabotaged Beothuk traps and caribou hunts; the Beothuk likewise stole traps back from the Europeans, an act for which the latter were obviously angry, and there was violence. Had the Beothuk fought back (which they apparently did not do), they would have been outnumbered and outweaponed, since the Europeans had guns—something the Beothuk didn't seem to want.

By the early nineteenth century, the Beothuk were isolated, few in number, and struggling to survive on the meager resources left in a small corner of Newfoundland; diseases for which they had no natural immunity may have also played a factor in their struggle. Several governors through the later eighteenth century tried to extend hands of friendship to try to assist the Beothuk, but the damage had been done.

In 1823, three weak, sick Beothuk women were taken to St. John's, Newfoundland, and nursed back to health with the intention of returning them to their homes and the few Beothuk that were left. Before that could happen, though, two of them died, leaving a woman named Shanawdithit who lived with the Europeans for years and who educated them on the Beothuk people. By the time Shanawdithit died of tuberculosis in June of 1829, the rest of her people were gone, too, and the Beothuk were officially extinct.

Or were they…?

Legend from the Mi'kmaq, a nearby First Native tribe, says that the Beothuk intermarried with some Europeans, possibly the Vikings—a legend that has turned out to have an astonishing nugget of truth to it.

In 2017, DNA was sequenced from the remains of prehistoric Beothuk individuals, and it was determined that they were genetically separate from all other First Nation peoples in the area four hundred years ago. That was a surprise; scientists had originally thought that the First Nation people from Newfoundland shared a single common ancestor, relatively recently. Instead, there were at least two distinct groups of genetically different humans who populated the area. A common ancestor between the Beothuk and others lived at least ten thousand years ago—maybe longer.

In early 2020, that information had been updated. It's now believed that the Beothuk are related to the Mi'kmaq; clear links show that Beothuk genes still reside in an anonymous individual who is distantly related to the family of Shanawdithit.

Stay tuned….

AMERICAN HISTORY: DID YOU KNOW?

ANNA HARRISON, THE wife of our ninth president, decided not to attend her husband's inauguration, figuring she'd be in the White House soon enough. She never made it there; William Henry Harrison died of pneumonia just thirty-one days after the ceremony and before she was able to move to Washington.

SARAH POLK WAS the only first lady (so far) to never give birth.

FORMER LOUISIANA GOVERNOR Bobby Jindal and former first lady Michelle Obama are both hard-core fans of TV's *The Brady Bunch*.

THOUGH IT BECAME a state in March of 1845, Florida was not much more than a frontier until the 1920s, when construction and population boomed. Just over 528,000 people lived in Florida in 1900. Thirty years later, that number had jumped to nearly 1.5 million folks who'd moved to the Sunshine State—which, incidentally, wasn't Florida's official state nickname until 1970.

TOILET PAPER FACT: seven out of ten people bunch toilet paper into a big wad of tissue before using; nine out of ten Europeans fold their TP neatly before completion.

Long after kissing Shirley Temple in the movie That Hagen Girl, *Ronald Reagan, as president, shook hands with Shirley Temple Black, who had been working as a diplomat.*

THE FIRST TWO monkeys that the United States sent into space were named Albert. The first Albert made it to 40 miles into space in June of 1948 but suffocated on the way back. A year later, Albert II shot more than 80 miles into the atmosphere but, alas, his parachute failed to open, and he died during his trip back to Earth.

WHEN FRED ROGERS (1928–2003), commonly known as Mr. Rogers, was a child, he was something of a piano prodigy and was said to have composed short songs before he was ten years old.

RONALD REAGAN (1911–2004) was the first Hollywood star to kiss Shirley Temple (1928–2014), which he did in the movie *That Hagen Girl*. Audiences reportedly hated it, perhaps in part because Temple was still a teenager and Reagan was thirty-six years old.

RED IS THE color used most on our states' flags. Red, white, and blue are, together, the most common color combination for flags of the world.

THE BARCODE WAS invented by Drexel graduate Bernard Silver (1924–1963) and was based on the Morse code he'd learned as a

HANDY FACT

"Dr. Charles R. Drew (1904–1950) was an African American surgeon who was a leading scientist in the study of blood and plasma. After graduating from Amherst College, Drew set his sights on medical school, after recuperating from a leg injury suffered during his star football career at Amherst. He finally found a place to admit him to medical school—McGill University Medical School in Montréal, Canada.

Dr. Charles R. Drew

"He later taught a pathology class at Howard University Medical School and then trained at Columbia University's Medical School. He set up the first blood bank at Presbyterian Hospital in New York."

—*The Handy History Answer Book,* 3rd edition

Alas, at that time, blood donations were sorted according to the race of the donor, a practice that Drew was very much against. In 1950, the year he died in a car accident, the Red Cross stopped segregating blood.

boy. He and friend Norman Joseph Woodland (1921–2012) knew that store owners had problems tracking merchandise, so they put their heads together and started working on a code that used something like the dots and dashes. The barcode system was patented in 1952, but no one was initially willing to try it. In 1973, grocery stores and tech companies devised the Uniform Product Code Council. The UPC code was based loosely on Silver and Woodland's creation.

ANIMALS: MY DINNER WITH FIDO

Meals in the seventeenth century could be a little challenging. Considering a lack of grocery store or farmer's market and that a butcher might be a day's ride away, most families raised, caught, or hunted their own dinner. One didn't want to waste a morsel, and cooking was becoming something that the average late-Renaissance wife took pride in doing or having done right.

Roasting meat in the 1600s was looked down upon as something one just didn't do; Renaissance palates knew better than to ruin the cut by putting it in an oven. Instead, cooks believed that the taste of meat was better if it was cooked over an open fire on a turnspit, which was basically a long iron rod upon which the meat was skewered. This was suspended on braces over the flames and turned, turned, turned constantly until the meat was cooked, a task that could be done manually for a short time. Imagine the exhaustion, though, not to mention the long-term effects of smoke and ash and the injuries of turning a handle or wheel for the *hours* it would take for the meat to cook. Early on, the chore was done by the lowest person on the kitchen totem pole, which was often a young boy. Later, it was done by a small dog.

The method was easy if not fiendish: on the wall of the kitchen, a large open cylinder was attached, with a loop of chain wound over and around it and to the handle of the spit. When it was time to put the meat over the fire, the spit dog (also known as a turnspit dog) was taken from a wooden cage and placed on the inside of the wheel, which was set up higher so that the dog could escape the majority of the heat. As soon as the dog's paws hit the cylinder, the animal was expected to run like a hamster in its exercise wheel. Some dogs were expected to run for hours.

An ingenious machine created during the Renaissance was a dog-powered turnspit. A small dog would turn a wheel (rather like a hamster wheel today), which would then turn a spit over a fire, cooking the meat slowly and evenly.

If you saw a drawing of a spit dog, you might think of a dachshund or a small spaniel. Spit dogs were said to be mongrelly and low-slung but extremely solid and hearty, with pendulous ears and a high, curly tail that wouldn't get caught in the mechanisms of the wheel. Some chroniclers wrote that the dogs had misshapen legs, which should come as no surprise. They appeared well fed, which is also no surprise, considering the necessity of their work. Like most dogs in the 1600s and 1700s, they were working dogs, but in the case of spit dogs, there was a twist: though this was a complete and separate *breed of dog*, they were mostly thought of as mere tools in the kitchen, no different than we might think of a spatula or a pair of tongs.

In wealthier kitchens, there might be multiple wheels and, thus, multiple spit dogs. Dogs may have worked in pairs so that one might rest a bit, but lazy dogs were not tolerated; those curs might be convinced to step quickly when a hot coal was tossed on the wheel by an impatient cook. Before you start frowning, know this: spit dogs were often given the Sabbath off from work in the kitchen.

And before you start frowning at the British, know *this*: the spit dog was also used for a time in American kitchens.

By the mid-1700s, one could find spit dogs in practically every kitchen in Great Britain and in many American kitchens, too, but a movement to stop the use of the pups was afoot: Henry Bergh, a New York City activist, was so bothered by the presence of and cruelty to spit dogs that he ultimately founded the American Society for the Prevention of Cruelty to Animals (ASPCA), which began efforts to take the

HANDY FACT

"A small female dog named Laika" was the first animal sent into orbit "aboard the Soviet *Sputnik 2*, launched November 3, 1957.... Laika was placed in a pressurized compartment within a capsule that weighed 1,103 pounds (500 kilograms). After a few days in orbit, she died, and *Sputnik 2* reentered Earth's atmosphere on April 14, 1958."

—*The Handy Science Answer Book*, 4th edition

A statue honoring Laika the dog cosmonaut stands in Moscow.

hot dog out of the kitchen. It's interesting to note that it may have happened anyhow; new inventions for turning a spit were made, and new methods of cooking were starting to appear in finer homes. At any rate, tens of thousands of spit dogs were in use in the mid-1700s, but a mere century later, one could hardly find *any* spit dogs in use.

But get this: by the turn of the twentieth century, spit dogs, as a breed, were *extinct*. Or were they?

While you won't find a turnspit dog at any Westminster or Crufts show and they're not listed on the American Kennel Club's (AKC) website, dog lovers say that corgis, basset hounds, dachshunds, and some breeds of terrier look a lot like surviving depictions of spit dogs. So, your little Fido may have ancestors who toiled in Renaissance kitchens....

HISTORYMAKERS: THE STORY OF CHARLEY PARKHURST

If there had been such a thing as a Bureau of Labor Statistics in the year 1870, you'd have found that the rate of unemployment was about the same as in 2019. There were jobs to be had in the mid- to late-nineteenth century—even if you had to invent one yourself—but not all of them were safe or desirable for someone who wanted a quiet life. For a soul looking for good money, adventure, a little danger, and maybe some fame, you couldn't beat driving a stagecoach, which was a sort of Uber of the Wild West.

Stagecoach drivers in the years before the Civil War were men of importance, often more admired than the wealthiest men and women they transported. "Whips," as they were sometimes known, had confidence, which was mandatory when roads were dust and gravel, at best, and were often located clinging to the sides of mountains. Drivers were fearless; they had to be to travel through open and openly hostile Indian territory, where death was a constant possibility and where a man had to keep passengers safe while also guarding money and/or gold. Transporting valuables and goods, in fact, was often fully half the job, since methods of commerce such as precious metals, gemstones, and cash were usually aboard the coach and were a major target of random thieves roaming the prairies and deserts of the untamed West.

A stagecoach driver was trusted, having sometimes been handpicked by the owner of the coach. Stable hands respected him, pas-

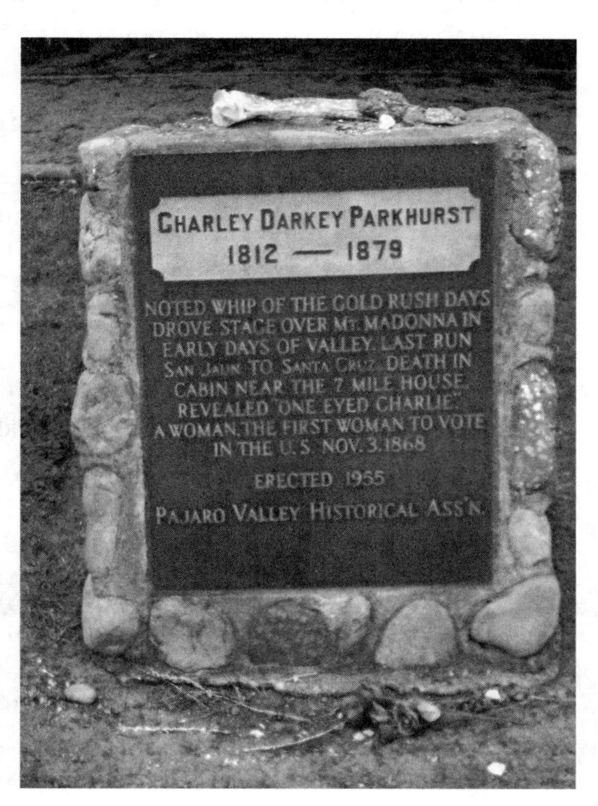

The grave of Charley Parkhurst is located in the Pioneer Cemetery in Watsonville, California.

sengers admired him, and, not always a young man anymore, he was at the top of his game.

And that's just where Charley Parkhurst liked to be.

Born in 1812 in New Hampshire, Charley Darkey Parkhurst wasn't much to look at: stocky, well under six feet tall, and nearly 200 pounds, Parkhurst was solid, with big arms and wide at the hips, but just a glance from one eye could chill the blood of most men—which was good because one eye was all Parkhurst had, having lost the other eye after an accident with a horse.

But we're getting ahead of ourselves....

In the beginning, Parkhurst was a twelve-year-old runaway from an orphanage, and that's about as far as accounts agree. Some say that a fellow named Ebenezer Balch promised to make a man of ol' Charley, and Balch lived up to that promise. Others say Parkhurst became a stable hand, or was escaping a marriage, or worked in an uncle's farm, or was running from something, or that Parkhurst traveled around while learning to work with horses and teams, or all of the above. It's all unknown and quite conflicting, but stories converge in about 1850, placing Parkhurst near or in San Francisco, then commencing to work as a stagecoach driver, often for the famed Wells Fargo organization.

Accounts say that Parkhurst loved to give candy to children, but he was mostly known to be tough: a rolled coach resulted in a few busted ribs, but a doctor was never consulted. Skilled with a six-horse team on rough terrain and fond of tobacco, hard liquor, and four-letter words, "Cock-Eyed Charley" was also good with a gun; bandits steered clear after word got around that one of them got a bullet in the chest for attempting to steal a fortune from Charley's coach. *Nobody* messed with Charley Parkhurst.

But all that jostling on a wooden seat and the wrangling of six horses took its toll, and by the early 1870s, Parkhurst was nearing sixty

HANDY FACT

The Handy History Book, 3rd edition, reminds us that the "Nineteenth Amendment (1920) granted women the right to vote." More than fifty years before that, though, Charley Parkhurst was listed in an official poll list for the election of 1868. There's no solid record of a ballot cast, but if she indeed did vote, Parkhurst was likely the first woman to do so in an American election.

years old, suffering from physical ailments, and was literally ready to step down and let someone else take the reins. By 1879, another kind of bandit came to call when cancer of the tongue killed "One-Eyed Charley," who died with close friend and business partner Frank Woodward at her side.

Her? Oh, didn't you know that Charley Parkhurst was likely born as *Charlotte?* Yes, and some say that she'd been someone's mother at one point. Still, for most of her adult life, few knew or were brave enough to admit that beneath the rough clothing of a courageous stagecoach driver was the body of a woman who apparently found it better (and probably more exciting then) to live as a man.

AMERICAN HISTORY: A PENNY SAVED, PART I

Money, as they say, makes the world go 'round. So, here's a basic and quick history of cash in America....

While it is true that some Native tribes assigned certain objects as worthy of trade substitutes, the first paper money issued by immigrants to the New World was issued as bills of credit by the Massachusetts Bay Company in 1690 to finance military expeditions. Perhaps not surprisingly, those were immediately followed by illegal counterfeit bills.

By the time America was involved in the Revolutionary War, each of the original thirteen colonies had its own version of paper money despite Great Britain having tried its mightiest to squash those upstarts and their cash; as for coinage, a big mix of British, French, German, and Spanish coins were used for purchase. In order to make sense of this mess and to raise money for the war, post-Revolutionary big shots introduced the first national paper currency in 1775. But because there was nothing to back it up, that colonial cash was quickly devalued. In the areas that the British occupied, it was banned entirely, and the currency was completely worthless by the war's end. This, and the rampant counterfeiting that had been taking place during colonial times, soured new Americans on the whole idea of paper money.

Benjamin Franklin devised a set of printing blocks to thwart crooks, but it didn't matter: folks were mistrustful of American cash, which spurred the first secretary of the U.S. Treasury, Alexander Hamilton, to found the Bank of the United States in 1791. Meant to stabilize the new economy and regain trust, the bank established credit for the government, issued private currencies, borrowed as needed, and likewise lent money.

Once the Constitution was signed and accepted as law, a new U.S. Mint was established in 1792 in Philadelphia by President George Washington. Coins were issued in half cent (discontinued in 1857), cent, half dime, dime, quarter, half-dollar, dollar, "quarter eagle" ($2.50), "half eagle" ($5), and "eagle" ($10) amounts. Still, the Mint couldn't keep up, and there weren't enough coins to support widespread use. So, while it was illegal to create and circulate one's own paper money, local banks and businesses did it anyhow, and counterfeiting was once again a problem.

The problems were in full bloom once the Civil War began. Congress had banned foreign coins as legal tender with the Coinage Act of 1857, but that probably mattered little in the War between the States in 1861. Both the North and South needed money for the war machine, and both created and issued it with impunity and without the gold or silver to back it; needless to say, things were chaotic and unstable, financially, for much of the country. Sometimes, there was a tax incurred for using the temporary money, which surely muddied the waters.

At the end of the war, all systems for both sides were folded into the National Banking Act of 1863 and 1864, which calmed the country—at least, on that front. For a time, "demand notes," issued during the Civil War, were exchangeable for gold or silver "on demand" from any one of seven banks within the United States. These were rather quickly replaced by dark green notes that were marked "legal tender" and were not exchangeable for bags of precious metals. Still, there

A quarter eagle minted in 1796 was composed of gold and was worth $2.50. If you happen upon this particular coin at a garage sale or flea market (or buried in a wall somewhere), the coin is now worth over $67,000.

HANDY FACT

"Paper money first appeared in China during the Middle Ages (500 to 1350). In the ninth century C.E., paper notes were used by Chinese merchants as certificates of exchange and, later, for paying taxes to the government. It was not until the eleventh century, also in China, that the notes were backed by deposits of silver and gold (called 'hard money')."

—*The Handy History Answer Book*, 3rd edtion

were a few issues with old Civil War-era money, including that various notes had been issued and were still in use. Americans still carried a lot of mistrust toward the banking system as a whole.

It didn't help when, in October of 1907, the New York Stock Exchange fell to half of what it had been the previous October, setting off what historians now call the Panic of 1907 (or the 1907 Bankers' Panic or the Knickerbocker Crisis), which caused an even further erosion of trust. If not for financier J. P. Morgan's cash infusion and his fast talk to fellow wealth holders, things would have been far worse. The Panic of 1907 led John D. Rockefeller's father-in-law, Senator Nelson W. Aldrich (1841–1915), to establish an oversight committee, which led to the Federal Reserve System. Fortunately, Federal Reserve Bank notes held up during the Great Depression.

For the most part, with a few alterations of appearance, we've enjoyed the same basic currencies since the end of the Depression era: the dollar bill, the two-dollar bill, the five-dollar bill, the ten-dollar bill, the twenty-dollar bill, the fifty-dollar bill, and the hundred-dollar bill. We also get pocketfuls of pennies, nickels, dimes, quarters, and the occasional half-dollar and gold dollar. Larger denominations in paper—bills of $500, $1,000, $5,000, and $10,000—were printed until 1945 and were discontinued in circulation on July 14, 1969; $100,000 bills, used for transactions between banks, were printed from 1934 to 1945. Those higher denominations have been quietly retired for public use (see History—American History: A Penny Saved, Part II).

HISTORYMAKERS: WHAT'S GOING ON OUT THERE?

If pressed, most people would admit that they can't go far from their cell phones. Computers are just about mandatory for today's life. It's hard to imagine life without either convenience but not if you never even knew those things existed....

In 1978, while searching for a safe place to land a group of geologists he was hired to transport, a Soviet helicopter pilot noticed something that surely shouldn't have been on the side of a remote hill in Siberia. He couldn't set down, so he couldn't be sure, but it looked man-made, like a garden, some 6,000 feet up the side of a mountain in the Republic of Khakassia. But that couldn't be—the furrows in the ground appeared some 150 miles from the closest known settlement. He reported it back to the geologists, who were in for a huge surprise.

It all started more than four decades before, when the brother of Karp Osipovich Lykov (1907?–1988) was killed by a member of the Bolsheviks, who were known to harass and threaten the Old Believers, a Russian Orthodox sect that had been persecuted in Russia for centuries. Upon the death of his brother, Lykov decided that it was time to literally run for the hills. He, his wife, Akulina, their nine-year-old son, Savin, and their two-year-old daughter, Natalia, escaped into a nearby forest, where it surely seemed safer for them.

They reportedly took a few things with them: seeds to plant, a spinning wheel for making cloth, books to read and learn by, and other small things they could carry. They went deeper and deeper into the wilderness until they came to the spot where they constructed a tiny, half-underground cabin from whatever materials they could find,

The Old Believers (also known as Lipovans), the group to which Lykov and family belonged, are still active today in eastern Europe.

and they settled in to stay. Akulina eventually gave birth to two more children—a male child in about 1940 and another daughter in 1943.

Through balefully cold Siberian winters and cooler summer growing seasons, the Lykovs kept to themselves. Clothing was said to have been patched until there was nothing left to patch, whereupon the family fashioned clothing from cloth they made from materials they grew. Cooking was difficult, since the utensils they brought with them had, through the years, fallen apart, so they cooked in vessels made of bark. The children were taught to read by their mother, who used the Bible and prayer books she'd carried with her in 1936; because of their isolation, they spoke a modified version of Russian to the geologists who arrived.

For the younger two Lykov children, that visit must've been a shock: those geologists were the first nonfamily humans they'd ever seen; they knew through their parents' tales that other people existed and that humans sometimes lived in cities, but they had no true frame of reference. They knew nothing about TV, telephones, or space flight, although they had noticed satellites in the skies at night and had seemed fascinated by the concept. When the technology was explained, Karp himself was excited by the idea. Despite their total isolation (or nearly total, because some say that locals might have occasionally made forays to the Lykov cabin), scientists were happy to note that the Lykovs were intelligent and had good senses of humor.

Still, although the area surrounding their cabin was absolutely beautiful and life was what they'd made it, it was undoubtedly hard. They'd brought no guns or even a bow or arrows with them when they fled the Bolsheviks, so meat had to be trapped. Other animals often ate crops that were planted before the Lykovs could harvest them. When the geologists arrived, the family's food had been reduced to little more than potatoes; often, they were near starvation. It was said, in fact, that in 1961, Akulina had died of starvation in order that her children would have enough to eat.

The presence of the scientists proved to be an irritation to Karp, as well as a blessing. He was happy to accept salt from visitors, but he didn't seem to want anything else. He tried to get his children to refuse modernity, too, but while the older Lykov kids seemed to agree with their father, the younger two children, Dmitry and Agafia, seemed open to what newcomers offered. Everything, that is, except television, which was seen as sinful but watched, nonetheless. Eventually, the family accepted some newfangled gifts, but not all by a long shot.

Three years after their forced re-emergence from hermitage, three of the Lykov children (now adults) were gone, possibly because

HANDY FACT

Artist Boris Kustodiev's 1920 dramatic depiction of a Bolshevik propagandizes the revolution.

"The Bolshevik Revolution was the culmination of a series of events in 1917. In March, with Russia still in the midst of World War I (1914–1918), the country faced hardship. Shortages of food and fuel made conditions miserable. The people had lost faith in the war effort and were loath to support it by sending any more young men into battle. In the Russian capital of Petrograd (which had been known as St. Petersburg until 1914), workers went on strike and rioting broke out. In the chaos (called the March Revolution), Tsar Nicholas II (1868–1918) ordered the legislative body, the Duma, to disband; instead, the representatives set up a provisional government. Having lost all political influence, Nicholas abdicated the throne on March 15. He and his family were imprisoned and are believed to have been killed in July the following year."

—*The Handy History Answer Book*, 3rd edition

Despite attempts to control them, the number of Bolsheviks continued to grow well beyond the end of World War I.

of their horrid diet, all within days of each other. In 1988, Karp died in his sleep and was buried not far from his cabin.

At the time of this writing, Agafia still lives in her mountain cabin.

AMERICAN HISTORY: YOU'VE GOT MY VOTE!

Imagine what it would be like if we still voted in the same manner as did our early American ancestors: until 1792, states could potentially hold presidential elections whenever they darn well pleased on any day of the week, any month, any time.

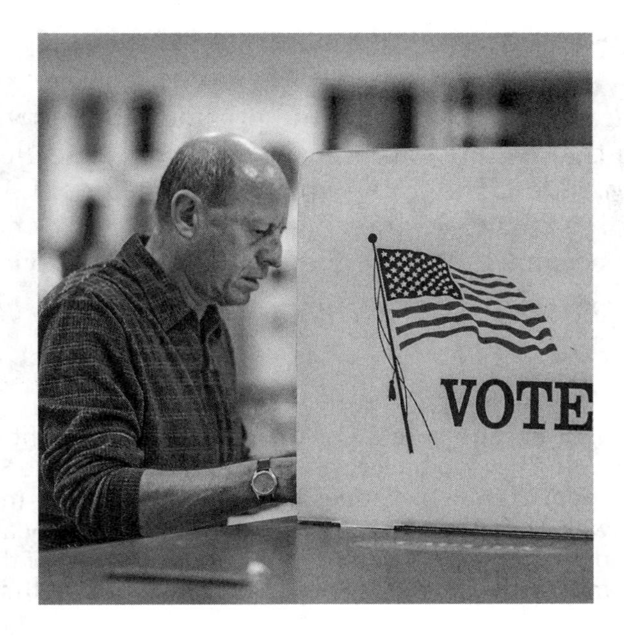

Why are American elections always held on Tuesdays? The answer has to do with religion.

That willy-nilly schedule ended somewhat in 1792, when a law was passed stating that the messy status quo could still happen as long as all state elections were completed within thirty-four days prior to the first Wednesday of December; on that day every four years, each state's qualified property-owning electors (read: white men of means) had to be present in Washington and ready to vote for president. That window of time, the thirty-four days, was most convenient for farmers who would have their harvesting finished by then and who would have still been able to travel easily to their respective polling places before winter set in.

Still, even this had its flaws.

If a candidate won in an early-voting state, his win (or his opponent's loss) could influence subsequent elections in neighboring states that might vote closer to the December date. While it's true that news didn't exactly travel at the speed of television back in 1840, and while it's true that there really weren't a lot of states or a huge population to contend with, this mishmash of election days could mean a clear (but unintended) imbalance of votes.

In 1845, Congress had enough of that and voted to make the Tuesday after the first Monday in November to be the day Americans in all states vote for their president. In 1875, that day became law for U.S. House of Representatives members, and it was law for senators in 1914.

But why Tuesday, and why not the *first* Tuesday, regardless?

HANDY FACT

"Twelfth Amendment (1804): Revised the presidential and vice presidential election rules such that members of the electoral college, called electors, vote for one person as president and for another as vice president. Prior to the passage of this amendment, the electors simply voted for two men—the one receiving more votes became president and the other became vice president."

—*The Handy History Answer Book*, 3rd edition

Because a Saturday or a Monday vote might mean that a voter would have to travel on a Sunday, which was the Sabbath and the day spent in church. Wednesday was market day for most farmers and, well, Friday was just out. And the whole Tuesday-after-Monday business was, in part, *because of* business: the first of the month was when most merchants closed out their previous month's bookkeeping.

CANADIAN HISTORY: A MOVING EVENT

Every adult who's ever done it knows one big, glaring thing: moving from one house to another really stinks. Pack up your things, take them to a different place, unpack them, place everything in new spots, and do it all over again sometime down the road. The average person moves more than eleven times in his or her lifetime ... so what would possess a citizen of Montréal to do it annually?

Call it tradition.

Every year on July 1, folks across Canada mark Canada Day, which is a reason for celebration, parades, cookouts, and, for hundreds of thousands of Quebecois and Montréalers, it's also a day to put their belongings in a truck, turn off the lights, and lock the door behind them.

Yep, it's Moving Day for them.

This mayhem all started in the eighteenth century by French officials who decreed that May 1 was the date for all legal agreements to commence, including leases. This exact date was partially designed to prevent landlords from putting tenants out when the weather was still cold. Rental documents, then, started May 1 and ended April 30, making Moving Day on the former date.

All well and good, right? Except that sometimes, it's still cold and snowy in Montréal on May 1. And except for parents, whose children were still in school at the beginning of May. With these little issues in mind, the Quebec government moved the moving date to July 1 in 1973, which also gave workers a break that year and beyond, since the day was already a holiday (Canada Day is July 1), and they didn't have to use a vacation day to haul their stuff.

Even so, though people surely know it's coming and have time to plan, Moving Day can be total chaos, with moving vans, trucks, motorcycles with sidecars, three-wheeled bicycles, cars packed to the hilt, specially made trailers, and even plain old wagons employed by those who are merely moving across the street all causing traffic jams and flared tempers. Statistics indicate that more than 60 percent of Montréal's citizens are renters—and that may mean up to 250,000 Quebecois who'll need more boxes this coming July.

It also means a headache for Montréal officials, who must deal with tons of garbage from discards and detritus that inevitably comes from moving; recycling is encouraged, but more than 50,000 tons of garbage have been collected in recent years. It also means an influx of homeless pets when new landlords won't accept pets in the new place. Sadly, the Montréal SPCA says that the number of surrendered pets more than doubles in the summer surrounding Moving Day.

Canada Day is celebrated every July 1 to honor the unification of Canada, Nova Scotia, and New Brunswick into a single dominion of the British Empire in 1867.

HANDY FACT

According to *The Handy History Answer Book*, 3rd edition, Norwegian-born Leif Eriksson (c. 970–1020) likely was the first European to set foot in North America when he and his crew of some thirty-five men stepped onto Baffin Island just north of the province of Quebec in the year 1001. Leif and his men "likely made it to Labrador, Newfoundland ... and later landed on the coast of what is today Nova Scotia."

Leif Eriksson Discovers America *by Hans Dahl.*

They did not, however, move there.

ANIMALS: ONE POTATO, TWO POTATOOOOOO

Although the sport still has its legions of fans, there have arguably been no others in history who loved horse racing more than did the folks in the eighteenth century.

Of course, prior to that, there's no doubt that people raced ponies and they likely bet on them, even in antiquity. But during the time of Queen Anne, who died in 1714, horse racing became a bona fide professional sport, and gambling on the racing outcomes got a little out of hand. In 1750, the Jockey Club was formed in Great Britain as a way to set rules and govern the sport.

When the British came to America, they brought horse racing with them, and they set up the first racetrack in 1665 in what is now Long Island, New York. The little chestnut colt that's the subject of our story, however, was born in Great Britain in 1773.

Even at—or, perhaps, especially at—that time, breeders of racehorses kept scrupulously detailed records of the pedigreed lineage of their equine athletes. That's how we know that the colt's breeder was Willoughby Bertie, the 4th Earl of Abingdon; his dam was Sportsmistress, the mother of other famous horses; and his sire was Eclipse, who was an undefeated racing legend himself. For the Earl, the birth of this particular foal must've been of some excitement.

No one knows for sure when the colt was named; today, a future racehorse's moniker usually leads back to those of his sire or dam or

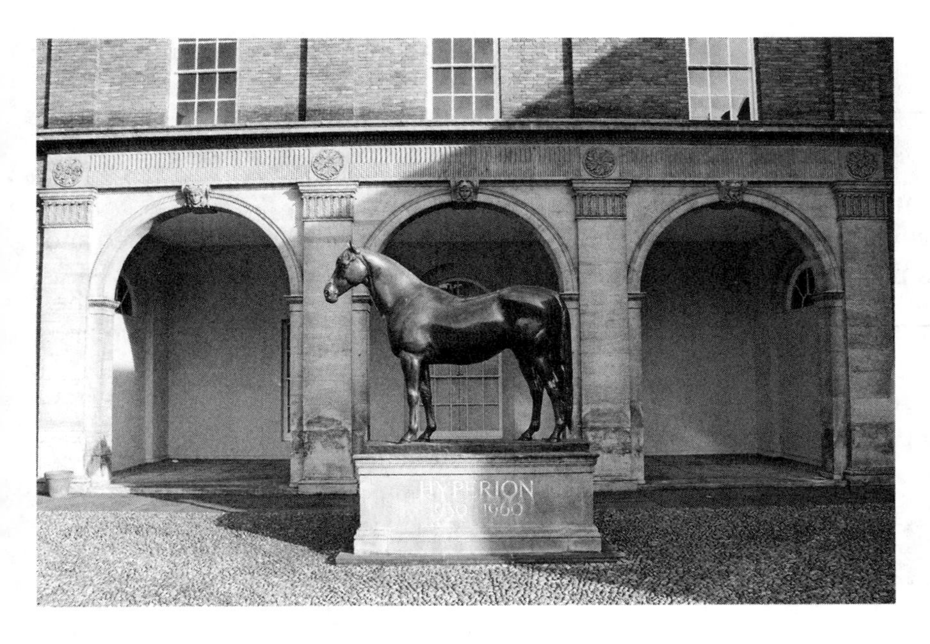

The Jockey Club in Newmarket, England, was established in 1750. It is no longer in charge of governing British horse races but is still heavily vested in the sport, owning 15 racetracks.

both, but this little chestnut colt ended up with an unusual one at some point in his life. Legend has it that a stable boy, wishing to personalize the horse's stall, asked the Earl what name he wished to call the animal and was told "Potatoes." Misunderstanding—and probably shrugging but doing as he was told—he painted "Potoooooooo" (Pot-eight-Os) on the stable door. The Earl was said to have been so tickled by it that he left the name just as it was, although it was sometimes officially shortened to include the numeric symbol.

The name never hurt Potoooooooo's career any: in the seven years he raced (1777–1783), he became known for his endurance on

HANDY FACT

The Handy Science Answer Book, 4th edition, says that Clydesdales, which are not at all considered agile enough for racing, were used as war horses "to carry the massively armored knights of the Middle Ages. These animals had to be strong enough to carry a man wearing as much as 100 pounds (45 kilograms) of armor as well as up to 80 pounds (36 kilograms) of armor on their own bodies. However, the invention of the musket quickly ended the use of Clydesdales and other Great Horses on the battlefield as speed and maneuverability became more important than strength."

The fastest racehorse ever clocked was Winning Brew, on May 14, 2008, at 44 miles per hour.

the four-mile track, and he racked up an astounding thirty-four wins. In 1784, Pot8Os retired and was put out to stud, and in the ensuing years, he sired more than one hundred winning racehorses.

Alas, in the fall of 1800, Pot8Os died and was buried in Hare Park in Great Britain. Today, the bones of the great racer are in the National Horse Racing Museum in Newmarket (see also History—Historymakers: Annnnd, They're Off!)

CULTURE: DON'T SMILE, DON'T SAY "CHEESE"

Imagine having no photographs of the people you love. In our all-phones-all-the-time existence, that's hard to, um, picture because worldwide, humans snap more than 14 trillion photos per year. That's a lot of scrapbooking and posting, but get this: before the invention of inexpensive photography, an individual might have one or two pictures taken *in a lifetime.*

Back in the early nineteenth century and prior, people who wanted a permanent image of themselves or their loved ones relied on skilled (or not-so-skilled) painters to craft as perfect a likeness as possible. Problem was, sitting for a portrait took awhile—possibly even months or years—and lots of things can change in that time, so a portrait might be started, but in the end, it could look very different from the real-life person it was meant to memorialize.

Enter Joseph Nicéphore Niépce and Louis-Jacques-Mandé Daguerre, who introduced their methods of photography in 1827 and 1839, respectively. They basically invented early methods of taking rudimentary photographs, which made it possible to capture someone's picture in much less time than it took to sit for a painted portrait.

Maybe you've seen photos taken with these methods: they look old because they *are* old, but they were also revolutionary, even though they had their drawbacks. Having your picture taken was expensive, for one, so only the upper-most of uppermost folks could

The daguerrotype was the first photographic process available for public use. This photo, taken in 1838 by inventor Louis-Jacques-Mandé Daguerre, is titled View of the Boulevard du Temple *and was one of the first images to include people in it.*

have paid to have it done more than once. Though a photo didn't take as long to get as did a painting, early photographs were still labor-intensive for the subject because having an image done might mean holding very still for interminably long minutes. The immobility was meant to avoid blurring the picture because of the way a negative was made—slowly and with a glass plate inside the camera, which was treated to expose the image with chemicals, some of which we know today are dangerous and could do long-term damage to the health of a photographer.

So, with this in mind, early Victorians called the guy with the funny camera—not always for themselves but often to take photos of their dearly departed.

Death, at that point in history, was tragic but especially common for infants and young children, who were especially vulnerable. One study showed that nearly half of all children in the 1800s died before they reached adulthood. Parents then didn't have any of the support parents get now when babies or children died, and it's natural to think that they'd want some sort of remembrance of their child. As for adults, well, they died, too, and who wouldn't like one last photo of Mom or Dad or a sibling—especially if the photographer was talented enough to make the deceased look lifelike, as though sleeping or taking a break?

Indeed, these *memento mori* (rough English translation: "re-member you must die") photos—which often also included the entire

HANDY FACT

The Handy Science Answer Book, 4th edition, says that there were 2,423,712 people who died in the United States in 2007. More than one-quarter of them died of heart disease; another 23.2 percent died of cancer.

family posed with the deceased or parents holding a child as if waiting for the little one to wake up—became precious possessions for grieving families. Sadly, in many cases, the *memento mori* was the only picture the family had together, mostly because of the cost but also because having a photographer nearby wasn't an option, especially in rural areas; in many cases, photographers went from city to city to ply their trade.

As *memento mori* became more widespread, devices were invented to allow the recently deceased to stay upright for a photograph. In other pictures, one can glimpse a grieving mother behind a curtain, holding her baby, who appears to be asleep; most *memento mori* are posed in a manner to make the deceased look healthy and alive or dozing, and many corpses were posed with belongings they enjoyed in life.

Look up *memento mori* online, and you'll see some of these heartbreaking pictures. Notice that you'll rarely see a coffin.

The Victorians' aim was not to be macabre but to remind people to meditate on the fact that death came for us all; to that end, our ancestors also used other items as *memento mori* and even gave them out to mourners, including specially made inscripted coins, jewelry, or watch chains made from the hair of the deceased, small baubles, rings, and other ephemera.

BRITISH HISTORY: THE ROYAL FAMILY, PART I

QUEEN VICTORIA IS known as the Grandmother of Europe, and here's why: She and Prince Albert had nine children and 42 grandchildren, most of whom married royalty or near royalty to forge bonds throughout Europe at the end of the nineteenth century. At one point, Victoria was the mother of a future king; grandmother of a future king; and grandmother of a king-emperor, an emperor, an empress consort, and four queens-consort.

VICTORIA WAS SAID to have hated pregnancy, and she thought newborn babies were ugly.

IN MEDIEVAL TIMES, British men of means wore underwear, but women did not.

THE QUEEN AND king of England do not sleep together in the same room. That they don't is a long-standing tradition and is said to be for "practical reasons" that have a lot to do with comfort and getting a good night's sleep.

QUEEN ELIZABETH ALWAYS takes her royal physician with her on her travels. If she goes to a place where the blood supply may have issues of safety, she also travels with her own blood supply.

MORE THAN FIFTY years ago, a documentary aired on the BBC showing what the royal family's day-to-day life was like. It was called *Royal Family*, and after it had aired, the queen approved, but she had some concerns: specifically, in a world growing increasingly tumultuous, it might not be such a great idea to show too many details on royal comings and goings and scheduling. Since then, the program has not been seen in its entirety anywhere.

An 1860 photo of Queen Victoria.

HANDY FACT

"England's royal houses are simply families, including ancestors, descendants, and kin. Since 1066 England's rules have come from a series of ten royal houses: Normandy (ruled 1066–1135), Blois (1135–1154), Plantagenet (1154–1399), Lancaster (1399–1471), York (1471–1485), Tudor (1485–1603), Stuart (1603–1649, restored 1660–1714), Hanover (1714–1901), Saxe-Coborg (1901–1910), and Windsor (1910–present)."

—*The Handy History Answer Book*, 3rd edition

DURING QUEEN VICTORIA's reign, London suffered under what was known as "The Great Stink." It happened in July and August of 1858, when the River Thames, which was London's main source of water, became clogged with sewage and other waste and the heat of the summer months exacerbated what had to have already been a horrid stench. Civil engineer Joseph Bazalgette (1819–1891) was hired to fix the situation—which he did, with a system that, as of 2016, was still partially in use.

HAD KING EDWARD VIII not abdicated, there's a real possibility that Queen Elizabeth II wouldn't have been queen (unless, of course, he'd died with no children. In that case, yes, she would have assumed the throne anyhow).

SINCE 1748, ALL British monarchs have had two birthdays. George II started the tradition because his birthday was in November, and the weather is usually rather dreadful then. There were no parades or outdoor events, either, so he moved his "birthday" to the second week of June. Ever since then, the person on the throne has a private family birthday on their real birthday and a public "birthday" with a Trooping the Colour parade, salutes, and great celebration in June. For the record, Queen Elizabeth's actual birthday is in April.

GOOD LUCK WITH adding a royal autograph to your collection. Absolutely no one in the royal family is allowed by rule to sign their names to anything that's not an official British document.

AMERICAN HISTORY: FLIPPIN' BURGERS

A century or so ago, if you were hungry and you were in Wichita, Kansas, you were in luck. Wichita had made American history by being the city where White Castle opened its first fast-food restaurant. Although that initial eatery and its fare weren't known then as "fast food," you have to wonder if they knew what they'd started....

Up until then, folks thought hamburgers were low-rung grub, and Upton Sinclair probably had something to do with that. In 1906, his novel *The Jungle* laid open a host of nasty, stomach-churning, repellent, little secrets about the Chicago stockyards and the meat that came from them then, and it's understandable why diners took a pass on a patty.

In 1921, cook Walter Anderson and Edgar "Billy" Ingram opened the White Castle in Wichita, and they decided to serve burgers. In response to the dining public's reticence to eat hamburger, they cooked their square meat patties in an open kitchen so that people could see the meat on the grill and watch their meal as it cooked. This was just before the Depression, and a hamburger could be had for a nickel.

Shortly after White Castle opened its doors in Wichita, a young man named Harland Sanders (1890–1980) had gotten an offer he

White Castle restaurants were the first fast-food chain to be established in the United States, not McDonald's, as many people believe.

The distinctive McDonald's building got its "Mansard" roof design for the first time in 1969. PlayPlaces were introduced at McDonald's in the 1970s.

didn't dare refuse in another part of the country. Coming out of the Great Depression, Shell Oil offered Sanders a gas station to run for a cut of the profits, which turned out to be a great deal for them: just off a main road, the station was the perfect place to serve food, so Sanders cooked chicken and other tasty dishes for hungry travelers, and he worked on devising recipes that would set him apart from the competition. On that note, Sanders's competition was eliminated in a shoot-out in the early 1930s; someone was killed, someone else went to prison, and Sanders went back to work. His restaurant eventually became known as Kentucky Fried Chicken.

Even though quick, tasty food was ... well, quick and tasty, such fare didn't really take off until people did, in automobiles. With the popularity of cars came a consumer need for the convenience of fast food, and Bob Wian took advantage of that in 1935 when he turned his tiny hamburger stand into a drive-in in Glendale, California, and called it Big Boy. Later, the McDonald brothers, Maurice and Richard, opened their first drive-up restaurant in San Bernardino, California, in 1940. It must've been a popular year for food: the first Dairy Queen also opened in Joliet, Illinois, in 1940, and also that year, Harland Sanders had ultimately settled on the final ingredients of his "Secret Recipe" for chicken, but he kept it to himself until almost 1953, when he franchised his idea.

The distinctive roof design of the early McDonald's restaurants had an arch on either side of the building like this one in Fresno, California.

HANDY FACT

The Handy Science Answer Book, 4th edition, says, "An empty stomach has a volume of only 0.05 quarts (50 milliliters). A full stomach expands to contain 1 to 1.5 quarts (a little less than 1 to 1.5 liters) of food in the process of being digested." If you can imagine a two-liter bottle of soda, roughly half full, you get the idea.

The coincidences keep coming: in 1953, the "golden arches" became a branding logo with the opening of a McDonald's restaurant in Phoenix, though it should be noted that "golden arches" were an architectural feature of other McDonald brothers' burger restaurants. A little over a year later, James McLamore and David Edgerton opened their first Burger King in Miami; a year after that, businessman Ray Kroc joined the McDonald's company and started quickly to put the wheels in motion to buy the small chain from the brothers.

Today, Harland Sanders's eateries have been rebranded as "KFC" restaurants. Today's Dairy Queen is no longer known for just ice cream treats. McDonald's feeds more than 68 million people a day worldwide. And the fast-food franchise with the most locations in America is Subway, the first of which opened in Bridgeport, Connecticut, in 1965.

CULTURE: GO WASH YOUR HANDS!

Right there. That's probably one of the earliest mandates you had when you were a child: go wash your hands. We listened to Mom so well that the average American washes their hands nine times a day, with women doing it slightly more than men.

It's not hard to imagine a time when that wasn't important—but don't imagine too far. Though animals seem to have an instinct for hygiene, humans as a whole likely didn't have a regular or ritualistic urge to be clean for centuries into our existence.

The earliest records of washing come from 4,000–3,000 B.C.E. People in the Zhou dynasty in China learned that oils could be removed from the skin when a certain kind of plant was reduced to ash. At roughly this same time, early Egyptian women were starting to experiment with beauty products made from ores and oils, and ancient Romans were busy building aqueducts to bring fresh water into their cities for bathing. It's possible that ancient Mesopotamians knew at this time that having or allowing poor hygiene could result in sickness and disease; they also may have been the first to use ritualistic hygiene in a religious context.

In about 600 B.C.E., the Greeks were able to enjoy public baths and everyone was welcome, especially after they'd exercised. The Greeks were also probably the first to regularly use chamber pots indoors. Other cultures had begun to use twigs and other natural materials with which to brush their teeth; the Greeks and Romans made rudimentary toothpaste from ashes and pumice. Soap was refined, but modern minds might not think so: some of the ingredients that the ancients used for soap were beans, urine, blood, ashes, and animal fat. Just thinking about that makes you want to wash your hands.

The Greeks seem to have been the ones who thought up "hygiene," a word with roots in mythology.

By the beginning of the Common Era, the Romans used cesspits to rid their cities of waste. The British had begun to promote better oral care, and if that advice wasn't heeded, then a trip to the *barber*

Ruins of a Greek bath are well preserved in this photograph. The Greeks were among the first to emphasize the importance of hygiene in their culture.

was scheduled because they were the guys who took care of that kind of thing. Hygiene was hardly modern, but it was better than what folks had in the Middle Ages.

Even so, it wasn't as bad then as you've been led to believe.

On a regular basis—perhaps even daily—people in the Middle Ages kept themselves clean; if someone was a person of means, they might have had a sort of bathing tub in their home and access to hot water—a mixed blessing if there were multiple dusty people in the household and the water had to be shared; in that case, men went first, followed by their sons, their wives, female children, and then babies. Poorer folks bathed in public baths, streams, or buckets, and they used soap, although with little to no access to hot water, baths might have been less frequent, and who would blame them?

As for waste, chamber pots were still in use in the Middle Ages—especially in average homes and especially at night—but seated toilets were a whole lot less awkward to use. Royalty had fancier toilets than did common folk (and royal bum-wipers, to boot), though people of minor means who had to "go" did it by sitting on a wooden or stone seat suspended over a cesspit or some other kind of hole in the ground. If they were fortunate enough to be in a castle when the urge hit, a toilet hole might have led to a channel, which allowed waste to drop into the moat. And now you know why smart folks stayed out of the moat.

Another good thing about the Middle Ages: the Chinese came through with the invention of toilet paper in the 1400s. They also invented the natural-bristle toothbrush.

In the early Renaissance years, though, hygiene fell apart.

For a time, doctors actually recommended that people *not* bathe because it was thought to be better for their health if they didn't. Skin, it was believed, was made to protect the body from toxins (which is sort of correct), but submerging it in water supposedly negated that protection. Instead, people were supposed to rely on their clothing to wick away the nasty things their bodies produced or came in contact with. To be seen wearing brilliant-white linens was to be clean, never mind what the nose noticed.

To say that this stunk is an understatement. King James VI of Scotland was said to have worn the same clothing day in and day out for months and months. King Louis XIV is said to have taken two baths in his entire life—both on the advice of his physician. Queen Elizabeth I, it's been claimed, only bathed once a month. Lower-class citizens

HANDY FACT

The Handy History Answer Book, 3rd edition, says that Sir John Harington (1561–1612) is credited with the invention of the modern flush toilet. Such a nice guy, he then set about installing a new potty in one of Queen Elizabeth I's palaces.

Sir John Harington

"Though he was a serious scholar and translator, Harington was also a rebel who wrote controversial satire, leading to his banishment. His invention of the so-called 'water closet' was not taken seriously in its day. But over the following two centuries, various inventors worked to improve it, ultimately developing the plumbed sanitary toilet, a flush commode that is connected to plumbing and sewers or septic tanks."

likely didn't bathe *at all, ever,* and the use of other hygiene was basically nonexistent: hand washing was rare, lice were common, and because the floors of homes were repositories for all sorts of waste, bacteria and vermin flourished. It was the latter that proved to be humanity's undoing when the Black Plague was spread by fleas and lice in the 1300s through the 1800s.

The tide started to turn in the latter 1700s when cleanliness began to take hold in certain places and societies. By the middle of the nineteenth century, we had hygiene (almost) down pat.

Baths still weren't common every day, but they weren't forbidden, either. Folks bathed and washed when they needed to if not every day, and on the warmest of days, genteel ladies carried nosegays, which were literally hand-sized floral arrangements to help avoid the smell of unwashed bodies. Even so, most people strove to take a bath at least weekly; showers were invented and widely available about midcentury.

Though chamber pots were still used when needed, indoor toilets and running water became more common—rural folks often still had to rely on outdoor privies and water hauled inside by hand—but the methods of getting rid of human waste were still not well considered. Except in cases where cesspits were still used (as with an outdoor privy), human waste was dumped on the ground or elsewhere outside, and it often landed in rivers and lakes and, occasionally, near consumable water sources. In England in the mid-1850s, a heat wave caused the River Thames to dry up enough to reveal raw sewage, the stench of which caused Parliament to take action through a better sewer system. At around the same time, it was discovered that cholera was caused by inefficient removal of fecal matter near drinking water.

In post–Civil War America, doctors understood that cleanliness was a good thing, and they began promoting better health practices via hygiene. Upper- and middle-class homeowners wanted bathtubs installed along with their running water, and better methods of washing clothes were developed (although, in less prosperous homes, having multiple changes of clothing was a luxury), but poor families still had to rely on less-effective methods. To answer that, one city took steps: New York City opened public bathhouses in the late 1800s and gave each attendee a free cake of soap when they arrived. The number of manufacturers of soap increased, as did the number of soap brands. Advertisers took note and began marketing products that would help rid delicate bodies of "B.O." and "halitosis," both of which are words coined to name maladies that weren't named before the early part of the twentieth century.

Today, the hygienic pendulum seems to have swung in the opposite way: scientists and physicians say that we take *too many* baths and showers and that the antibacterial products we use in our quest to be clean will backfire as bacteria develop resistance to our efforts to eliminate it—if it already hasn't.

HISTORYMAKERS: FUNERAL FOR A LEADER

Here's something sobering: every person to ascend to the office of president of the United States has an unpleasant task to do almost immediately after taking office: they must imagine what their funeral should be like.

It's almost definitely not fun, but it does allow him (or her) one final chance to leave behind reminders of their legacy. The funeral can be simple or grand, it can include any number of flourishes, and it should include things that were important to him or her personally and while in office. Time and locale of lying in state (if that's the kind of ceremony chosen) can vary, motorcades can be ordered, pallbearers (other than military members) can be named, even the color of the funeral drapery is chosen. No detail is too small, and few things are impossible.

State funerals for presidents commonly include official notifications to each branch of the military; public viewing in the individual's home state and in the National Cathedral; a funeral procession along Constitution Avenue to the Capitol; and a closed-casket period of lying beneath the Rotunda before the deceased is interred in the place he or she (or the family) has chosen.

In the last fifty years (and at the time of this writing), America has seen the death of six of its presidents: Harry Truman (1884–1972), Lyndon Johnson (1908–1973), Richard Nixon (1913–1994), Gerald Ford (1913–2006), Ronald Reagan (1911–2004), and George H. W. Bush (1924–2018).

In the days after his death, Harry Truman's family opted to have a private funeral at the Harry S. Truman Presidential Library and Museum in Independence, Missouri, which is where former president Truman is buried next to his beloved wife, Bess.

Richard Nixon's family, reportedly fearful that the controversy of his presidency might taint the respect he was due, decided likewise to hold a private funeral at the Richard Nixon Presidential Library and Museum in Yorba Linda, California, where he is buried next to his wife, Pat.

Johnson, Reagan, Ford, and Bush all opted to have state funerals, which are official funeral rites conducted by the federal government. State funerals are only offered to deceased presidents, a president-elect, and anyone else who has been deemed to have given significant service to the country. The commanding general of the U.S. Army Military of Washington administers a state funeral and all its pomp, but the basic planning of such is done by the family of the deceased.

Barely a month after Harry Truman died, Lyndon Johnson suffered a heart attack and died at his Texas ranch. His body was taken to Washington, D.C., where he lay for two days in the Capitol Rotunda. Flags were lowered for thirty days at half-staff following the thirty-day period meant for Truman, and after a funeral at the National City Christian Church, Johnson's body was flown back to his ranch in Stonewall, Texas, where he is buried in the family cemetery near his wife, Claudia (Lady Bird).

411

The first state funeral was held for William Henry Harrison (1773–1841), who was also the first president to die in office—a mere 32 days after taking the oath of office.

When Ronald Reagan died in June of 2004, thousands of people filed past his casket as he lay in state beneath the Capitol Rotunda. During the funeral at the National Cathedral, George H. W. Bush (Reagan's vice

The casket containing the remains of President Ronald Reagan is shown here in 2004. Lying in state at the U.S. Capitol Rotunda is an honor reserved for presidents and a select few other important figures.

president), former Canadian prime minister Brian Mulroney, and former British prime minister Margaret Thatcher all lauded Reagan as dozens of world leaders listened from the pews. Not surprisingly, and perhaps because it was one of the first large international events to be held following September 11, 2001, Reagan's funeral was one of the first former presidents' ceremonies to feature an incredibly large and powerful security force at the site. In this case, the Department of Homeland Security assumed control of Reagan's funeral, superseding all other plans.

Former president Gerald Ford apparently put very careful thought into his funeral: he lay in state in three places before his interment: Palm Desert, California; Washington, D.C., at the Capitol; and Grand Rapids, Michigan, where he is buried at the Gerald R. Ford Presidential Museum. Ford requested that he lie in state in both the House and the Senate, but he eschewed a funeral procession along Constitution Avenue, instead asking for a motorcade through Alexandria, Virginia, and through the National Mall near the World War II memorial. He requested that his casket be placed on the historic bier that was constructed for Lincoln's funeral while he was in Washington.

When Lincoln was carried across the street from Ford's Theatre and undressed to facilitate treatment, it was found that he carried a pocketknife, a wallet containing a few newspaper clippings, some Confederate money, eyeglasses, a watch fob, a cufflink, and a handkerchief.

Following his death in late November 2018, former president George H. W. Bush's body was flown to Washington, D.C. Almost immediately after his arrival, a short ceremony was held for him, members of Congress, the Supreme Court, and his family before the public viewing in the Capitol Rotunda. His body was moved to the National Cathedral on December 5, which was declared a National Day of Mourning; the funeral was attended by the sitting president, Donald Trump; three former presidents, Jimmy Carter, Bill Clinton, and Barack Obama; and many world leaders. After the funeral in Washington, Bush's remains were flown back to Texas for a private funeral service, then to College Station, Texas, where he is buried at the George H. W. Bush Presidential Library.

The first nonpresidential state funeral was held for Thaddeus Stevens (1792–1868), a Republican member of the House of Representatives from Pennsylvania and a fierce champion of African Americans.

Thaddeus Stevens

HANDY FACT

According to *The Handy History Answer Book*, 3rd edition, there have been four American presidents assassinated while in office: Abraham Lincoln (1809–1865); James Garfield (1831–1881); William McKinley (1843–1901); and John F. Kennedy (1917–1963). Each of them had a state funeral.

CULTURE: IT'S NO BARREL OF LAUGHS

For most people, it seems like the highest of follies: to step off perfectly good soil and into a rickety craft in order to tumble, head over heels over head, down onto rocks below. And yet, dozens of people have tried to go over Niagara Falls in a barrel—and some even lived to talk about it.

Niagara Falls is the collective name for three waterfalls that run through Niagara Gorge from Lake Erie into Lake Ontario, along the border between Ontario, Canada, and New York in the United States. The largest of the three, Horseshoe Falls, straddles the international border between the two countries; it is also known as Canadian Falls. The two smaller falls, American Falls and Bridal Veil Falls, are both geographically located in the United States. During the daytime, more than

Niagara Falls State Park is America's oldest state park. It's open year-round, all day and all night. Visit www.niagarafallsusa.com for information.

Annie Edson Taylor is shown here with the barrel she used to take the falls challenge in 1901. She survived the plunge, though with some injuries.

six million cubic feet of water go over the falls; at night, water flow is reduced to take advantage of the falls as a power source. Collectively, Niagara Falls was created by glacial activity at the end of the last Ice Age.

Nobody seems to know for sure how Niagara Falls came to be named, but it definitely astounded early European explorers when they first came to North America. Word went back home to France and Europe that the falls were something to see ... and so visitors came. By 1801, Niagara Falls was a tourist attraction and a honeymoon destination for the rich. A hundred years later, it became a destination for daredevils.

The first person to go over Niagara Falls in a barrel was 63-year-old schoolteacher Annie Edson Taylor (1838–1921), who jumped in an airtight, cushioned barrel on October 24, 1901. She was no dummy: before she jumped in the barrel herself, Taylor sent a cat in a barrel over the falls first; when Kitty lived, Taylor knew she could do this thing.

As it happened, October 24 was also her birthday, and maybe she thought it was a lucky day: Taylor was a widow, and she needed the fame and fortune that would surely come with such a stunt. It took just twenty minutes to go from a rowboat above the falls to the very bottom, and while Taylor emerged bloody and with a head wound, she also said, "If it was with my dying breath, I would caution anyone against attempting the feat."

Alas, Annie Edson Taylor's intended fame and fortune never happened. She died near penniless and in eventual obscurity. What she did achieve, however, was a legacy that was copied—and still is.

Before we go any further, you might be asking yourself: "Why a *barrel*?" The answer seems to be because wooden barrels were cheap and easy to get back in the earliest years of this fad, because barrels floated (at least for a while if they remained intact), because open boats obviously weren't going to work so well, and because the most common materials that we'd use as cushions today hadn't been invented yet. Having said that, as you'll see by continuing to read, though a barrel was the object that caught the public's fancy the most, other crafts were eventually used for the ride.

411

While the exact number is unknown, it's estimated that more than 5,000 people since the mid-1800s have gone over Niagara Falls and were killed by it. Some fell by accident; most are believed to have been suicides.

In 1920, Englishman Charles Stephens (1862–1920) went over the falls with an anvil at the bottom of the barrel but was killed.

It took nearly a decade after Taylor's successful tumble for someone to figure out that a wooden barrel probably wasn't the safest vessel. Daredevil Bobby Leach (1858–1926) became the first man to survive the journey in a custom-made steel barrel, but he fared worse than did Taylor: Leach suffered two broken kneecaps and a busted jaw. It's ironic that Leach died of an infection resulting from a leg injury after having slipped on a banana peel.

In 1928, Canadian American Jean Lussier (1891–1971) devised a steel-and-rubber sphere with air bladders and oxygen tubes that took him over the falls successfully. In order not to have a repeat of Annie Edson Taylor's poverty-stricken post-trip, Lussier dismantled his craft and sold pieces of it to tourists and souvenir hunters.

Two years later, Greek immigrant and chef George Stathakis (1894?–1930) went over the falls in a vessel and survived the fall but not the journey: his barrel got trapped behind the falls with just eight hours of oxygen. Unfortunately, it was 18 hours before rescuers could reach him, and Stathakis was dead by then. He'd reportedly brought his pet turtle along for the ride; the turtle survived.

After that, it seemed that nobody had the stomach for going over the falls intentionally for many years. William "Red" Hill Sr. (1888–1942) had spent his life studying the Falls and working to rescue those who had gone over, whether intentionally or unintentionally. In 1951, his son, William "Red" Hill Jr. (1913?–1951), decided to do what his father had always dreamed of doing by going down the Falls in a vessel he made of inner tubes and netting. He did not survive.

And that was that. Because of the outcry that came from a needless, some said foolish, death like Hill's, Ontario premier Leslie Frost asked the Niagara Police to arrest anyone who tried to go over the falls in a barrel. Lawsuits notwithstanding, trying the stunt on a whim

HANDY FACT

According to *The Handy Science Answer Book*, 4th edition, "The water dropping over Niagara Falls digs great plunge pools at the base, undermining the shale cliff and causing the hard limestone cap to cave in. Niagara has eaten itself 7 miles (11 kilometers) upstream since it formed 10,000 years ago. At this rate, it will disappear into Lake Erie in 22,800 years."

See it now, while there's still time.

Niagara Falls

could cost you thousands of dollars in fines and possible jail time today because going over Niagara Falls in a barrel or any kind of vessel is still illegal.

BRITISH AND AMERICAN HISTORY: I LOVE YOU, I HATE YOU

Nothing says "I love you" quite like a gorgeous Valentine. It doesn't matter whether it's a fistful of flowers plucked from the backyard or a sparkly diamond—either can set your heart aflutter. In 2018, Americans spent nearly $20 million to wow their paramour.

'Twasn't always so, though. Once upon a time, it was just as common to send hate mail to your enemy as it was to send love letters to your sweetie on Valentine's Day.

Now, when you think about the Victorians, you might think of people who possessed gentility and manners—folks for whom protocol was mandatory and who demanded chastity. They were, after all, the ones who covered up piano legs lest anyone see them because, well, the word "legs" and all, you know.

So, you may be surprised to hear that the Victorians were all about sending vinegar Valentines.

Though Shakespeare mentioned Valentine's Day in *Hamlet*, sending paper cards to a beloved wasn't a fad until the eighteenth century. Paper cards weren't factory-made then, so what a prospective sweetheart could expect was a fancy, handmade card, perhaps embellished and including a poem or thoughts of love. Preprinted cards were available starting in the late 1700s and were commonly sent in the early 1800s.

Who knows why the whole thing began, but by the 1840s, insulting Valentine cards became available in both the United States and in Great Britain. And by "insulting," well, whoever got one got a great big In Your Face.

Made cheaply and mostly illustrated by artist Charles Howard, vinegar Valentines were basically plain postcards (open-faced,

A circa 1912 postcard depicts the Municipal Baths and Beach on Coney Island. All this one is missing is a "Wish You Were Here" caption.

so others could see what you got). They featured a mocking caricature or nasty drawing on one side, along with a poem or two-line verse that was meant to hurt or demean. Almost always, the cards were sent anonymously and, in many cases, or at least for the first few years of the fad, the recipient of the message had to pay for postal delivery.

Talk about adding insult to injury!

You might think something so crass might come from the lower classes, but vinegar Valentines were sent across all classes to both men and women of all ages. While it's true that some of them might have been lighthearted and sent as a joke or tease, most were meant to mock traits or habits—someone's looks, occupation, or even their intelligence might be insulted. They were handed out to barmaids and store clerks. They were given to unwanted suitors or inept admirers. Some were sent to soldiers. Neighbors sent them to neighbors. Outspoken folks and late-nineteenth-century activists (particularly suffragettes) were particular targets. Reports of violence as a result of a vinegar Valentine began to dot newspapers and police ledgers.

By the late 1840s, vinegar Valentines sold at the same rate as normal hearts-and-flowers-I-love-you cards, but the tide was turning, albeit slowly. In 1895, teachers began to see the cards as detrimental to young minds, and they tried turning the Valentine's Day holiday to thoughts of love again. By 1905, things had gotten way out of hand, and many of the vinegar Valentines were held up in a Chicago postal

HANDY FACT

"The French called the tomato *pommes d'amore* (meaning 'apples of love'). This ... name may have referred to the fact that tomatoes were thought to have aphrodisiac powers...."

—*The Handy Science Answer Book*, 4th edition

A circa 1900 "vinegar Valentine."

hub, unsent because they were deemed too vulgar for delivery. Even so, sending vinegar Valentines continued to be somewhat common into the World War II years, and you could still purchase them up until the early 1970s.

CULTURE: BRUSH UP ON THIS HOBBY

What hangs on your wall? When you were a kid, you might've had posters hanging in your bedroom, but now you've likely hung a framed print or a collage of photos that have meaning to you.

If you were a particularly lucky kid in the 1920s, though, you might have proudly tacked up your very own work, done with a very basic fill-in-the-block paint kit that Mom gave you. In the kit was a fragmented picture outlined in blue on some sort of cardboard with each small bit of the picture labeled with a letter of the alphabet that corresponded to one of the paint colors. It was foolproof for a kid just goofing around.

And then the Great Depression hit, followed by World War II. Nobody had the money, let alone the time, for hobbies, and resources were often funneled toward the war. At the end of the war, though, and in the final years of the 1940s, Americans felt free, fresh, and optimistic about the future. It was a new age, and they had access to more leisure than ever before, thanks to recently developed "time-saving" inventions and newfangled products. Coincidentally, at this same time, an idea for a fun new hobby that could tap into this new leisure was brewing at two different companies.

Paint-by-numbers kits are still available today and enjoyed by young and old hobbyists alike.

Dan Robbins (1925–2019) was working for one of the two corporations, the Palmer Paint Company in Detroit, Michigan, and he knew that his boss, Max S. Klein, was looking for new ideas. Robbins recalled his days of being an art student prior to World War II and the things he studied while in school; he later said that he also remembered Leonardo da Vinci's supposed practice of teaching his students to paint by allowing them to fill in numbered backgrounds in his work. This all led Robbins to play around with painting and to create a paint-by-number kit with the goal in mind of making his employer money by bringing art to the masses.

At first, the kits created a marketer's nightmare: sold under the brand name Craft Master, available only at department store S.S. Kresge and only in Detroit, the paint pots for two different paint-by-number kits were inadvertently swapped in their boxes, resulting in some pretty weird (and very definitely wrong) paintings. Buyers weren't happy at all and demanded refunds. By the spring of 1951, though, the bugs had been worked out and, with the help of a few friends who acted like excited buyers at the New York Toy Fair and helped create a buzz, the kits finally took off. Original sets were priced at less than $3.

Obviously happy with the final results and excellent sales, Robbins and Klein began to plan for the future and for a wider variety of kits. They hired a team of more than sixty Detroit-area artists, including a man who'd survived the Holocaust because of his talent, to work on the canvases; with guidance from Robbins and Klein, the artists created landscapes, figures, flowers, religious images, animals, wildlife canvases, buildings, and, later, personalized (and more expensive) kits

HANDY FACT

So, you think you're a good portrait painter? *The Handy History Answer Book*, 3rd edition, says that Titian (Tiziano Vecelli) is known as "the father of modern painting" because "during Titian's time (1488 or 1490–1576), artists began painting on canvas rather than on wood panels. A master of color, the Venetian painter was both popular and prolific. His work was so sought-after that even with the help of numerous assistants, he could not keep up with demand."

Titian

made from the photograph of a loved one. If there was any question of the possibility that a picture might be created or a color used, Klein had the final say.

Artists started with a painting, which was then reduced to the color blocks, which were then matched with the proper Palmer paint color. The kit was then tested by multiple people to ensure that the results would be reasonably uniform; later, kits were made for various levels of hobbyist interest and ability, including kits for kids that were nearly foolproof.

Despite that, art critics sniffed about the tasteless, lowbrow existence of paint-by-number pictures, and although home decorators were horrified that such pictures were framed and hung in living rooms everywhere, the fad took off. President Dwight Eisenhower was a big fan, and his work was hung in a gallery in the White House, along with paint-by-number works from other dignitaries, including FBI director J. Edgar Hoover, former president Herbert Hoover, and New York governor Nelson A. Rockefeller. Hollywood stars painted. Famous singers painted. Everyday dads and CEOs and housewives painted. Even Andy Warhol painted a painting of a paint-by-number kit. At the height of the fad, Palmer Paint Company was making some 50,000 paint-by-number kits each day; by 1955, obsessed Americans had painted nearly twenty million masterpieces. The most popular kit was *The Last Supper;* overall, we didn't like paint-by-number kits of abstracts or mere shapes.

Alas, like most fads, the craze for paint-by-numbers kits had pretty much wound itself down by the late 1950s. The market became saturated by copycat companies, Max Klein sold his share of Palmer-Pann (renamed from Palmer Paint in 1955), and the whole operation was moved from Detroit to Toledo, Ohio. General Mills purchased the company in the late 1960s. Dan Robbins died in 2019 at age ninety-three.

With the onslaught of the COVID-19 pandemic, paint-by-number kits found new customers in 2020.

And yes, they're still fun to do.

AMERICAN HISTORY: CAN YOU HEAR ME OKAY?

Chances are that when you got up this morning, you started your day with the news or weather from your favorite TV or radio station, or maybe both. Though they may call themselves Mix96 or Cowboy105 or The News Ten Team, the place you got your news from is really officially known by four letters, starting with either a "W" or a "K." So, what's the deal?

It all started over a hundred years ago when dozens of countries met for the International Radiotelegraph Convention. What they discussed, among other things, was the need to identify each country's radio signals. The United States was assigned several letters, but the "W" and the "K" were set aside by the government for commercial broadcast stations. In 1923, it was decided then that all stations east of the Mississippi River would use "W" and those west of the Mississippi River would use a "K" as the first of four call letters. Stations that were already using an opposite letter were allowed as exceptions.

At first, just three letters were required, total. That's why you probably know about WGN (Chicago), WRR (Dallas), WHO (Des Moines), and KOA (Denver). It didn't take long to realize that a mere

An app icon for WGN in Chicago. Older stations only needed three letters in their call letters, but most stations now have four.

HANDY FACT

According to *The Handy Science Answer Book*, 4th edition, Nikola Tesla had "over one hundred patents, among which are patents for alternating current and the seminal patents for radio. Tesla was responsible for many other innovations, including the Tesla coil, radio-controlled boats, and neon and fluorescent lighting."

Nikola Tesla

three letters would exhaust the possibilities too quickly, so the government changed the rules in April of 1922, requiring four letters, including the "K" or "W." The last three-letter assignation went to a radio station in North Carolina in 1930.

Although most of the first assigned letters were randomly selected, then, as now, letters after the "W" or the "K" could have meaning. For instance, a new station owner could ask to be assigned prefix letters to indicate a news network, the owner's initials, the city or state in which the station is licensed, or even letters to indicate the kind of things they plan to air. Call letters WLS, for instance, originally stood for "World's Largest Store" because the station was founded by department store chain Sears, Roebuck and Company. Other stations might just let the Federal Communications Commission (FCC) choose call letters at random and see what happens.

Like the individuality of fingerprints, no two radio stations can have the same call letters, as assigned by the FCC. Low-power or web-only stations are not regulated by the FCC and can therefore get away with copying other low-power stations' call letters. To further complicate the understanding of a panoply of regulations, a radio station and a TV station can share the same call letters, as can an AM station and an FM station, but the important distinction is generally made by "-FM" or "-TV."

Of course, there are always a mess of exceptions, permissions, and allowances to all of the above, so if you're thinking about throwing a noninternet, over-the-air radio or TV station up in the basement of your house, it's best to check the laws.

CULTURE: SUCH A CARD!

Hey, it's your birthday, and although you rarely get much in the mail on a normal day, today was a big day for your mailman. You got a handful of cards today, each one wishing you a good birthday and making you smile. So, whose big idea was it to send a stiff piece of paper through the mail in the first place, huh?

Like a lot of fun things, it all started with the Chinese and Egyptians.

In ancient times, they were the ones who thoughtfully sent good wishes on paper from one household to another at the New Year. At roughly the same time in history, the Egyptians sent papyrus scrolls filled with salutations from friend to friend. And that's about all she wrote for hundreds of years until the Germans began sending handmade, wood-cut New Year's greetings in around the year 1400. A few years after that, other Europeans began sending Valentine missives at the appropriate time.

During the Renaissance, when literacy increased and art and literature flowered, the sending of paper greetings grew, too. By the later 1600s, books were created to help the uninspired to get words flowing, artists were in demand to decorate the message with images of love and happiness, and adventuresome folks played with materials other than paper with which to craft their message. This, of course, meant that the card *was* the gift: one-of-a-kind, unique, well-made, and worth keeping for as long as possible.

Those paper gifts, as you might have guessed, were not cheap. Fortunately, however, things were changing: by the early 1800s, the sending of Valentine cards became more common for Europeans, in part because the cost of purchasing them and the price to send them was low. Even so, it took nearly half a century before the first recorded Christmas card was made in London, created thanks to Sir Henry Cole (1808–1882) and his artist friend, John Calcott Horsley (1817–1903). In about 1850, Esther Howland (1828–1904) began making and selling Valentine cards that were lovely and not at all pricey for U.S. consumers. By 1880, Louis Prang (1824–1909) of Baltimore began holding annual competitions for artwork for his iconic Christmas cards.

When soldiers left home to fight in World War I, the use of greeting cards surged in the United States. As Americans moved around during the Great Depression, cards were a cheap way to keep in touch

An 1889 advertisement for Prang Valentine's Day cards. While his Valentine's Day cards were popular, he was even more famous for his Christmas cards.

with far-flung loved ones. World War II saw another surge of card use for our boys overseas; it was during the war that greeting card maker Hallmark officially sealed the card in our culture with the slogan "When You Care Enough to Send the Very Best."

We've never looked back.

Today, even though messages and letters are often sent electronically, there's still something about getting a card from someone. Canadians send some 40 million Valentines each year; 90 percent of all American households buy Christmas cards to send, with an average of 30 cards per household.

Eight out of ten greeting cards are purchased by women.

BRITISH HISTORY: STICK A FORK IN IT

So, as it turns out, your camp counselor or camping buddy was right: fingers really *were* invented before forks. So were knives. So were.... Well, read on.

Watch any small child eat, and you'll get a glimpse of history: we've eaten with our fingers in our food for millennia. Sometimes, our ancestors had strict rules for *which* fingers were allowed or which hand one could bring to one's mouth (which is still true in some countries). Sometimes, ritual washing was required before communal dining (imagine if not!), and in time, other expectations for eating were set, including the use of napkins, tablecloths, and finger bowls. Today, in many cultures (sometimes, even our own!), eating with one's fingers is the only way to go.

For a long time, scientists have said that spoons were our oldest utensils: first, perhaps, spoons of bread to sop up a meal, followed by shells or horns that were shaped as needed for the cook more than for an individual. As early as 1000 B.C.E., the Egyptians added handles to make spoons more usable. Later, spoons were made from wood; the word "spoon" comes from a Germanic word that means "shaving."

Once people learned to make things from ores such as copper and iron, they began making spoons and knives specifically for the individual. It was common, then as now, to make the utensils with some sort of decoration; spoons might sport an engraving to inspire the diner. The first recorded use of a manufactured spoon comes from the mid-thirteenth century when spoons were for dining but were also meant to show off one's wealth and power.

Could it be possible that knives predated spoons? Yes, if you consider flint as a knife; it's not much of a stretch to imagine an ancestor pulling a hot piece of meat from the fire with a utensil that cut the meat up in the first place. Like spoons, the manufacture of knives took evolutionary leaps once copper, iron, and bronze began to be worked, but most of them were meant as weapons rather than for eating. The Egyptians and the Romans both manufactured spoons with sharply pointed handle parts, which may have been an early attempt at making knifelike skewers. By King Edward I's time (roughly the latter 1200s), being a knifemaker was an honored profession.

And yet, these things didn't entirely filter down to the masses. If, say, you were invited to dinner at the neighbor's house in the Middle

Ages, you would have sat down to a table without much more than food on it. Individual spoons were not yet widely used. Knives were for cutting and stabbing—and besides, you brought your fingers, didn't you? Your meal would have been put in a large kettle in the middle of the table (or servants might have circulated with platters, depending on the type of host you had), and you would have been given a trencher, or a thick piece of hard, crusty bread, to use as a rudimentary plate. When you were nearly done with dinner, you might've eaten the trencher or, more likely, your host would have collected everyone's trenchers from the table and given them to the poor.

By the fifteenth century, table knives were common on English tables, although guests and travelers usually brought their own knives with them. By that time, too, spoons became hot gifts for certain times of a person's life: grooms made them of wood and gave them to their brides on their wedding day, and Apostle spoons depicted tiny statues of the twelve Apostles on the end of each respective spoon's handle. Apostle spoons were often given as gifts to babies at birth. Knives were handy things to have, but spoons, as a whole, were so valuable that they were specifically mentioned in the wills of wealthier people.

Nothing, however, says Johnny-come-lately to a meal than a fork.

The ancient Greeks, Romans, and Egyptians surely used a pronged instrument to flip and cook meat over the fire, probably beginning in the seventh century if not earlier. Back then, they were massive and long in order that the user could lift large slabs of meat and not be singed doing it. Legend has it that a Middle Eastern princess made quite a splash, then, when Europeans learned that part of her dowry consisted of miniature versions of the cooking utensil.

The first time a human used chopsticks is a historical mystery, but it might have happened as early as 3,000 B.C.E., according to some sources; other sources say that chopsticks didn't appear until about 1200 B.C.E. Both of those were likely used for cooking. For sure, by about 400 B.C.E., folks in China widely used chopsticks for eating meals. The earliest chopsticks were likely made from bronze or some other metal, although today, they're traditionally made with unfinished bamboo. The exceptions are Korean chopsticks, which are made of metal.

What do you call your evening meal? "Dinner" is generally the most-used term to indicate your final meal of the day, but calling it "supper" might indicate something about your ancestry. Long ago, most Americans ate their largest meal during midday, but when we started working away from home from morning till night, our largest meal shifted to be eaten later at night. Thus, the "dinner" meal, in certain places, moved to a later hour—but not always; farmers still ate their big meal at midday, and they had a light supper in the evening. More evidence: look it up. Dinner doesn't indicate the time of day so much as it indicates the meal; "supper" comes from old French and English, indicating the last meal of the day. Today, if you have "supper," you're probably from or have family in the Midwest and Plains areas.

Alas, the fork didn't take off; the legend goes on to say that the princess later died exactly due to her fork since it obviously insulted God. Indeed, a fork, then, with its two tines, looked a lot like the devil's trident. And so, until about the late 1300s, forks at the table were largely unknown.

Starting in the mid- to late 1300s, table forks began to show up on inventory lists. The Italians embraced the fork as a table utensil, and forks made their way to France a little over a century later. By the early seventeenth century, the British had decided that it was perfectly okay to stab their food to take a bite—starting, of course, with royalty and ultimately filtering down to the commoners.

In 1669, King Louis XIV of France decreed that pointed knives were illegal at tables in order to discourage violence (what interesting times they were, eh?), and so knives for meals were rounded off from then on in British homes and castles. By that time, dinner plates had been introduced and were regularly used at court. For the wealthy, napkins were once again de rigueur at mealtime.

By the 1800s, everyone in Britain and Europe was on board with three main eating utensils and had even created new kinds: there were short forks for salad and forks for shellfish, butter knives and meat knives, and there were even special spoons for use with special foods. In America, though, little of that had crossed the ocean: here, early Americans had forks for cooking, but using a fork to eat was some highbrow stuff. Most homes had one kind of knife that did double duty on and off the table. To top it all off, fingers were still used just as much as were utensils by everyday Americans.

HANDY FACT

The real beginning of eating utensils came about during the Iron Age. *The Handy History Answer Book*, 3rd edition, says that "the real advent of the Iron Age came not with the discovery of metal (in about 2500 B.C.E.) but with the invention of the process of casing or steeling it, probably about 1500 B.C.E. This happened when it was learned that by repeatedly reheating wrought iron in a charcoal fire and then hammering it, it not only became harder than bronze but kept its hardness long after use.

"The next technological improvement, which again meant a further hardening of the metal, was the process of quenching it, which involved repeatedly plunging the hot iron into cold water. It was only after this series of discoveries and inventions that the significant impact of iron on culture and civilization was appreciably felt."

WORLD CULTURE: PICKLES AND CHRISTMAS THINGS, PART I

As late as 1840, most Americans refused to have a Christmas tree in their home because the tree was considered to be a pagan thing. Christmas itself was thought to be a serious, somber holiday, and outward joy was forbidden, never mind being "merry."

With this as background, no one knows for sure which U.S. president first brought a Christmas tree to the White House. Some say that James Buchanan did in the late 1850s; others say that Franklin Pierce brought a tree to the president's home in 1853. Benjamin Harrison had the first recorded tree in the White House in 1889; five years later, Grover Cleveland was the first to enjoy electric lights on a tree.

Early Christmas tree fans surely loved the twinkle as much as modern fans do. But by the time Teddy Roosevelt stepped into the White House, a growing American anti-Christmas tree voice that started during the William McKinley Administration was getting louder.

The White House Christmas tree is traditionally set up in the Blue Room (President John F. Kennedy and First Lady Jacqueline Kennedy are pictured here in 1961).

People hated the idea that a living, rooted tree was being cut down for the sake of a few weeks' time in a bowl of water in someone's home. The impact on forests and land was debated, and many folks called for this desecration to end. Some even advocated for artificial trees.

One of the people in the no-tree camp was Roosevelt himself. TR, a famous and robust outdoorsman, had never allowed a Christmas tree in his home but, with six children in the house and with the idea that Christmas trees were especially for small kids, the expectation was that he might bend his rules.

Not so much—at least, at first.

In 1902, son Archie Roosevelt supposedly hid a small tree in a closet and put presents beneath it; his father knew but didn't comment. He might have done it again the following year and, again, TR

HANDY FACT

According to *The Handy Science Answer Book*, 4th edition, it's not that hard to tell what kind of Christmas tree you have: "An easy way to identify the various trees is to gently reach out and 'shake hands' with a branch, remembering that the pine needles come in packages, spruces are sharp and single, and firs are flat and friendly."

knew about it but didn't make a fuss. Perhaps that nonreaction felt like tacit permission because, in subsequent years, the Roosevelt children had a small, lighted tree or two in the White House, and TR learned to admire the trees.

CULTURE: TOMATO, TOMAHTO

Imagine that you've just been given the fruit of a vine that most people believe would kill you. You wonder if you might die, just holding this piece of . . . well, what would you call it? It's not food if it kills you. But *will* it kill you? You take a bite. . . .

For decades, legend has it that this all happened to Robert Gibbon Johnson in the fall of 1820. Johnson, it's said, had promised his neighbors that he was going to bite into the rounded object, and so they gathered with breaths held in anticipation. Johnson, however, knew full well that the fruit he had in his hand wasn't going to spell the end of his life; in fact, he knew it was going to be delicious. And so, he took a big bite of a tomato to the horror of onlookers in Salem, Massachusetts.

Great story. Except, it's probably not true. And to further the misinformation, there have been hundreds of similar stories floating around through the years, including one that attributes the introduc-

Europeans once believed that tomatoes were poisonous and should not be eaten.

tion of the tomato to Thomas Jefferson (which isn't true but, incidentally, Jefferson did actually introduce America to many of his favorite foods, including ice cream and French fries).

The truth is that the tomato took a very long journey to your table.

Says Andrew F. Smith in his book *The Tomato in America*, the first European mention of a tomato happened in 1544, when Pietro Andrea Mattioli referred to "golden apples" because the tomato he knew was yellow. As a very early botanist, he classified tomatoes with the mandrake in the nightshade family but, unlike others in that classification, he never claimed that it was poisonous. Instead, Mattioli thought that maybe the tomato was an aphrodisiac, and he offered tips on cooking and serving it.

Eleven years later, tomatoes were known to be red (at least sometimes) and were called "love apples." In an attempt at further explanation (and, perhaps, exoticness adding romance and a good reason to take a bite), Renaissance herbalists declared that the fruit came from any one of several less-traveled areas—an argument that, by the way, wasn't settled completely until the middle of the twentieth century. Renaissance scientists didn't know it then, but tomatoes first appeared millennia ago on the west side of South America in the New World and on the Galapagos Islands. Of course, it was natural that early people living in what is now Mexico would find them.

If you've ever bitten into a juicy, ripe tomato, you'll understand why the Aztecs loved this fruit, naming it "xitomatl," which differentiated it from the fruit they called "tomatl," which Smith says were small and bitter but looked like the tomatoes we recognize. The Aztecs introduced the xitomatl to the Spanish during the Conquest; they happily took the seeds home with them to Europe, cultivated the plants, and shared them all around. By the late sixteenth century, tomatoes were being grown in England for food; in the earliest years of the seventeenth century, records show that tomatoes were used for cooking in Italy and Spain. A century later, they were also used for medicinal purposes.

HANDY FACT

"Thomas Jefferson (1743–1826), author and signatory of the Declaration of Independence and third president of the United States, was a self-proclaimed Epicurean."

—*The Handy History Answer Book*, 3rd edition

So, how did the tomato go from medicine to poison?

In 1728, Smith says that botany professor Richard Bradley claimed that the tomato plant was pleasant to see but that most tomato fruits were "dangerous." There might have been an element of truth to this, but it wasn't the fault of the tomato: in the 1700s, wealthy diners ate from pewter tableware, which had a high lead content; the acid in the juice from tomatoes might have caused the lead to leach out at high volume, thus poisoning those who ate the fruit. Tomatoes, in other words, got a bad rap.

Other scientists, however, very much disagreed with the tomatoes-as-poison theory; people were still consuming tomatoes in various ways and nobody died from it, including one Jewish physician who took tomatoes to the colony of Virginia from his home in England in the middle of the eighteenth century. Since there were, by then, several scientists who claimed to have seen tomatoes growing in various places in the New World, it's possible that tomatoes came up from Jamaica, Barbados, or South America, or here by another ship or ships from Europe. One thing's for certain, though: by the close of the eighteenth century, Bradley had been proven wrong, and tomatoes were seen in all kinds of dishes, both in America and abroad.

CULTURE: COOKING WITH COLOR

Back a couple centuries ago, making a meal was quite the undertaking: the cook (or an assistant, if she had one; it was almost always a "she") brought in the firewood, banked the wood, and lit the fire. For a while, meals were cooked on open flames and then on some sort of stove that still required wood and tending. Making sure the food was cooked was dicey, and it required close monitoring—not long enough in the fire, and it could make someone sick; too long, and you had a ruined meal.

And then came electricity.

Having an electric stove in the house meant more uniform meals since they could be boiled in a pan on the cooktop or baked in an oven that stayed at a regulated temperature. No more guesswork on cook time. Not only that, but keeping food uniformly (and safely) cold

By the 1950s, kitchen appliances such as this stove were available in an interesting array of colors, including turquoise, yellow, brown, blue, and pink.

was possible, too, with an electric refrigerator. No more melting ice mess from an icebox. Electric stoves were first patented in the late 1800s and the first fridge in 1913; neither was widely in use until the 1920s, and cooks must've rejoiced then at the clean lines of their modern, new, all-white appliances.

Following the austerity and rationing of World War II, Americans were better off financially and looking at the future. That, of course, meant a promise to housewives that new appliances would change their lives, and in 1955, General Electric introduced the beginning of the change with their "Mix-or-Match Colors." First Lady Mamie Eisenhower was said to have truly liked petal pink (pink was Mrs. Eisenhower's favorite color); other colors were canary yellow, turquoise green, cadet blue, and wood-tone brown. Sherwood green was also available and, of course, the owner of a newly remodeled, nicely updated kitchen could still have white appliances ... but why would any cook want white when she could have bold, new color in her kitchen?

As the 1950s morphed into the hip 1960s, kitchen colors shifted but only just slightly. If you still loved the bright colors, your appliances could still sport pink or turquoise, or your contractor could get them to match the cabinets in your room, but the cook who wanted the most up-to-date kitchens were more apt to pick appliances (including dishwashers) in harvest gold, avocado, or coppertone, the first two being the ones most people remember with fondness. Those three colors stuck around in the 1970s and were joined by poppy red, coffee, and orange (or "sunflower," if it was a Kohler model) appliances, often matched with darker brown cabinets and bright, floral, red-yellow-orange wallpaper on the walls.

HANDY FACT

If you're like many cooks, your family and friends gather in the kitchen. According to *The Handy History Answer Book*, 3rd edition, the Kitchen Cabinet "was the name given to President Andrew Jackson's unofficial group of advisers, who reportedly met with him in the White House kitchen.

"The Kitchen Cabinet was influential in formulating policy during Jackson's first term (1829–1833), many believe because the president's real cabinet, which he convened infrequently, had proved ineffective."

By the late 1980s, though, color was passé in new kitchens, although it seemed to leave with a whimper, as evidenced by the appearance of bisque and off-white appliances in new kitchens then. Neutral was where things were heading; black appliances made an appearance, and white made a comeback then. This trend continued through the 1990s and was followed by the sleek, ultramodern look of stainless-steel appliances at the dawn of the twenty-first century.

Today, you can still find some places that will build or make color appliances if you aren't able to find one that's vintage. If you're dying for a pink kitchen, it might be worth looking....

BRITISH AND AMERICAN HISTORY: THE WRITE STUFF

Arguably, one of mankind's most important inventions was the writing or drawing utensil, without which we would have no way of communicating with anyone now or in the future and no way of knowing what our forebears once said, felt, and thought. Today, such instruments are so commonplace that you likely have one within a few feet of where you are at this very minute.

The very earliest writing tools were probably fingers, sticks, or rocks, made to etch or scratch images into harder surfaces. At some time in prehistory, early humans learned to make a sort of rudimentary red or black pigment made of coal or iron oxide paint, and so they used their fingers or brushes made of reeds on the walls of caves to depict action scenes of hunting prey. Crayons, you see, hadn't yet been invented.

Ancient Babylonians and Sumerians (and, possibly, the Chinese) discovered that they could etch symbols into soft clay and then bake the clay into tablets that would last if you were careful not to break them with everyday use. The Romans took that one step further and created wax tablets with which to use a stylus, which made the messages erasable but also, alas, too meltable. The Egyptians finally got it right when they realized that they could put ink inside a stylus made of a reed with an appropriately shaped end, but reeds didn't last very long, so in the seventeenth century, feathers—quills—were employed as writing instruments.

And thus it was for centuries.

PENCILS

In about 1565, a miner in England discovered a vein of graphite and, though there didn't seem to be a lot of uses for it, farmers found it handy for marking their livestock. Imagine the aha! moment when someone realized that the graphite—which was said to be sturdy enough to cut with a saw—would work as a dandy writing material that wouldn't ruin hard-to-come-by paper. The word chosen for this handy instrument was "pincel," which is Middle English meaning "little brush," coming from the Latin word that means "little tail."

At first, pencils were made of thin, handmade tubes of graphite wrapped in leather for a better grip and to protect a user's fingers from

Graphite was first used as a writing medium in the sixteenth century in England, and it was the Italians who came up with the idea to wrap the graphite in wood.

smudging. Later, the Italians created pencils made of graphite wrapped in wood; early pencils were made flat, so they didn't roll away from busy professionals such as carpenters. In 1795, French inventor Nicolas-Jacques Conté (1755–1805) added clay to the graphite to extend a dwindling supply of graphite due to an economic blockade from Great Britain, thus making pencils of various hardness grades possible and also thus inventing the pencil as we know it. Mechanical pencils—the kind with refillable lead—were patented in 1822 by Sampson Mordan (1790–1843) and John Isaac Hawkins (1772–1855).

Pencils, by the way, never did contain lead inside, not in the part that leaves a mark. If a pencil did contain lead, it was *on* the instrument, not *in* it: the lead came from the pencil's paint. In 1978, lead was banned as a paint ingredient in the United States.

It should be noted that erasers specifically meant to erase pencil marks were invented in 1770 by Edward Nairne (1726–1806) and Joseph Priestly (1733–1804); attaching an eraser to a pencil didn't happen until 1858 when Hymen Lipman (1817–1893) patented the famous addition.

According to folks who study this kind of thing, there are four basic methods of holding a writing instrument. The Dynamic Tripod is the one your teacher probably made you use: the thumb and first finger pinch the instrument, which is at a lesser angle and supported by the middle finger; the ring and pinkie fingers trail along as stabilizers. The Lateral Tripod is similar to the Dynamic Tripod in that the instrument is gripped by the thumb and first two fingers, but the thumb extends over the instrument more. The Dynamic Quadropod looks rather like you're pinching the pencil with the thumb and three fingers; the pinky finger stays next to the ring finger. The Lateral Quadropod is similar to the Dynamic Quadropod, but the thumb is crossed over the instrument. These four basic methods don't include the Underhand Grip, which is mostly used for artwork or sketching.

There is, of course, no "right way," and individuals will vary in any method of holding; just keep in mind that some psychologists think they can tell a lot about you by how you hold your writing instrument.

PENS

At about the time that graphite was discovered in a mine in England, the idea of using feathers as writing instruments began to take off. It helped that writing and reading were both becoming more important for more people.

Early quills were made most often from the largest goose feathers; if a writer could afford it and could find it, a quill from a swan

Quill pens were an early form of the writing instrument. A large feather from most any big bird could be used (pictured is a quill from a falcon).

or crow was preferred. In a pinch, one could use a feather from another large bird like an eagle or turkey, too. Making a quill could take time and effort, fire and stone … or it only required a sharp knife with which to sharpen or square off the feather's point and make a slit to hold the ink. Once that was done, one merely dipped the business tip of the feather in ink from a bottle or inkwell and scratch a few words on parchment before having to dip the pen again.

It was quite a process; in fact, to write a lengthy (or even not-so-lengthy) letter or a book took considerable patience using a quill pen because quills didn't last very long, and they constantly needed freshening. Still, that was what sufficed for several centuries. Let's just say, though, that they were good enough to have been used to write shipping logs, to order supplies, to sign the Magna Carta and the Declaration of Independence, and to write the original articles of the U.S. Constitution. Quill pens are so embedded in human history that today, they are set out at each place at the counsel table when the U.S. Supreme Court is in session.

In 1822, manufacturer John Mitchell of Birmingham, England, created a pen with a machine-made steel tip. Six years later, Josiah

Mason (1795–1881) created a pen tip that was replaceable; this invention, coupled with Mitchell's efforts, made Birmingham world famous for pen tips. (It sounds quaint now, but there it is.)

This pen-and-ink thing was great but not perfect: writing meant that one had to carry a quill and a bottle of ink around, which was unwieldy. In the mid-1800s, various inventors came up with a method of holding ink inside the pen via a small well; in 1884, inventor–manufacturer L. E. Waterman (1837–1901) came up with a better way to feed the ink to the pen's nib without a mess—a fountain of ink, if you will. A fountain pen.

In 1888, John J. Loud (1844–1916) received a patent for a pen that would write on rough surfaces using a ball-and-socket mechanism. The method was basically good, but the ink was not, and the pen was prone to clogging, and it was not suitable for letter-writing (which didn't seem to be Loud's intention anyhow). Still, because the idea was sound, newspaper editor László Biró asked his chemist brother to come up with an ink that would make the best use of the ball-and-socket point. Biró filed a patent in Great Britain for his old-new invention, then moved to Argentina with his brother and a friend, filed a new patent in 1943, and commenced selling pens to the British RAF. Following World War II, inventors around the world began to tinker with ballpoint pen designs.

Bear in mind that a pen with continuous ink that flowed just by writing, that didn't leave ink stains on one's hands, and that you didn't have to fiddle with was, for postwar folks, absolutely *revolutionary*. The first successful company in the United States, the Reynolds International Pen Company, founded by Milton Reynolds (1892–1976), introduced its new ballpoint pen at Gimbels Department Store in New York City on October 29, 1945, to great fanfare. The pen sold for $12.50 (over $185 today), and thousands were sold in a week's time.

And then the market folded. The fancy ballpoint pen was a disappointment—for a while.

In the 1950s, however, the ink was once again reformulated, and the points were made better with new kinds of manufactured steel. Prices went down considerably, and so did the need to keep a pen forever. Today, we throw away more than 1.6 billion pens that end up in landfills and litter.

You can still buy ballpoint pens; in fact, ballpoint pens are the most common types around. You can buy rollerball pens (similar to ballpoint but with different ink); gel pens (also a different kind of ink); markers (often with felt tips); stylus pens (with two different tips on each end); if you really want a throwback, you can buy fountain pens from specialty makers; or, you can always make your own pen if you have feather and a knife handy.

CRAYONS

Hate to break it to you, but the unique smell of crayons that you remember from your childhood? The scent that's listed as one of the Top Most Memorable Scents? That's from stearic acid derived from beef tallow—yes, from bovine animals, moo cows, cattle—used to give crayons a waxy feel.

The first crayons were made a long time ago, if you can employ the word (coined in 1644) loosely. Ancient Egyptians, Romans, and Greeks made art by heating pigmented beeswax, then allowing it to cool on the surface of stones—which was not exactly child's play. Later, crayons were made of charcoal and oils first, then dry pigment and oils, then dry pigment and melted wax.

In 1864, Joseph Binney founded the Peekskill Chemical Company, which created paints and other chemicals related to pigmentation. In about 1885, Joseph's son, Edwin Binney (1866–1934), and Joseph's nephew, C. Harold Smith (1860–1931), joined the company, and Joseph retired. The cousins didn't let any green grass grow beneath their feet: Edwin continued to tinker with chemicals and,

An assortment of crayons from Binney and Smith's company from the years 1903 to 1910 are shown here.

through experiment, developed the world's first dustless chalk, which made teachers all over the world awfully happy.

In 1902, Binney & Smith took their business to the next level through incorporation and, later that same year, introduced a marking crayon for industrial use. Extra to the business, Binney began working with his wife to develop colored crayons for children and others; they introduced Crayola crayons in 1903. The name "Crayola" famously came from the French word for "chalk" (*craie*), and -ola, or oily.

The first box of Crayola crayons was sold door to door in 1903 for a nickel (about a buck and a half today) and came with crayons in black, brown, red, green, yellow, purple, orange, and blue. Each crayon came in a paper wrapper that was hand-rolled around the wax cylinder because machinery to do it hadn't yet been invented.

Two years later, the product line had expanded to include boxes of crayons in eighteen different amounts and five sizes of crayon; only two of the latter exist today, a "standard" crayon that's most common and the chunkier version for tiny hands.

In 1958, a box of sixty-four crayons was introduced and included a built-in sharpener, both of which were coveted by every American third-grader for a decade. In 1993, parents could buy their child a box of ninety-six Crayola crayons, complete with new colors to celebrate the company's ninetieth year. Other Crayola products have come and gone through the years as well.

HANDY FACT

On the work of Rembrandt van Rijn (1606–1669), *The Handy History Answer Book*, 3rd edition, says, "Some thought it too personal or too eccentric. An Italian biographer asserted that Rembrandt's works were concerned with the ugly, and he described the artist as a tasteless painter. Rembrandt's subjects included lower-class people, the events of everyday life and everyday business, as well as the humanity and humility of Christ...." Rembrandt "was also known to use the butt end of his brush to apply paint. Thus, he strayed outside the accepted limits of great art at the time." And it's a safe bet that he never used a crayon to do it.

Rembrandt van Rijn

As for the colors themselves, there's a surprising argument behind many of them. Plain old blue has changed, says one collector. Colors disappear and reappear without much hoopla, while others are introduced with great fanfare. Fans are sometimes allowed to name colors—or, in some cases, rename them—and some colors have sported names that Old Masters artists would recommend. Nobody who grew up in the time of the civil rights movement can forget the kerfuffle over the "flesh" crayon. And who ever totally understood the difference between blue-green and green-blue, anyhow?

CULTURE: THIS PART'S GOT LEGS

In the earliest days of humankind, it's a pretty safe bet that nobody was concerned about keeping warm. Nobody until, that is, people started to migrate into colder climates, and then the concern to keep skin covered became an increasingly urgent issue. It was then that early humans covered their legs with animal skins, and though they probably didn't care a bit, the outfit looked more like furry breeches than stockings.

The first stockings were worn by the Egyptians, Greeks, and Romans, the people who seemed to do everything in ancient times. They wore socklike garments made of woven grasses or lightweight animal skins on their feet, garments that rather resembled boots and were intended more for warmth and protection than for modesty's sake.

The modesty came about in the fifth century B.C.E. when early Christians wore leg wraps to signify piety and pure hearts. By then, the Vikings, too, were wearing stockinglike leg warmers that were knitted, but purity was probably not on their minds; the Vikings were more concerned with warmth.

Spartan soldiers reportedly wore only cloaks to protect themselves from bad weather—no coats for them! They were almost always barefooted.

In around the tenth century, stockings became more than just utilitarian; they were a fashion item and a status symbol. Linen and silk stockings displayed the wealth of the wearer, especially since the lower classes used cheaper wool for their stockings. To keep the stockings from falling down (as annoying then as it is now!), garters were used,

These Egyptian socks—the oldest socks yet found by archaeologists—date back to between the fourth and sixth centuries C.E. They are made of wool and were designed to be worn with sandals.

or the stockings were attached to a pantslike garment with ties or fastened with sewn-in ties. Stockings were made in various lengths, they were getting longer, and they often sported embroidery as embellishment.

It should be noted that stockings (or hose, as they were starting to be called) were men's clothing then. Women, it was thought, didn't need any sort of leg covering because their leg skin was covered up by voluminous skirts and, by the way, stockings on women were somewhat scandalous. Medieval women, in fact, didn't wear undergarments at all, aside from a chemise (basically, a type of slip meant more to protect the garment).

By the late fifteenth century, hose had become more fancy pants, less sock. Men like Henry VIII (1491–1547) wore their hose long: stockings were joined in the back at the waist but were open in front. This left a bit of a problem because tunics were shorter and pantaloons were all but nonexistent; for coverage, then, a codpiece was worn over the genitalia. Over time, the codpiece was padded, decorated, and emphasized until everyone came to their senses and, by the Elizabethan period (late sixteenth and early seventeenth centuries), breeches were longer and stockings for men were more like socks.

It was Elizabeth I (1533–1603), in fact, who first popularized stockings for women. In about 1560, someone gave her a pair of knee-high, knitted, silk stockings, and she was said to have been very taken with them. This royal approval was all the encouragement needed for entrepreneurs to open knitting schools and factories, and in 1589, an English chap named William Lee (1563–1614) invented the stocking frame knitting machine.

William Lee, inventor of the stocking frame knitting machine, tried twice to get his device patented. Both times, Queen Elizabeth I turned him down, fearing that it would hurt the hand-knitting industry that had been making most of Britain's stockings. Alas, her denial meant the end of Lee's business in Britain, so he went to France and was successful there until the assassination of Henry IV of France in 1610—a blow from which Lee never recovered. After Lee's death, his brother took the machines back to England and tried to establish the system again, but the hand-knitting industry had control of the industry and fought to keep it so. This fight spilled over into the earliest part of the nineteenth century when a war between Britain and France led to financial unrest in Britain and protest from hand-knitters, who ultimately destroyed all the machines they could find. The protesters became known as Luddites, after the fictional weaver Ned Ludd, who was the leader of such mayhem; their efforts were seen as so alarming that in February 1812, Parliament made the destruction of knitting machines a capital crime. All because of stockings.

For the first time, it was okay for women to wear stockings for more than just warmth, and while most women wore stockings that closely resembled those that men wore, upper-class women were able to find stockings made with patterns or designs on them or, much later, topped by lace and held up with garters. Garters themselves were improved once elastic was invented in 1820, although few would know this since skirts were still worn long, and the sight of a feminine ankle was scandalous to some.

Alas, toward the end of the nineteenth century, men's trousers styles meant that they ultimately stopped wearing longer stockings in favor of shorter socks.

And that's the way it was until the turn of the twentieth century, when hemlines crept upward (for more on hemlines in history, see History—Culture: Who Wears Short Shorts), but it was still not quite acceptable to show bare leg. Hosiery hadn't lengthened much at this time and garters were still used to prevent slippage, but flappers took obvious delight in flaunting designs and size of the netting. It scandalized their elders; what could be better?

As it had been for some time, silk, rayon, and wool were preferred materials for stockings (the latter for everyday; the former for evening wear), but simmering tensions in Japan in the late 1930s and a U.S. boycott on Japanese products a few years later sent scientists to their laboratories in search of a better material to use for ladies' hosiery. The search didn't take long: in 1935, Wallace Hume Carothers (1896–1937), a DuPont chemist, had invented a synthetic material that showed incredible promise but, sadly, he'd suffered from depression and committed suicide before he could learn that nylon could be used for myriad things—and that included ladies' stockings.

Nylon stockings were first modeled at the 1939 World's Fair in New York, and by the

The brilliant chemist Wallace Carothers invented nylon but, sadly, committed suicide before his creation took off.

time they were officially released for sale to the public, word had spread, and millions of pairs of stockings were sold in a matter of days. Every woman needed not just one pair but several, and no well-dressed woman left the house without wearing nylons, usually held up with a garter, garter belt, or girdle.

While women—particularly teens—were content to wear rolled-down socks or anklets, nylons were so popular that in the early days of World War II, when rumors swirled that the material might become scarce or be rationed due to the war efforts and the making of tents and parachutes, there was a run on nylon stockings, and they soon became difficult, if not impossible, to find. Not a problem: women just got creative, going bare-legged but dyeing their legs with light brown stain made of household products or cosmetics made specifically for this need, then drawing stocking "seams" up the backs of their legs to fool the eye. Once World War II ended, DuPont announced that nylons would soon be widely available, but it took a year to catch up with the demand. Women were so hungry for nylons that, in 1945 and 1946, America saw "nylon riots," and it was common for women to stand in line all day to buy hosiery.

Though some Hollywood designers began connecting panties and stockings in the Golden Age of Film, pantyhose became available

HANDY FACT

"The process of silkmaking starts with raising silkworms on diets of mulberry leaves for five weeks until they spin their cocoons. Then, the cocoons are treated with heat to kill the silkworms inside (otherwise, when the moths emerge, they would break the long, silk filaments). After the cocoons are soaked in hot water, the filaments of five to ten cocoons are unwound in the reeling process and twisted into a single, thicker filament; still too fine for weaving, these twisted filaments are twisted again into a thread that can be woven."

—*The Handy Science Answer Book*, 4th edition

in 1959 and seemed something like a revolution. Pantyhose left no sags, were smoother under clothing, were easier to put on and take off, and, best of all, pantyhose allowed hemlines to rise and rise and rise again. Ultimately, they were more affordable, too.

More than half a century later, fashion has done an about-face on the subject of stockings once again. Unless there's an office rule against bare-leggedness or a religious or medical reason for wearing them, many women forgo hosiery because wearing stockings or hose is not as fashion-mandatory as it once was. And to add a twist to this saga, men have since discovered that pantyhose beneath jeans are warmer than not.

Stay tuned....

HISTORYMAKERS: STAGECOACH MARY

Here's a lesson for anyone who thinks it's "too late" in life to make a change: Mary Fields (1832?–1914) came into this world, historians believe, in Hickman County, Tennessee, around 1832—nobody is positive of the date or even the place because Fields was born a slave. By at least one indication, "Mary" might not have even been her first name.

Because of her status, nothing is known about Mary's childhood or her early adulthood; it's presumed that she was a slave until the end of the Civil War, when she headed north like so many freed slaves did, working her way up the Mississippi River on the *Robert E. Lee* as a laundress or maid. Some say that it was there that she met Judge Edmund Dunne. In the late 1870s, Mary found a job in an out-of-the-ordinary place: an Ursuline convent in Toledo, Ohio, where she found a deep connection with Mother Amadeus; it's possible, say some sources, that Mother Amadeus might have been related to or knew one of Mary's former owners and offered the woman shelter and employment.

But where the friendship started isn't as important as the fact that it strengthened. For the nuns at the convent, six-foot-tall, 200-pound Mary became handywoman, gardener, fix-it person, fetcher, hauler, and laundress. She did the work of a man but also helped the nuns inside the convent, as needed. When Mother Amadeus traveled to Montana with the intention of setting up a mission school there, Mary stayed behind in Ohio for a time, but when Mother Amadeus sent for her in about 1885, Mary happily packed up and headed for the frontier.

Doing basically the same work that she'd done in Ohio, Mary lived and ate her meals with the nuns at the new mission, but she also reportedly spent her evenings at the local saloon, drinking and

Mary Fields, the U.S. Postal Service's first black female star-route carrier, worked from 1895 to 1903.

arguing with the menfolk. Often dressing in men's clothing and smoking a cheap cigar, by all accounts, she was a brawler and took no nonsense from any man with or without a gun. Local men learned that one didn't mess with Mary lightly. Even many of the nuns trod carefully.

This led to complaints that reached the bishop's ear, and in about 1894, he banned Mary Fields from the convent. Once again on her own, Mary did what Mary had done probably all her life: she took in laundry and did odd jobs to get by. Some accounts also say that she was the proprietress of several different restaurants that all lost money because Mary gave away the food to those who needed it rather than sell it.

This reputation for self-sufficiency paid off when Mary—who was in her sixties then and quite literate—applied for and received a contract from the Postal Service to become a star-route carrier in 1895, making her the second woman and first African American to work for the Postal Service. From then until 1903, Mary never missed a day of work, carrying the mail between Cascade, Montana, and St. Peter's Mission. Her reliability gained her the nickname "Stagecoach Mary" because she was always on time, taking her mail via stage-coach over rocky roads in summer and snow in weather. It was said that if the snow was too deep for horses to get through, Mary delivered the mail on foot, wearing snowshoes. In that time, thieves were a common problem with mail delivery, but Mary's reputation preceded her in that respect, although she was reportedly inordinately kind to children.

In 1903, age and the hard conditions of the job got to Mary, and she finally retired to Cascade, where she made ends meet by taking in laundry. She died in Great Falls, Montana, in 1914 and, despite that she undoubtedly still felt the pain of racism every now and then, they say that her funeral was one of the largest the town had ever seen.

HANDY FACT

Delivering the mail on foot would have not been easy, no matter how strong Mary was: an average year in Cascade, Montana, sees 48 inches of snow. *The Handy Answer Book for Kids (and Parents)* says, "Ordinarily, 10 inches of snow has about the same amount of water as one inch of rain. But temperature affects this general rule."

CULTURE: AND THE KANGAROOS WIN IT!

Sports fans are known to be, well, *fan*atical about their teams' statistics; for some fans, for instance, one of the main benefits of attending a baseball game in person is keeping stats, and what fan can't name the last time their team went to the Super Bowl?

Yep, there's no doubt about it: fans love their teams, but why are sports teams named after animals or mythical beings rather than just known by their home? Why do sports teams even have mascots? What's it like to be inside one of those goofy costumes? And what kind of trophies can Bucky win, anyhow?

To begin, let's look at the root word: "mascot" comes from an old French word indicating charm or sorcery, also talisman. Literally, a "mascot" is a good-luck charm.

It's hard to say who (or what) was the first mascot, but one source indicates that it was an actual person, a small boy who col-

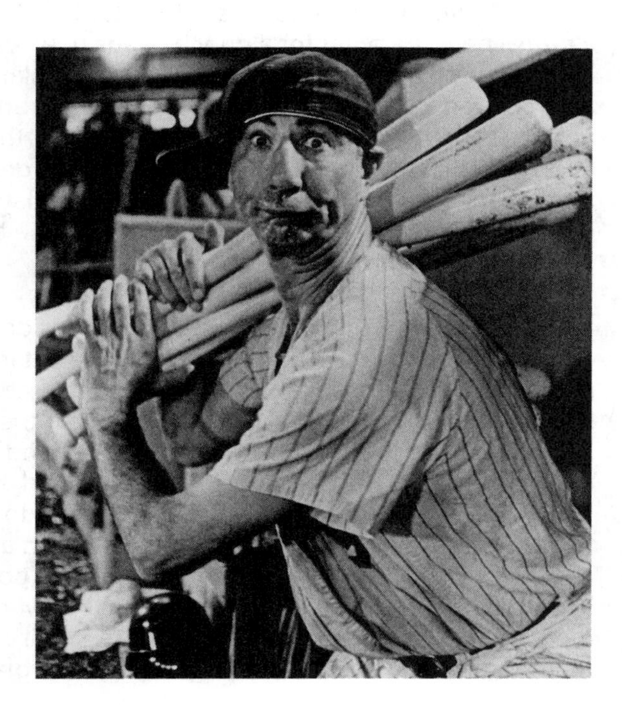

Max Patkin, the "Clown Prince of Baseball," entertained fans by joking around and horseplaying with Joe DiMaggio.

It should come as no surprise that animals are the number-one mascots (bulldogs are the most popular, then tigers), followed by humans of some super-sort or another.

lected bats and balls for a nineteenth-century baseball team. There are at least two references to "mascottes" from 1884; one of them was a live goat. This continued at least through the early part of the last century: the vast majority of mascots were children or animals, usually dogs.

During World War II, the idea of having a "mascot" to entertain the crowds began to take hold, starting with stadium-pleasing shenanigans that regular baseball team players would do during a game. The first "professional" mascot is said to be pitcher Max Patkin (1920–1999), whose silly horseplay with Joe DiMaggio (1914–1999) during an exhibition game was said to have launched Patkin's career as the Clown Prince of Baseball. Patkin didn't have a costume, per se; the first costumed mascots appeared in 1964: Mr. Met, for baseball, and Brutus Buckeye, for Ohio State University's football team.

It may come as a surprise to know that the widespread practice of having professional mascots—those larger-than-life, goofy-looking guys in costume—at a game is a relatively new thing. And in the beginning, it had little to do with sports....

In 1974, San Diego radio station KGB hired a cartoonist to create a funny, new character for a bunch of ads they were making; the guy came up with a chicken. The ads were so popular with listeners that they created a costume to match the drawing, and they hired a journalism student named Ted Giannoulas to wear the costume, dance, hand out prizes, meet and greet listeners, and generally have a good time. Giannoulas did such a great job his first day that he was asked to come to a Padres game to do the same with the team's fans—and he did it for forty more years.

Now, you might wonder what it's like inside one of those mascot costumes. In a word: HOT. It can get *mighty* hot in there. Sweaty is another word, possibly pretty smelly, and hydration is key if you're inside all that fur and foam. Most mascots wear the costume for a limited amount of time, and they usually have handlers, in fact, to ensure they have water nearby at all times, to ensure the human inside can see (because the mesh part of the head often doesn't correspond to the line of sight of the wearer, and there's often a lot of padding to a costume), to help get the costume on and off (potty breaks, you know), and to ensure that fans don't get too overzealous when they meet their beloved mascot. On that note, small children don't know that there's a human inside that grinning, fiberglass head, and they demand an extra caution: they can get hurt as easily as they can accidentally hurt the person inside a costume.

At any rate, though Giannoulas was a tough act to follow, it didn't take long for other baseball teams to want their own fun, fan-favorite mascots. By the 1980s, many college and pro football, baseball, and basketball teams had their own mascots. Hockey teams sometimes had them. Even the Olympics started to choose mascots. There's even a Mascot Hall of Fame. Mascots often—but not always—are named after their teams, but some mascots refer to local landmarks, history, or culture, or they sport a name that fans might be able to figure out.

Overall, though, the reason sports teams even have mascots can be summed up in one word: merchandise. A lot of fans will buy a lot of stuff to support their team, but they'll buy even more if their team's mascot is on it.

Prior to about the year 2000, ethnic groups were often used as mascots (for instance, the Red Raiders might have a Native American mascot, which was usually depicted in some sort of cartoon or caricature). Slowly, starting at the middle- and high-school level, schools began to be aware of the offensiveness of that, and they changed. At the time of this writing, renaming had trickled upward to pro sports and, in the year 2020, there were just a few major pro sports teams left who hadn't gone with a name and/or mascot change.

HANDY FACT

"The Heisman Trophy is the top individual award in college football. Named after former college football coach John Heisman (1869–1936), the award was created by the Manhattan-based Downtown Athletic Club in 1935, where Heisman was its director at the time. When Heisman died in 1939, the award was renamed in his honor."

—The Handy American History Answer Book

This Heisman Trophy was won by Oklahoma Sooners quarterback Kyler Murray in 2018.

HISTORYMAKERS: ONE LUMP OR TWO?

Few things can be more thrilling to horse lovers than the sight of a herd of wild mustangs running across a mesa. To a miner in the Arizona Territory in the late 1800s, however, the sight of an inordinately large, reddish horse was certainly terrifying.

Reports then of this animal that ran across the sand on cloven hooves that were bigger than a man's hand began to filter down to the news media. At first, few believed the reports—who didn't make up yarns now and again around the campfire?—but when rancher Cyrus Hamblin saw the creature, the mystery was solved.

The creature was a camel, the result of a failed U.S. government experiment. But this camel had human remains tied to its back, and though it escaped capture, locals were able to retrieve skeletal parts of some unfortunate victim of "The Red Ghost."

Old West Mythology? Or true?

What's true for sure is that camels were once an imagined part of settling the West. Back in the earlier part of the nineteenth century when America consisted of both states and territories, methods of binding the West (which was largely unexplored by white men) to the East (where the bulk of the population resided) via easy roads or routes were being blocked by many things, including distance, ease of travel, the fact that horses don't move well on sand, hostile natives,

A camel is seen in this photo taken at the Drum Barracks in San Pedro, California, c. 1863.

and a lack of interest from Eastern politicians to see the urgent need for a transcontinental railroad. In early 1853, Congress was very ready for someone—anyone!—to find an easily accessible new route, in part because of the gold miners, who were badly in need of an efficient way to get supplies to their outposts and camps.

When camels were first mentioned as a possibly efficient way to tackle desert sand, it didn't take much persuasion: Jefferson Davis (1808–1889) was already interested in using camels to help in the War Department. He'd known about employing camels for the troops since 1836, when Major George H. Grossman (1799–1882) used them during the Seminole War in Florida. After other experts weighed in on the use of camels and their advantages, Davis petitioned Congress in 1883 for the money for "a sufficient number" of two different types of camels (the Bactrian camel, with two humps; and the Arabian camel, with one hump) to see which would make a better military machine. In 1855, he received $30,000 for his experiment.

By late summer, the British, who'd heard of the herd-to-be, had purchased several thousand camels to use with their armed forces, and the Americans were the proud owners of 21 beasts.

Arriving in the spring of 1856, the camels were housed in Texas and were joined not quite a year later by forty-one more animals. Turkish and Egyptian handlers were hired, and in the summer of 1857, experiments were conducted to see how the camels would handle southwestern American climate, soil, and environmental situations. While the animals came through with flying colors, some Army officers had already been giving the whole project the side-eye, concerned that the camels wouldn't be properly cared for or

Bulgarian forces are shown here with a camel caravan during the First Balkan War in 1912. While successfully used in warfare before, camels were dumped by the U.S. Army by the time of the Civil War.

that the whole endeavor would be abruptly and unceremoniously scrapped should the then-current administration that approved them be removed due to the brewing War between the States.

Still, Edward Fitzgerald Beale (1822–1893) was appointed to take a platoon of soldiers, camels, handlers, and wagons from the New Mexico territory to California in the summer of 1857 to see how the animals reacted in real time. The problems getting there were many, but they weren't from the camels: the camels were sure-footed, unfazed by cacti and rocks, and easily swam in swollen rivers. The problems were with the handlers and other humans who liked to drink and carouse a bit too much along the trail.

Once in California, however, issues with the camels became apparent. Camels weren't exactly floating in perfume; in fact, they smelled *awful*. Other domestic animals were often extremely startled by the size and odor of a camel. It cost a lot to feed a camel, and the beasts weren't fit for quick trips. What's worse: they were noisy, recalcitrant, and could have unpleasant, often violent, tempers.

Although Secretary of War John B. Floyd (1806–1863) had recommended the government order a thousand more camels for military use in 1858, after the Civil War began, the military was abruptly no longer interested in camels as war machines. In the fall of 1863, Union secretary of war Edwin Stanton (1814–1869) demanded that the camels be put up for sale for whoever wanted them.

New owners, however, had the same issues with the camels as did the army, namely, that they weren't cheap to keep. For sure, several of them were set free to roam.

HANDY FACT

"The hump or humps [of camels] do not store water since they are fat reservoirs. The ability to go long periods without drinking water, up to ten months if plenty of green vegetation and dew is around to feed on, results from a number of physiological adaptations. One major factor is that camels can lose up to 40 percent of their body weight with no ill effects. A camel can also withstand a variation of its body temperature by as much as 14 degrees. A camel can drink 30 gallons of water in 10 minutes and up to 50 gallons over several hours. A one-humped camel is called a dromedary or Arabian camel; a Bactrian camel has two humps and lives in the wild in the Gobi Desert. Today, the Bactrian is confined to Asia, while most of the Arabian camels are on African soil."

—*The Handy Science Answer Book*, 5th edition

The last confirmed spotting of a feral camel was in Arizona in 1931, although rumors of wild camel sightings still swirl today....

HISTORYMAKERS: GOING TO THE CHAPEL

Every first-time bride knows that The Dress is absolutely necessary for the fairy-tale wedding of her dreams. They say "you'll know" when The Dress is on. They say The Dress will find you. There are even TV shows about The Dress.

When Jacqueline Bouvier (1929–1994) got engaged to John Fitzgerald Kennedy (1917–1963) in June of 1953, she no doubt began, as every woman with a sparkly diamond on her ring finger does, to dream about her Dress.

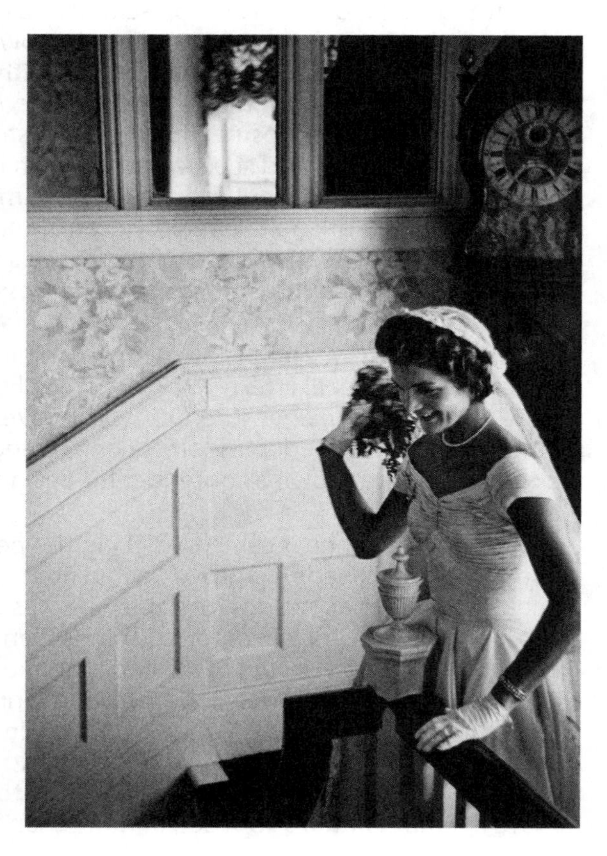

Jackie Kennedy is pictured here in her wedding dress designed by Ann Lowe.

Designer Ann Lowe (1898–1981) was not Jacqueline's first choice. Lowe probably wasn't even Jacqueline's second choice—but she was always in the running because Lowe was then very highly in demand by all of New York's best-dressed society ladies. Every woman who'd spent her existence wearing haute courier for the previous twenty-some years had wanted at least one of her designs.

Looking back, it almost seemed as though Lowe had been born with a needle and thread in her hand: her mother and grandmother had both been seamstresses for Alabama's wealthy, and Lowe had fond childhood memories of sitting on the cutting-room floor, amusing herself with cloth scraps. Lowe was only sixteen when her mother died, and there was an order of ball gowns outstanding and unfinished, so she did what seemed right: she finished her late mother's work and made the deliveries on time. Still, life was not easy: when Lowe showed up for class at an elite New York fashion school, administrators immediately segregated her from other students. They'd allowed her to attend the school by mistake, not knowing at the time of admittance that Lowe was black.

After graduation from fashion school and having worked and further honed her craft for a time in Florida, Lowe opened a studio in Harlem, where she quickly became popular with the likes of Oscar winners, society ladies, debutantes, Roosevelts, Rockefellers, and DuPonts. Wanting more exposure, she opened a store on Fifth Avenue in New York City in 1950, and she started her own label. Known as one of New York society's "best kept secrets," Lowe admitted to *Ebony* magazine in 1966 that working for wealthy clients made her "an awful snob" because, after a taste of being in demand, she simply was not interested in seeing her couture on the bodies of less wealthy folk.

That last part was no problem for Jacqueline Bouvier, daughter of Black Jack Bouvier III (1891–1957) and socialite Janet Norton Lee (1907–1989). Bouvier, equestrienne, world traveler, and a socialite like her mother, had enjoyed a casual working relationship with Lowe for some time. But was Lowe a contender for designer of The Dress?

Reportedly, when it came to picking the person who would make Jacqueline's wedding gown, Jacqueline's mother had in mind someone different as the designer, but she was overruled by the bride. Given Jacqueline's love for All Things France, it should come as no surprise that she very much desired a simple gown of French design, but her father-in-law-to-be, the domineering and controlling Joseph Kennedy Sr. (1888–1969), had a final say in the entire wedding, right down to the clothing, and he put a stop to any notion of an overseas gown. And so, Lowe was ultimately hired to transform more than 50 yards of taffeta (that's 150 feet, or more than three times the length

HANDY FACT

According to *The Handy History Answer Book*, 3rd edition, the Kennedy presidency was called "Camelot" because "shortly after President John F. Kennedy was assasinated (November 22, 1963), the former first lady was talking with a journalist when she described her husband's presidency as an American Camelot, and she asked that his memory be preserved. Camelot refers, of course, to the time of King Arthur and the Knights of the Round Table and has come to refer to a place or time of idyllic happiness. John Fitzgerald Kennedy's widow, who had with fortitude and grace guided her family and the country through the sorrow and anguish of the president's funeral, quite naturally held sway over the American public. So, when she suggested that the shining moments of her husband's presidency were reminiscent of the legends of Camelot, journalists picked up on it. Despite subsequent revelations that there were difficulties in the Kennedy marriage, public opinion polls indicate that the image of Camelot—albeit somewhat tarnished—has prevailed."

After the assassination of JFK, the first lady described the brief period of American history as a kind of idealistic Camelot.

of an older ranch-style home) into the wedding gown alone, not to mention what was required of the other gowns that Lowe and her assistants made.

It would be a wedding that brides still talk about. For an African American woman in pre-civil-rights-era America, the project was a major thing.

But with ten days to go until the wedding day, the unthinkable happened: a water pipe in Lowe's studio burst, sending water all over two-thirds of the gowns, including the wedding dress itself. Working day and night, Lowe and her assistants re-created and remade each of the ten ruined dresses at Lowe's expense, which meant, in the end, that she lost money on what could have potentially been her most influential contract.

And then it got worse.

Always angling for publicity, Joseph Kennedy had invited reporters from the major news outlets of the day to attend the wedding, and they dutifully wrote extensively about every detail of the wedding, the reception, the food, the bride and groom, the guests, and the stunning gown. When asked about the latter, Jacqueline, who was said to have disliked The Dress because it clearly wasn't her choice of style, downplayed the designer. Some sources claim that she also offhandedly uttered something racially hurtful; others vow that that *never* happened and never would. Regardless, in the end, just *one reporter* mentioned Ann Lowe as the creative mind behind the silk taffeta.

It is to Lowe's everlasting credit that she managed to forgive and to work with the new Mrs. Kennedy again some time later, although it would apparently be many years before Jacqueline learned fully what had happened at her wedding in September 1953. Still, with the Swinging Sixties in full force, Lowe's creations began to fall out of demand shortly after John Kennedy was assassinated, and she struggled, a victim of fickle fashion and her own underpricing of her couture.

More than a decade after the wedding, with age a factor and her business in decline, Lowe was reportedly in serious financial trouble when an anonymous benefactor paid off her outstanding IRS tax bill.

Friends claimed that Lowe always secretly thought the benefactor was Mrs. Kennedy....

WORLD CULTURE: PICKLES AND CHRISTMAS THINGS, PART II

Other fun things to know about Christmas....

LUCK AND HAPPINESS are guaranteed for the Catalonian household that displays a Caganer. Traditionally, the Caganer is a small, crouched figure with a barretina on his head, a pipe in his mouth, and his pants down around his thighs. Appearing to be in the middle of defecation, he's placed near the nativity during the holidays. Depending on the legend, the Caganer symbolizes fertile fortunes, or he reminds us that we are all human. In many houses, the Caganer is joined by Tió, a log that must be fed until Christmas Day, upon which it is beaten and sung to in the hopes that it defecates gifts to the kiddies.

Celebrants haul a Yule log into their home as part of the traditional pagan celebration of Yuletide that was later co-opted by the Christian Church.

WHY DO WE use pine trees for decoration at Christmas? The answer's easy: because long before Christians began celebrating Christmas, pagans used pine boughs to decorate their homes during the winter solstice and the Romans used pine boughs to decorate for Saturnalia. Besides, a Christmas maple just isn't the same....

RELIGIOUS LEADERS KEPT Christmas somber and serious through about the mid-1800s, but tinsel as a decoration has been around for more than 400 years. Meant more for decorating statues than trees initially, tinsel was made of super-thinly hammered silver by Nuremberg silversmiths in 1610, and then it was cut by hand into thin strips; needless to say, it was very expensive, affordable only for the very affluent. Later, when it first became a tree decoration (possibly around Queen Victoria's time), it was noticed that silver tinsel turned black when placed near the candles used to light trees. Never mind the chance of fire; that blackness ruined the sparkly effect, so people began to experiment with ways to make tinsel. By the early 1900s, tinsel was made of copper and aluminum; it disappeared during World War I due to the war effort, but it reappeared as a decoration made of lead. In 1972, lead tinsel was deemed unsafe, but by then, a tree without tinsel was a terrible thought, so the lead product was discontinued in favor of other material, such as cheap aluminum. Today's tinsel is made of plastic.

UNTIL ABOUT A century ago, candy canes were entirely white.

ACCORDING TO THE *Washington Post*, assuming that on Christmas Eve, Santa visits every Santa-believing child under the age of 14, *and* assuming that he doesn't dilly-dally or take even a minute to lay any digit aside of his nose, *and* assuming that he doesn't visit nonbelievers or really bad kids, *and* assuming that he honors alternate holidays in other countries and religions, he will have roughly 20 hours to bring toys to millions and millions of kids worldwide. Unless he's a master of conjuring, the toys on his sled would weigh nearly as much as a loaded freight ship, and if he eats just a half a cookie and takes a tiny sip of milk at each household he visits, he would consume millions of calories on that one single night.

BAD KIDS BEWARE: many cultures tell stories of a creepy Christmas creature that punishes those who don't behave themselves at Yuletide. In Germany, bad kids may want to watch out for Hans Trapp, a scarecrow-looking dude who helps Santa by eating miscreants; in Switzerland, Germany, and France, Belsnickel supposedly whips children who greedily snap up goodies; in France, Père Fouettard is said to kidnap and whip kids who are bad; Joulupukki is a Finnish creature that looks like a goat and goes around scaring children; and Krampus is a demon-goatlike being that comes from central Europe, also to frighten children into behaving.

PRIOR TO THE 1920s, Christmas gifts were wrapped in plain, brown paper or in white tissue paper. The emergence of preprinted, mass-produced paper made it possible for gift-givers to add a festive look to the presents they put beneath the tree.

HANDY FACT

According to *The Handy Answer Book for Kids (and Parents)*, it's important to recycle your nonmetallic wrapping paper "so new paper can be made from" it. "During the recycling process, wastepaper is chopped into very fine pieces, mixed with water, boiled, and made into a thick soup called pulp. This pulp may be washed and bleached to make it white or left a natural color. Pulp is then spread out onto large screens that allow water to drip through. Next, giant rollers squeeze the pulp, removing more water. Once the new paper is dry, it is wound onto huge spools and later cut into sheets of paper. Other paper products like cardboard boxes are made in similar ways. Wood fibers and other plant materials are also used in paper-making."

CULTURE: GET A ROOM!

Every traveler in the world knows one unalterable fact: at the end of a long road or a long flight, the idea of laying your head down in a quiet room is what gets you through. Experts say it's been this way since the beginning of mankind: hosts welcomed trippers at rudimentary hotels some tens of thousands of years ago.

The title for the very first vacation hot spot is up for grabs, but in ancient times, the Greeks and Romans both had resting places for weary travelers within their city limits, generally in the form of small inns, baths, and eateries to encourage travelers to stay awhile. One could argue, though, that those weren't *hotels*—at least not as we know them—nor were any of the religious orders that opened convents and abbeys for travelers in need, nor were the homes of everyday citizens opened as inns. No, hotels as we know them weren't really a thing until medieval times. That was about when Nishiyama Onsen Keiunkan opened in Japan and Zum Roten Bären opened in Germany.

Both of those, by the way, are still open for business.

To various governments, all these travelers and foreigners coming and going must have been rather chaotic, and in the early fif-

Founded in 705 C.E. by Fujiwara Mahito (the building in the photo is obviously not the original structure!), the Nishiyama Onsen Keiunkan hot spring hotel in Hayakawa, Japan, is recognized by Guinness World Records *as the world's oldest hotel still in operation.*

teenth century, the governments of France, Great Britain, and Spain enacted rules for hotels and inns, including a need for a registry. By this time, inns and hotels began to see that more amenities could mean more staying guests and, therefore, more money: some offered kitchens for guests to use and shelter for their carriage horses. In Louis XIV's time, a hotel was constructed that also consisted of small shops for the traveler to visit; by the late 1700s, nicer restaurants had been added to professional hotels, and the idea of making some rooms better than others (for a fee) began to take root; by the early 1800s, some hotels specifically catered exclusively to the wealthier traveler.

The idea of vacations for the rich had taken hold by then in Europe, and fancier hotels sprung up in picturesque areas of France and Italy. America, however, was still mostly frontier, and travel could be dicey; still, folks away from home had to have a place to stay, and inns were generally where they sheltered. Yes, the City Hotel had opened in New York City in 1792 and the Tremont House could be found in Boston by 1829, but neither of those helped in less-populated areas. Instead, tavern keepers in wilder areas of America opened their facilities to men (almost always men) who were on the road—the caution being that such shelter often meant sharing a bed with a stranger or two. Some innkeepers copied the "Penny Sit-Up" mode of shelter used in New York City for its homeless population: for a single penny, a traveler could reserve a bench to sit on and a taut rope to lean against while they slept. Rest assured, though, that as American cities grew, so did the need for places for folks to stay while just passing through, although many of those were considered luxury venues. By the beginning of the twentieth century, American hotels had elevators (New York, 1957), were built with fire resistance in mind (Chicago, 1870), and had electricity in all rooms (New York State, 1880).

One might think that the Great Depression would put a halt to the construction of hotels, but that wasn't true. Not even a worldwide economic collapse stopped the building of the Waldorf Astoria in New York or in many of Florida's booming areas. However, motels (a word coined by adding motor + hotel = motel) began springing up along American roadways in 1925, and they struggled during the Depression.

Still, the motels did what they had to do, and some that survived during the 1930s did so with the help of gimmicks, such as nearby roadside attractions, amenities for children, and specialty-shaped buildings that

411

If you've got the money, honey, you can spend one single night in each of the hotel rooms in Las Vegas. But you'll have to have a lot of time: it would take you nearly three hundred years. It would take you the rest of your life to spend a single night in each of the rooms at Disney World in Orlando.

A circa 1880 photo of Tremont House in Boston. This elegant hotel, which hosted such luminaries as Charles Dickens, Ralph Waldo Emerson, Alexis de Tocqueville, Daniel Webster, Abraham Lincoln, Andrew Jackson, and Davy Crockett, was open from 1829 to 1895.

housed the motel itself. Seriously, if you could afford a few pennies, how could you ever resist saying that you slept in a teepee or a dog's hind leg?

By 1958, hotels were rated up to five stars. They'd appeared in guidebooks for decades—at that point, the guidebooks were for wealthy travelers, road-trippers, and African Americans. Hotel and motel chains had already taken hold, and loyalty programs were instituted in the early 1960s. Amenities and freebies changed to reflect what travelers might need: sewing kits were given out in the late 1950s, vending rooms were added (free ice!), wheelchair-accessible rooms became available in the 1960s, and minibars were added in the early 1970s. Also, in the 1970s, business-class hotels began to take reservations.

HANDY FACT

How many everyday, working-class people were able to stay in a hotel in the early part of the last century? Not a lot because few had regular vacations: according to *The Handy History Answer Book*, 3rd edition, workers' unions made "important gains during World War II (1939–1945)" for member employees in the United States, which helped guarantee that they received health insurance and "paid vacations and holidays" from their employers.

Today, no matter what your interest or why you're going somewhere, you can find a hotel there—even if it's on a desert, on an island, or yes, even in Antarctica.

CULTURE: GAMES PEOPLE PLAY

A lot of kids of a "certain age" will remember when their mom used to tell them to go outside and "find something to do." Chances are, if there were other kids around, that something ended up being a game that everybody could play—but did you know that most of those games had their roots in antiquity?

HIDE-AND-SEEK (or, in some areas, hide-and-go-seek) was first described by a second-century Greek writer, so you can bet it's old. Literally, it's a game in which one child covers his or her eyes and counts to a mutually agreed-upon number while the other children hide. The child chosen as "it" must seek them. Hide-and-seek is played throughout the world under several different names.

RED ROVER FIRST appeared in the early part of the nineteenth century. In this game, an equal number of children form a line parallel to each other and tightly hold hands. When a child is

The children's game of jacks dates back to ancient Greece and perhaps even before then.

chosen, they must run at full tilt toward the opposite line and try to break through; if they succeed, they remain with their original team. If not, they become a member of the opposing team.

The game of blindman's bluff (in which one child covers his or her eyes or is blindfolded and attempts to find the other kids) was passed down from the ancient Romans of 1000 B.C.E.

THE GAME OF jacks was described in mythology and the *Iliad* and the *Odyssey* and was almost certainly played by grown women as well as by older children. Called knucklebones then, it was played with the foot bones of sheep or goats; in its earliest days, it was a counting game of toss-and-catch, in which the bones were tossed up and caught on the back of the player's hand or tossed in a hole to land in a certain way. Modern games are played with a rubber ball and ten jacks; like most children's games, the game of jacks has variants around the world.

TAG LIKEWISE HAS many variations and is known by many names. Nearly all versions are chase games, in which "it" chases the other players and tries to touch them in some way so that the tagged person then becomes "it." Note that because things can get out of hand and kids can get hurt or feel uncomfortable being tagged, this game has been banned in many schools and playgrounds.

HANDY FACT

Here's another blast from childhood, courtesy of *The Handy Answer Book for Kids (and Parents)*: "'Please' and 'thank you' are not magic words like 'abracadabra,' which a magician says as he pulls a rabbit out of a hat. But they are special words because they make dealing with other people go more smoothly. Saying 'please' shows that your request also comes with respect for the person you are asking.... Someone whose actions are appreciated will be more likely to help out or be generous again. So, you can see how being polite can help people get things done. The words 'please' and 'thank you' make the world a more thoughtful and generous place."

WORLD CULTURE: FACTS ABOUT FOOD

GROUND BEEF AS we know it was not around until the nineteenth century, after the meat grinder was invented. They were first available for sale in the mid-1800s, but once the meat grinder was featured at the 1876 Philadelphia Centennial Exposition, every home cook wanted one.

IF YOU SEE "sweetbreads" on a restaurant menu, try not to think about doughy, yeasty goodness slathered with butter. Sweetbreads are actually meats from the thymus gland or pancreas of a calf or lamb or, possibly, another edible organ meat from a different animal.

DESPITE WHAT YOU learned by watching Bugs Bunny, Welsh rarebit has absolutely nothing to do with rabbit. It's actually, at its very basic, toast that's covered in cheese or cheese sauce.

AMERICA'S PRESIDENTS PAY for their own and their family's food, plus everything that would be consumed by private guests. This includes birthday cake, holiday treats, and the Thanksgiving feast.

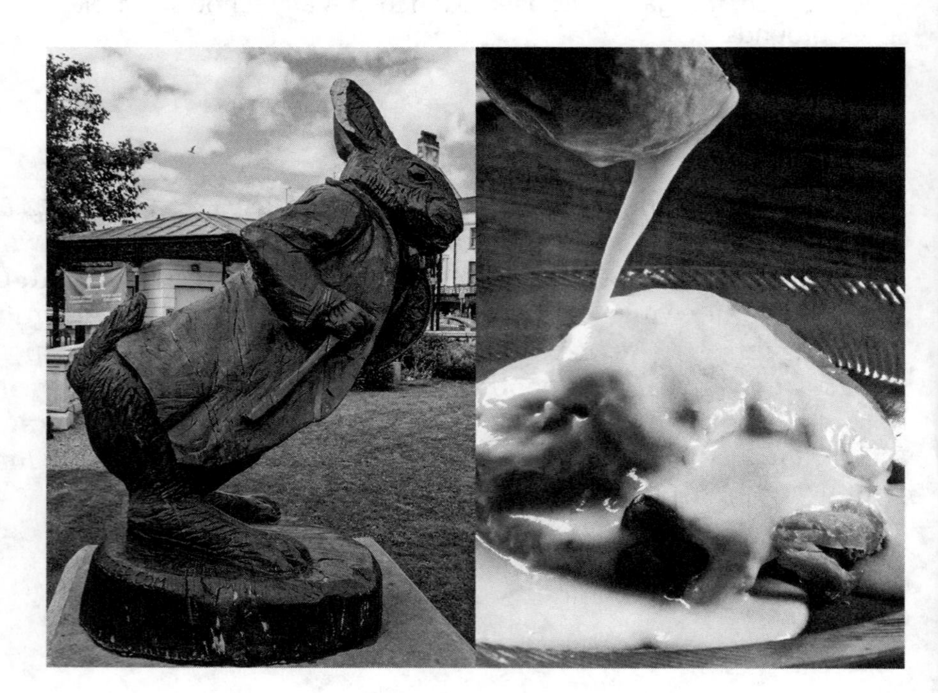

At left is a Welsh rabbit (a statue of the White Rabbit from Alice in Wonderland*, located in Llandudno, Wales), and at right is a dish called Welsh rarebit. Don't confuse them!*

THOUGH IT MIGHT sound like it, black pudding contains no chocolate. Nope, black pudding is a kind of sausage made from pork fat and pork blood, possibly some beef suet, and some kind of grainlike oats, oatmeal, or barley stuffed in a sausage casing. Also, colloquially, it's called "blood pudding."

A POPE'S NOSE (also known as a sultan's nose) is part of a plucked chicken's behind. It looks like the chicken's tail because it *was* a tail back when the chicken had feathers. Historically, the pope's nose was the part the cook saved for herself before she served the rest of the bird to her employers.

THE EAR OF corn you had at your last summer barbecue had an even number of rows, generally between 8 and 22 but averaging 16 rows on the cob. That's because the cob's spikelets (a botanical arrangement found on grass flowers) always appear in pairs.

RICE IS THE most eaten food in the world. It's a dietary staple for more than 3.5 billion people around the planet. Corn comes in second place. As for animals eaten, pork is the top meat, on 36 percent of the world's plates; chicken is a very close second at 35 percent.

NUTRITIONISTS SAY THAT if you're looking to gain weight, you should eat lots of bananas. One medium-sized banana has more than 100 calories and lots of other vitamins and minerals you need.

ONCE UPON A time, making a pound cake was easy. The recipe went like this: take a pound of butter, a pound of eggs, a pound of flour, and a pound of sugar, mix, and bake.

HANDY FACT

The Handy History Answer Book, 3rd edition, 3rd edition says that Mary Mallon, also known as Typhoid Mary, was "the first known carrier of typhoid fever in the United States. Though Mallon had recovered from the disease, as a cook in New York City area restaurants, she continued to spread typhoid fever germs to others, infecting more than fifty people between 1900 and 1915. The New York State Sanitation Department connected her to at least six typhoid fever outbreaks there. Officials finally—and permanently—instutionalized her in 1914 to prevent further spread of the acute infectious disease."

ANIMALS: THE BEAR FACTS ABOUT FIRE

If you were a child at any time in the last seventy years or so, you grew up with an icon that helped you be a good citizen. You might have known him as a cartoon character, but Smokey Bear (please note: never Smokey THE Bear) was a real, actual bear, and his story is a good one. It started nine years before the *real* Smokey was born.

During World War II, the country's need for wood was huge—not only was it needed for military use, but some Americans heated their homes with wood, and some even cooked over wood stoves. By 1941, the U.S. Forest Service estimated that 90 percent of all forest fires were caused by humans, probably accidentally.

But because so many of the country's firefighters were away at war and the heavy machinery used to keep forest fires at a minimum was likewise employed, it became necessary to organize an all-volunteer team to combat what looked to be a growing problem in the nation's forests. In 1942, the Forest Service put together the Cooperative Forest Fire Prevention Campaign, which encouraged Americans to be vigilant personally and on watch for flames in local forests. The Wartime Advertising Council chipped in, urging everyone to do their part with ads that featured a trio of wide-eyed forest creatures, none of them bears.

The original Smokey Bear (above) was injured in a fire as a cub and then brought to the National Zoo in Washington, D.C., where he was a popular sight in the 1950s. He has been immortalized in the Smokey Bear fire safety mascot (pictured) ever since.

That oversight was too much for bear lovers and, in 1944, the Forest Fire Prevention Campaign first described the new spokesbear. He was given a name, a jaunty hat, sturdy britches, and intelligent eyes. Because "Smokey" was a popular nickname for firefighters, he instantly gained a name. He was such a hit that the artist who originally drew Smokey, Albert Staehle (1899–1974), was asked to do other Smokey posters.

By 1947, World War II was over, the Wartime Advertising Council became just the Advertising Council, Inc., and Smokey had a new tagline: "Only You Can Prevent Forest Fires," a slogan he's used ever since. By that time, artist Rudy Wendelin (1910–2000) had taken over as Smokey's official artist. Radio announcer Jackson Weaver (1920–1992) became Smokey's voice for radio. Bringing on a live spokesbear was discussed, but the logistics must've been insurmountable.

In the spring of 1950, in the Lincoln National Forest in New Mexico, a fire tower operator spotted smoke and notified the local authorities, who investigated and, determining that it was a pretty major fire, called in regional- and state-based fire departments to help put the conflagration out as quickly as possible. In fighting the blaze, firefighters noticed a small bear cub alone by the edge of the burned area, and they hoped he'd find his mother. Once the blaze was extinguished, they knew he hadn't and that he was too small to live by himself in the now destroyed forest. Climbing a tree on which the bear was clinging, one of the firefighters, Ray Bell (1911–2000), caught the little guy, and they took him to a local veterinarian, who tended to the bear's burns.

Of course, the bear was immediately called "Smokey" (seriously, would *you* call him anything else?), and when word got out about the

HANDY FACT

To answer a question you've no doubt had, *The Handy Answer Book for Kids (and Parents)* says: Burning things smoke because "during a fire, the air around it becomes heated. The heated air sweeps up water vapor (molecules of water that float in the air) and tiny specks of the fuel (the material being burned) into a dark cloud of smoke. The more incompletely something burns, the more smoke it produces because more particles are left to be swept up into the air. Smoke gradually spreads out and drifts away, with gravity pulling the heaviest bits back to the ground. When a fire first starts to burn, there is usually a lot of smoke, which decreases as more of the fuel is burned completely."

fire and Smokey's rescue, the newswires went a little wild. Shortly afterward, Bell took his four-year-old daughter to check on Smokey's progress, and the little girl begged to take him home—and Smokey had a new, temporary home.

But little bears become big bears, and Smokey began acting like a bear that takes no guff from anyone. He reportedly began nipping Bell, and it was apparent that the cute, little cub was no housebear. While several states wanted him to move behind their borders, it was ultimately decided that Smokey would go to live at the National Zoo in Washington, D.C., a journey he took in the mid-1950s.

The original Smokey Bear died at the zoo on November 9, 1976, but before he passed, he was so popular that American children played with stuffed likenesses of him, read books about and "by" him, and saw countless posters of him (to hear the famous song about him, look for Eddie Arnold's classic rendition of "Smokey the Bear" online).

And remember: Only YOU can prevent forest fires.

CULTURE: TURN YOUR RADIO ON

Imagine this: you're in the back yard, and you want something different to listen to, something that'll give you a variety of music but that you don't have to tend. You want, in other words, to have your radio on, but imagine that it sounds very unlike that which you are used to enjoying....

The beginning of radio programming can best be described as a free-for-all. Some stations ran soap operas, others aired music programs, some ran dramas or talk shows, and some played a combination of everything available. Some stations were on the air just sometimes, depending on the person owning it and the person behind the mic—which were often one and the same. Still, radio was the dominant electronic entertainment—at least until TV appeared in the late 1940s—and then, radio began to struggle.

By 1949, local music programming had mostly replaced soap operas and live dramas on the radio. But when you touched that dial, you never knew what you could expect—and when rock and roll began hitting the airwaves, it could cause trouble for people who

Radio programs became popular reasons for families to gather during the 1920s and 1930s.

either loved it or hated it. It's probably coincidence that at about this time, two radio station owners decided to upend the old way of doing radio in favor of a new way that would endure for decades, and a new electronic device would deliver it.

For a number of years, Gordon McLendon (1921–1986) had enjoyed success with his Liberty Radio Network, which broadcast Major League Baseball and other programming, with the help of Western Union. This was live baseball, not re-created stuff, and it was a moneymaker for Liberty and for McLendon. This allowed a bit of expansion, and in 1947, McLendon and his father founded KLIF in Dallas, Texas.

Some months later, up in Oklahoma, Todd Storz (1924–1964) and his father purchased KOWH-AM and KOAD, an FM companion station (FM radio wasn't a "thing" just yet; until the 1960s, FM radio was often the simulcast of an AM parent station). To say that KOWH and KOAD struggled is an understatement: in the Storzes' first couple years of own-

Some of your favorite singers did "crossover" music in the earliest part of their careers. Fans know that Elvis Presley sang gospel songs long before he sang "You're the Devil in Disguise." Conway Twitty was a rock and roll singer at the beginning of his career. Carl Perkins dabbled in country music, and Johnny Cash did the same in rock and roll. Classically trained tenor Mario Lanza even had a crossover hit that went to number one on the Billboard charts.

HANDY FACT

From *The Handy Physics Answer Book*, 2nd edition: "In 1895, Guglielmo Marconi (1874–1937), a twenty-year-old Italian inventor, created a device that transmitted and received electromagnetic waves over a 1-kilometer (3,280-foot) distance.... As a result of Marconi's work on radio transmitters and receivers, he was the co-winner of the 1909 Nobel Prize in physics."

ership, polls showed that just 4 percent of Omaha households listened to either station in any given week.

But Todd Storz was watching, and he noticed that listenership rose when music was aired and was much lower during the talk-show programs the station ran. By late 1953, KOWH had switched to an all-music format with in-house "talent" and locally produced commercials. McLendon took Storz's idea and made it better, with jingles and traffic reports for the Dallas area.

By the mid-1950s, a small handful of Top 40 music stations could be found sprinkled in American cities. Within a short time, every major city (and not just a few small towns) had their Top 40 music stations with on-air personalities.

And you can thank three men and Bell Laboratories for your ability to take your tunes outside: in 1947, William Shockley (1910–1989), John Bardeen (1908–1991), and Walter Brattain (1902–1987) invented the technology to create transistor radios.

AMERICAN HISTORY: A PENNY SAVED, PART II

ACCORDING TO THE U.S. Mint, our currency still shows a lot of ancient symbolism: the torch on the back of a dime signifies liberty, while the oak branch (to the right of the torch) signifies "strength and independence." On the back of a dollar bill, the arrows the eagle holds indicate war, while the olive branch means peace. The wreath on the oldest Lincoln pennies signifies victory.

IF A BILL is misprinted for whatever reason, it is usually caught before being circulated and destroyed. The rare misprint that ends up in circulation is often worth many times its face value; likewise, a bill with a very low serial number may fetch more than its face value. For details, check with a numismatist (someone who specializes in coins and paper money).

IN SOME PLACES in the United States, it used to be illegal to take a check as payment for any goods purchased on a Sunday.

HUNDREDS OF THOUSANDS of people in America live on less than $100 a month. If, in fact, you have $10 in your wallet and zero debt, you are richer than nearly 3 out of 10 of your fellow Americans.

THE PAPER USED to print this book is very different from the paper used to print money. This book was published with mostly wood-pulp paper. U.S. currency is printed with paper that is 75 percent cotton and 25 percent linen. Paper money, in other words, really isn't *paper* money, it's cloth.

THE AVERAGE ONE-DOLLAR bill lasts less than two years before it's returned to the U.S. Mint and shredded. Ninety percent of worn-out money is now recycled.

YOU CAN PROBABLY imagine that money is dirty, but recent studies show that nearly *all* money, both paper and coinage, is filthy and can carry more than 100 different bacteria and viruses. Bottom line: wash your hands after paying.

THE U.S. BUREAU of Printing and Engraving prints between $500 and $600 million a year, with most of it going to replace worn-

The dollar bills in your wallet or purse are not actually paper; they are cloth made from a combination of cotton and linen.

out bills. The game maker Hasbro prints some $30 billion dollars a year in "Monopoly" money.

MORE THAN 85 percent of the world's money is digital only. It exists because a computer says someone has it.

FOR SHEER SILLINESS, check this out: lay out a mile of pennies, and it'd cost you $844.80. A mile-high stack of dollar bills would net you not quite $15 million. A million bucks in $1 bills weighs just over a ton. The better bet would be to get your cool million in $100 bills: it would weigh just about 22 pounds.

WITHIN THE UNITED States, the $20 is the most counterfeited bill. Outside the United States, our $100 bills are the most frequently counterfeited.

GRAB YOUR MAGNIFYING glass: on the back of the $5 bill, all 50 states are listed along the roofline of the Lincoln Memorial. Squint. You'll see 'em.

IN 1957, THE U.S. one-dollar bill became the first bill to carry the motto "In God We Trust." It had been on U.S. coins since 1864.

THE LAST SILVER Certificates were printed by the U.S. Mint in 1963; later that same year, the Mint made minor changes to the front of the $1 bill, in anticipation of the end of Silver Certificate redemption (an exchange of a Silver Certificate for silver coinage or bullion), which would happen in 1964 (coins) and 1968 (bullion). Aside from those small upgrades and a switch from Latin to English on the Treasury seal in 1969, the dollar bill has remained the same all these years.

KEEP AN EYE on your cash and watch for a "star note": you'll know you've got one when you spot a "star" at the end of a bill's serial number. The star indicates that the bill is a replacement for one

HANDY FACT

The Handy Answer Book for Kids (and Parents) says: "After new money is printed at the Bureau of Engraving and Printing, the U.S. Treasury Department ships it to the 12 Federal Reserve Banks that are spread throughout the country. The Reserve Banks then distribute the cash to commercial banks and other institutions where people keep their money. Customers withdraw cash from banks and spend it on gas, food, books, and so on. Eventually, the stores deposit the bills back at the bank, and the process begins again."

that was misprinted or otherwise damaged in printing. The serial number is identical to the one that was printed wrong (and which was destroyed); the Bureau of Engraving and Printing just adds the symbol at the end. Alas, your misprint is probably only worth its face value, but you can always check with an expert to be sure.

NEARLY HALF OF the money the U.S. government prints is in the form of $1 bills.

DURING WORLD WAR II, when copper and zinc were rationed, the government had to figure out another material from which to make pennies. They settled on zinc-coated steel, which worked just fine for about eleven months in 1943 (thus saving enough copper to make over a million ammunition shells for the war). When problems with the zinc-coated steel penny became apparent—it rusted and it looked too much like a dime—a different amalgamation was devised and used in 1944. By 1946, the pennies as we know them were back in production.

SPEAKING OF WAR pennies, copper-clad pennies made in 1943 are considered some of the rarest and most valuable coins, selling for many millions of times their face value. Only 40 of them are known to exist, and those are all in collectors' hands, so your chances of finding one in your change is zero.

FINALLY, KEEP YOUR eyes on the ground: Americans throw away more than $60 million a year, most of it in coins.

AMERICAN HISTORY: CHUGGA-CHUGGA-CHOO-CHOO

Back before there was TV, your ancestors had far fewer options for entertainment: there were books and stories, visiting, singing, parlor games, and … well, not much else. Except train crashes. They had train crashes.

The whole thing seems to have started in about 1895 by Ohioan A. L. Streeter (or Streeters, depending on your source), who sold railroad equipment to the growing railroad companies. He got the bright marketing idea to have an organized train wreck, literally: he got a length of track of a mile or so, laid it down, and had two locomotives brought in and placed on either end, facing one another. When

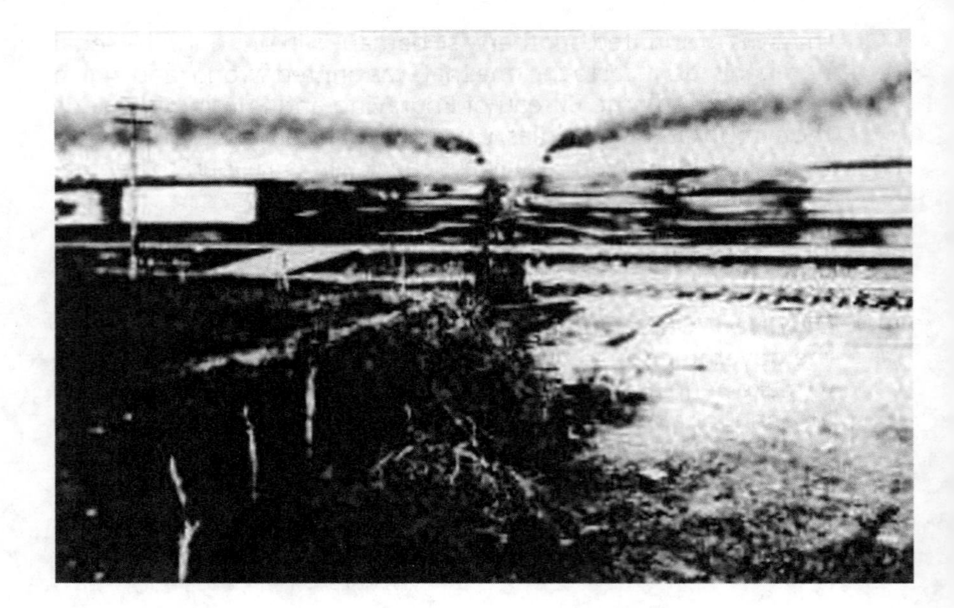

The train crash stunt in Crush, Texas, was captured on film, including the moment of impact between the two locomotives.

Streeter gave the "go," the two engineers brought their engines up to full throttle and went full steam ahead (no pun intended) at one another. Just before the BOOM!, they leaped from their respective locomotive cabs to safety, to the delight of the few thousand who paid a couple of dollars to see the spectacle.

Of course, if you're a fan of stock car races or demolition derbies, you can well imagine the popularity of such a thing. In the year after Streeter's great PR stunt, several other locomotive crashes were held to the delight of the crowds. The biggest, perhaps most famous, one was held in the tiny town of Crush, Texas, just north of Waco.

No, Crush wasn't named after the result of two trains colliding. Crush, Texas, was a temporary town cheekily named after William George Crush (1865–1943), a passenger agent who was tasked with trying to help the Missouri-Kansas-Texas Railroad (known as "The Katy") with its struggling bottom line. It's been said that Crush was a pal of P. T. Barnum (1810–1891), which is a good indication that he understood the average person's love of a show, especially if that show was a good locomotive demolition; with this idea in mind, he conceived the Crush crash. But a crash is nothing without a crush of crowd, so every person in Texas was offered a ride to Crush on The Katy for the princely sum of two bucks.

On September 15, 1896, two locomotives, one painted red, the other painted green, both pulling a small line of ad-poster-covered boxcars, faced one another on a railroad track surrounded by hillsides that, as it turned out, were great places for spectators. Crush had thought ahead: he had refreshments, and he'd erected a large tent for some 40,000 revelers to relax in. It was, by all reports, a festive, carnival-like atmosphere; in fact, the event itself was delayed because Crush didn't expect that many spectators, and crowd control became a brief issue.

At just a little after 5 P.M., Crush rode his horse to a point where both engineers could see him. He turned his mount, dramatically paused, tugged his hat from his head, and waved it. The engineers, who'd already started their locomotives, started toward one another, gaining speed until they were traveling at up to 60 miles per hour.

That doesn't seem fast, but in 1896, it was *incredible*.

The locomotive-on-locomotive collision occurred with a spectacular roar of metal-on-metal, timber, and screeching. Both engines raised up, as if they were lions fighting to the death before falling on their sides—and then there was a pause before the engines' respective boilers exploded, almost simultaneously, sending hot, ragged debris flying through the air.

Train-crashing events were so popular in the early part of the twentieth century that one man, Joe "Head-On Joe" Connolly (1859–1948), made it a career. In the years between 1896 and 1932, Connolly crashed nearly 150 locomotives in more than 65 staged events. Connolly said once that he became a locomotive demolition driver because, as a boy, he always wondered what would happen if two steam engines collided; his first chance to see that happened in Des Moines, Iowa, a week before the Crush Crash. He walked away with a pile of money in his pocket, and a star was born.

The crowd, needless to say, wasn't expecting such a disaster; people panicked and ran, screaming in fright, and several were very

HANDY FACT

You have to wonder what William Crush would think of a MAGLEV train. According to *The Handy Physics Answer Book*, 2nd edition, by Paul W. Zitzewitz, PhD, "MAGLEV, or magnetically levitated trains, are different from conventional trains in that they use electromagnetic forces to lift the cars off the track and propel them along thin magnetic tracks. Some demonstration trains have reached speeds of 500 kilometers per hour (300 miles per hour). Although the United States has no MAGLEV train nor an active research program in this technology, Germany and Japan have conducted a great deal of research in the field."

badly injured. Two (or some reports say three) people died, one man lost an eye, others lost limbs, were burned, or were otherwise hurt in the melee. Unbelievably, some immediately returned to grab smoking souvenirs to take home as proof of having been to see the Crush crash.

Officials, who'd been assured that this event would be completely safe for spectators to watch, immediately fired William Crush. Later, upon learning that most of the paying customers had a great time and that the publicity they received was invaluable, they reached out to Crush the very next day and hired him back to work.

CULTURE: SEVEN PLACES YOU CAN'T (OR SHOULDN'T) GO

You know what you need? A vacation! Right now, you need someplace to go relax, get away from the stresses of work and home. You need a vacation… but, in random order, just not to here:

7. Ramree Island, just off the coast of Myanmar, is home to thousands of crocodiles—big ones that will literally eat you. Evidence: in February of 1945, during the Burma campaign, a thousand Japanese soldiers reportedly entered a swamp on Ramree Island. Reports say that there were a few shots fired, followed by the screams of men and the sounds of alligators. Of the thousand soldiers who went on Ramree Island, only twenty survived. Today, Ramree Island is home to considerably fewer crocs and a gas pipeline, mosquitoes, and scorpions; aside from that, though, it's said that the danger is over, but if reports were ever true—and many sources today have cast doubts—would you still want to go there?

6. Dallol, Ethiopia, is located in Africa's horn, and you don't want to go there. First of all, it's hot—average temperatures are in the mid-90s Fahrenheit, and it's nothing for the temp to reach 10 degrees on top of that. The area boasts a volcano, and once upon a time, Dallol was of interest as a trade route with a port and for geological exploration; by the mid- to late 1960s, however, the place was abandoned, structures were left to the elements, and roads disappeared. Today, you can travel to Dallol by camel or via an armed caravan, but neither is really recommended since local tribesmen have been known to attack and kill any visitors who dare to try.

5. Ilha da Queimada Granda, an island just off the coast of Brazil, sure sounds like paradise—until you know its name in English: Snake Island. You don't want to go there because the place lives up to its name: It's estimated that between 1 and 5 snakes per square meter live on the island, and most of them are the endangered golden lancehead snakes, which are exquisitely poisonous. How the snakes got there is interesting: rising sea levels eons ago cut snakes off from returning to the mainland; over time, they evolved to the two-plus-feet-long monsters they are. How *you* get there is equally interesting: you don't. The Brazilian Navy only allows a very small handful of scientists to go to the island, to protect the snakes as much as to protect people.

4. North Sentinel Island, off the coast of India, is another one of those places that seem like a paradise. White, sandy beaches, swaying palms, men with arrows.... Indeed, it wasn't so long ago that the Sentinelese caught an American missionary who thought he might want to get to know them, and they killed him. The Sentinelese have been on the island for tens of thousands of years, and they're not about to get off it. You are not welcome there, not at all. In fact, approaching to within five nautical miles of those pristine, lovely beaches is illegal.

3. You *can* go to Aokigahara Forest in Japan. Go ahead, but be warned that it's not just a lovely, serene woodland area. Known also as "Suicide Forest," this beautiful (but extraordinarily creepy) 30-square-kilometer bit of land on the northwest side of Mt. Kilimanjaro is dark and quiet and rumored to be haunted by demons, and for good reason: for at least the last 50 years or so, despondent Japanese citizens have chosen the forest as a place to commit suicide. As if that isn't reason enough to just stay away, there's this: attempts to remove bodies are made just about once a month, which means that some bodies aren't immediately found—if at all—and for others, animals reach the bodies before authorities do. If you dare to visit Aokigahara Forest, the likelihood that you might find a dead body is relatively high.

2. If you have the patience to fill out reams and reams of paperwork, you can visit Poveglia, Italy. You won't find much there, though, except history: the island, which is bisected by a narrow canal, was first mentioned in historical records dating back nearly 600 years the island

HANDY FACT

According to *The Handy Science Answer Book*, 4th edition, rattlesnakes (including the sidewinder), moccasins (including the copperhead), and coral snakes are the most venomous snakes that are native to the United States. Snakes in the rattler category can reach lengths that exceed six feet.

was populated until the fourteenth century when its residents were chased off by warring factions. Photos of Poveglia show beautiful, grand buildings, vast forests, and verdant lawns, but the truth is that later, Poveglia was used as a place of quarantine, first for victims of the Plague in the 1700s and then for the mentally ill in the nineteenth and twentieth centuries. Rumors were that unbelievable cruelty happened during that latter time in Poveglia's history, hence its reputation for being one of Europe's most haunted areas. In 1968, the mental health hospital shut its doors, and Poveglia has been uninhabited ever since.

1. You can want to go to Mezhgorye, nestled in Russia's Ural Mountains. You'll want forever, though, because you can't go there. First built some fifty years ago, Mezhgorye is a "closed administrative-territorial formation, or a closed town," which means that the Russian government lets in exactly who they want—and that doesn't include you. What it does include are several thousand folks who apparently work there on something the outside world isn't sure about. It's thought that Mezhgorye holds military forces, possibly something having to do with nuclear missiles, maybe underground facilities, perhaps storage or bunkers, but only the Russians know for sure … and they're not telling.

Ironically, the town of Mezhgorye has a large "Welcome" sign outside the city limits, even though the government only allows residents there.

AMERICAN HISTORY: READ ALL ABOUT IT!

Without a doubt, you are a reader. When it comes to reading, there's never enough time, and if you're like most, you even have authors you follow and favorite genres to read. You know what you like—but do you know these things about the printed word?

THE HEAVIEST NEWSPAPER ever printed was the *New York Times*, published on September 14, 1987. It weighed in at a little over twelve pounds and was 1,617 pages.

AMERICA'S FIRST NEWSPAPER was *Publick Occurrences Both Forreign and Domestick,* printed in Boston in September of 1690.

OUR NATION'S OLDEST newspaper is the *Hartford Courant*, which has been in continuous print since 1764.

THE PHILADELPHIA TRIBUNE is the oldest African American newspaper in the United States, having started its run in 1884.

AT THE TIME of this writing, the *Guinness World Records* says that the thickest book ever published was *The Complete Miss Marple*

Numb. 1.

PUBLICK OCCURRENCES

Both FORREIGN and DOMESTICK

Boston, Thursday Sept. 25th 1690.

The first newspaper published in what is now the United States was Publick Occurrences Both Forreign and Domestick, *first printed in Boston, Massachusetts, on September 25, 1690.*

by Agatha Christie, consisting of every one of Christie's Miss Marple novels in one volume. Printed only in the U.K. in May of 2009, it measured nearly 13 inches high when laid on its back cover and weighed some 25 pounds.

BECAUSE, WELL, IT'S some kind of challenge, the record for the world's smallest book is absolutely changeable, depending on who has the better technology. At the time of this writing, it's a story etched on a bit of film that lays softly on the side of a poppy seed. No, this is for real: the story, created by Russian Vladimir M. Aniskin, was made in a gallery in St. Petersburg and measures 70 by 90 micrometers. Forget about reading it to your kids at bedtime; this book is too small to be read with the naked eye.

CZECH AUTHOR BOHUMIL Hrabal wrote *Dancing Lessons for the Advanced in Age*, a book that consists of one very, very long sentence. Not anywhere near that, author Jonathan Coe's *The Rotter's Club* includes a 13,955-word sentence.

YOU SHOULDN'T BE surprised at this: The Bible is the world's best-selling book. It's also one of the books most stolen from public libraries (the *Guinness Book of World Records* is the other most pilfered book).

MERCY OTIS WARREN (1728–1814) was the first woman to register a book title in America, which she did in 1790, right after Congress passed a new copyright law to protect Americans from British rip-offs. Warren's book was *Poems, Dramatic and Miscellaneous*, featuring poetry and two plays she'd written.

HISTORYMAKERS: HE'S MY BROTHER

To volunteer for service to your country is such a proud and outstanding thing to do that countless soldiers have followed in the footsteps of a parent, grandparent, aunt or uncle, or an older sibling. Look it up online, and you'll find countless stories of extended family members in service to the U.S. armed forces. But for the most tragic of reasons, during World War II, five brothers changed the way the military stations soldiers from the same family.

In the thick of World War II, patriotism ran high. Men of all ages (and not a small number of underage boys) enlisted to fight overseas, and the Sullivan brothers of Waterloo, Iowa, were no exception.

George (1914–1942), Frank (1916–1942), Joe (1918–1942), Matt (1919–1942), and Al (1922–1942) ranged in age from about nineteen to twenty-seven years old when they enlisted in early January 1942, an enlistment that came about, in part, because they learned of the death of a family friend at Pearl Harbor. The two oldest brothers, George and Frank, had served in the Navy during peacetime until until about a year before Pearl Harbor, and they likely still had buddies in the service. It's probable that the entire family did, so the brothers decided to enlist together; later, at the advice of her brothers, their sister, Genevieve (1917–1975), joined Women Accepted for Volunteer Emergency Service, better known as the WAVES (then the Navy's all-female arm).

It's been said that the family had a rough existence. The brothers were reportedly troublemakers. Their parents were lenient, to say it mildly. Still, patriotism won out, and the brothers went to their local Iowa naval recruiting office.

At that time, the Navy had in place a rule that no two members of a family could serve together on the same vessel, but it was obviously only loosely enforced: the Sullivan brothers enlisted with the condition that they remain together, and all five young men received special permission from the secretary of the Navy to do so. Shortly after enlisting, they were together assigned to the light cruiser USS *Ju-*

From left to right, Joseph, Francis, Albert, Madison and George Sullivan. All five brothers were killed while serving on the USS Juneau *in 1942. As a result of the tragedy, the U.S. Department of Defense implemented the Sole Survivor Policy in 1948 that prevented people from being drafted or serving on active duty if they had previously lost a family member.*

neau. By August 1942, the *Juneau* and its crew were overseas at the Guadalcanal Campaign, where it participated in several engagements, along with other U.S. ships.

At some time in the early hours of November 13, 1942, the *Juneau* took a hit from a Japanese torpedo and withdrew from battle. Later that same day, as it was on its way from the Solomon Islands, the *Juneau* was hit again. The cruiser split in half and almost immediately sank, taking Matt, Joe, and Frank down with it. Reports were that they were killed instantly. But some of the *Juneau*'s crew survived the second torpedo and lived—for at least a while. Survivors said that Al drowned the day after the *Juneau* sank; George survived but died less than a week later, by exposure or possibly by shark attack.

Astoundingly (at least by modern standards), it took almost exactly two months for the Sullivan brothers' parents to be notified of their deaths, mostly because of reasons of national security and safety.

In the aftermath of the Sullivan brothers' deaths, their parents, Thomas (1883–1965) and Alleta (1895–1972), threw themselves into war efforts on the homefront through volunteer work at USO outlets and by visiting factories and shipyards to boost the morale of workers there. Alleta, especially, was busy: It's estimated that she reached several millions of soldiers and homefront employees, either in person or on the radio, but it didn't come at further cost: some made accusations that the Navy "used" the Sullivans and their five sons as a sort of

HANDY FACT

Other heroes in World War II, according to *The Handy History Answer Book*, 3rd edition: Rosie the Riveter "referred to the American women who worked factory jobs as part of the war effort on the home front, where auto plants and other industrial facilities were converted into defense plants to manufacture airplanes, ships, and weapons. As World War II (1939–1945) wore on, more and more men went overseas to fight, resulting in a shortage of civilian male workers. And so, women pitched in. However, at the end of the war, many of these women were displaced as the men returned home to their jobs and civilian life. Nevertheless, the contribution of all the Rosie the Riveters was instrumental to the war effort."

The famous poster of Rosie the Riveter was an effective morale booster during World War II.

propaganda device. Such accusations remain, nearly eighty years after their deaths.

The cruiser on which three of the Sullivan brothers died was discovered on March 17, 2018, lying just over 2.5 miles underwater, exactly where it sank off near the Solomon Islands.

Because of the Sullivan family's sacrifice, the U.S. military instituted the Sole Survivor Policy (DoD Directive 1315.15) in 1948, a directive that protects families of military personnel from ever suffering such a devastating loss again.

CULTURE: I'LL DRINK TO THAT!

For sure, ever since our ancestors discovered that something really yummy could be made from fermented grapes, humans have relied on alcohol to celebrate, mark, praise, and numb. America, however, has had several unique relationships with alcohol over the last 400 years.

It's absolutely true that the Europeans who came from the Old World to the shores of what would eventually become America were already accomplished tipplers: while the Puritans have had a stern and joyless reputation since they arrived here in 1620, the truth is that they very much liked alcohol. Experts say that the ships that brought the Puritans from Europe to the shores of the New World probably carried more beer than they did water. They then brought their fermenting knowledge to the New World.

The Puritans' love of drink didn't mean that they tolerated drunks, though: folks who drank to excess were punished soundly and with no sympathy; still, drinking alcohol was seen as something that was good for a body, and even children were given beer. Alcohol was so widely approved that some of the Founding Fathers were involved in booze-producing family businesses and, by the late 1700s, the average colonist drank up to six gallons of pure alcohol annually.

By the way, six gallons of pure alcohol doesn't sound like much, but that's *pure alcohol*, and that's a lot—especially considering that today's average is just about two gallons of pure alcohol per year per person.

Just three decades into the nineteenth century, those six gallons had increased to just over seven gallons of pure alcohol per American per year and, looking back, you can see problems looming. Though the Native people had access to mind-altering substances and were probably already familiar with fermented drinks, they embraced the colonists' habits of social drinking—at least until the sale of alcohol to Indians was outlawed in 1832. In 1838, Massachusetts banned the sale of alcohol except in bulk, but that law was easily ignored. In other areas of the fledgling country, alcoholism (also quaintly known as "dipsomania") was becoming a major problem for both white and Native families and, though they weren't discussed then as much as they are now, domestic abuse rates soared. Even government agencies had to deal with the problem: The Navy once ensured rations of a half pint of rum per soldier per day, but that was abolished in 1862.

Even so, America still had quite a big love affair with alcohol, as evidenced by the plethora of saloons in western states at this time.

The first saloon in America was probably a hole in the wall in Brown's Hole, Wyoming, opened in 1822 with the intention to serve the fur trappers who lived and traded in the area. As the West was settled, the first to arrive needed a place to wet their whistles, and saloons were happy to help them do it. Not twenty years after the Navy stopped sailors' rations of rum, the saloon had become an important business in Old West towns; it was, in fact, quite common for a sizable prairie town to have more than one or two or a dozen busy saloons in their town limits. Regional breweries began to have strong ties with local saloons, and politicians understood their social importance. The old cliché of a madam and a saloon full of "girls" was more than just a Hollywood trope.

And here's where the glass started to become half full: by 1893, the Anti-Saloon League began protesting strongly against saloons and taverns. Within two years, a good number of Americans practiced temperance, and the Anti-Saloon League had grown into a strong organization, much stronger than the Woman's Christian Temperance Union or the Prohibition Party. Before the century was over, a huge voice in favor of total control of alcohol could be heard across the United States nearly everywhere but the saloons.

411

The first state to ban the sale of alcohol was Maine, in 1846.

On January 16, 1919, the United States outlawed alcohol through the Eighteenth Amendment. Prohibition totally banned any manufacture, importation, transportation, and sale of alcohol completely—but, of course, there are always those who know they can get around the laws.

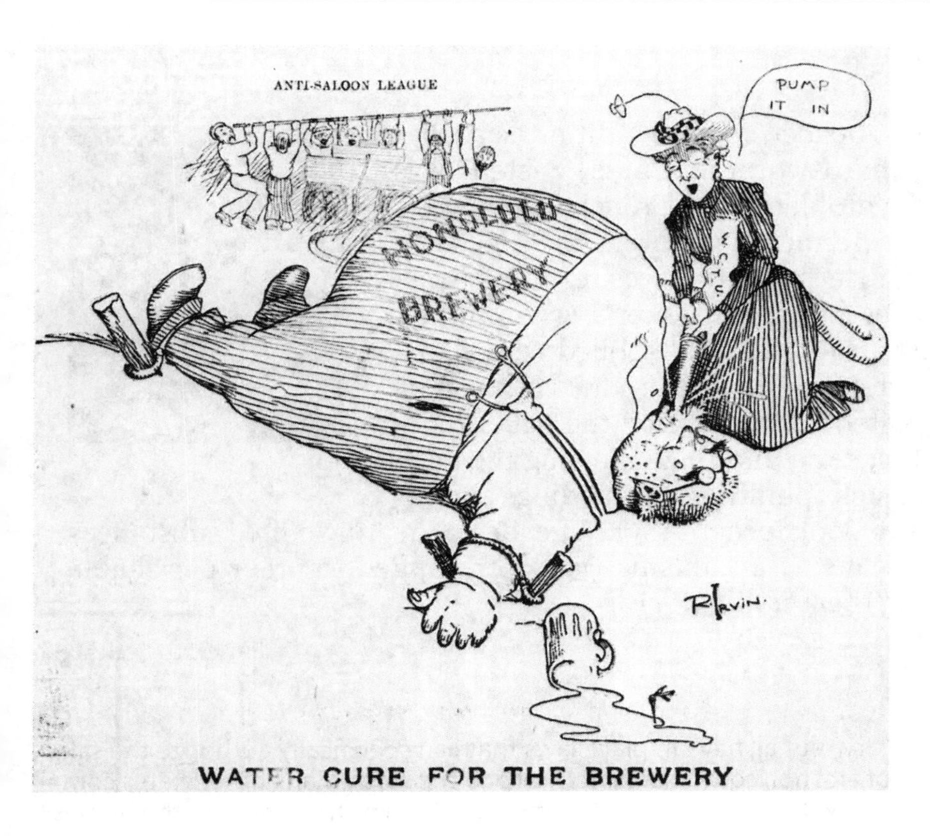

A cartoon in the 1902 issue of the Hawaiian Gazette *pokes fun at the Anti-Saloon League and the Woman's Christian Temperance Union.*

Moonshiners in rural areas made their own liquor and, because the nature of their business meant outrunning federal agents intent on arresting them, started America's love of stock car racing. If you wanted some company while you illegally imbibed, you could find a speakeasy (a bar) if you knew where to look or who to ask. You could make your own booze at home if you could find a recipe (although a fair number of people made a fair number of mistakes, and a wrong recipe resulted in death or blindness). Organized crime ran their very own underground alcohol distribution businesses, and gangsters like Al Capone, Bugs Moran, and Machine Gun Kelly rose to power, at least in part because of Prohibition.

A little over a decade after it began, though, Prohibition was on the wane. Many came to understand that the production and sale of alcohol was a great way to collect taxes, not to mention jobs at a time when jobs were scarce. On February 16, 1933, the Twenty-first Amendment did away with Prohibition.

Today—at the time of this writing—three states in the United States are "dry," meaning that the sale of alcohol is prohibited but the consumption is still legal. Kansas, Mississippi, and Tennessee all are

HANDY FACT

The Handy History Answer Book, 3rd edition, says that "in the 1860s, the hard-working [Louis] Pasteur was asked to investigate problems that French winemakers were having with the fermentation process: Spoilage of wine and beer during fermentation was resulting in serious economic losses for France. Observing wine under a microscope, Pasteur noticed that spoiled wine had a proliferation of bacterial cells that produce lactic acid. The chemist suggested gently heating the wine to destroy the harmful bacteria and then allowing the wine to age naturally. Pasteur published his findings and his recommendations in book form in 1866. The idea of heating edible substances to destroy disease-causing organisms was later applied to other perishable fluids—chief among them milk.

Louis Pasteur

"dry," but all have in place laws that can specifically authorize the sale of alcohol, control of which is up to the state. Many other states contain within their borders counties and towns that are "dry," the highest number of which are mostly in the South.

AMERICAN HISTORY: JIGSAW HOUSES

Watch one of those home-makeover, home-buying TV shows for almost any length of time, and it won't take long before you hear someone ask for a "Craftsman-style home." Here's how those houses got their iconic name....

Launched in 1893 by Richard Warren Sears and Alvah Curtis Roebuck, Sears, Roebuck and Company was the perfect answer to a problem that Americans were just learning they had: namely, that they had lacked access to goods in a timely manner. At its inception, the business best known as just plain Sears was a mail-order catalog that sold only watches and jewelry in its first catalog. Later, when Sears started selling more items to a mostly rural America, all a person had to do to get clothing, shoes, tools, bicycles, toys, or even furniture was look through a well-illustrated catalog, send the order through the mail, and wait a few weeks or a month or so for home delivery.

By 1894, the Sears catalog was well over 500 pages of goods and products. By 1897, Sears was a $750,000 business (the equivalent of $23 million today), which was a considerable amount nearly 125 years ago.

After weathering a few storms, both business-wise and financially, Sears stepped into what was already a growing industry: they began selling homes and house plans. From the outset, there was competition: three other businesses had already been selling home plans and blueprints by mail order since 1895, and one of them was "Wardway Homes" from the Montgomery Ward company, Sears' biggest rival.

You can bet that never stopped the Sears corporation.

The first Sears Modern Home catalog went out in 1908 and featured designs based on a then-popular movement called "American Craftsman." Sears offered a buyer's choice of 22 different styles and

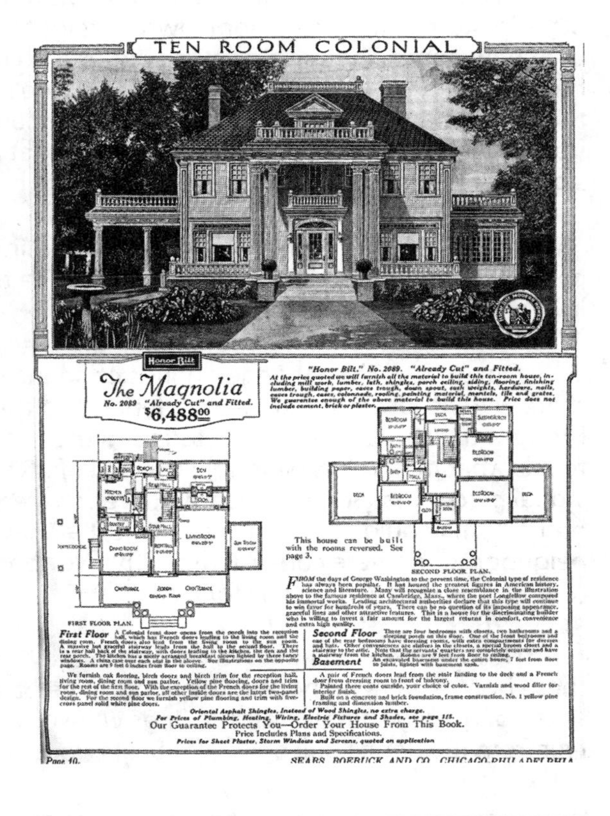

A Sears catalog home for a colonial-style house boasting ten rooms for under $6,500 in 1921.

sizes of homes ranging in price from $650 to $2,500 (prices later went up considerably, but they were always under $10,000). That price included plans and instructions, joists, walls, fixtures, nails—just about everything the homeowner would need to build. The kit came with an estimation for the labor, so a homeowner shouldn't have any surprises. At first, homes were sold without financing, but starting in 1911, Sears offered loans to their customers, and by 1918, nearly all homes were available through credit with loans of up to 15 years. At that time, Sears also offered kit hospitals to the Red Cross.

By 1925, some 30,000 homes had been sold by Sears; just five years later, the number had grown to almost 50,000 homes, and nine years after that, more than 100,000 mail-order homes had been shipped across the country.

Here's how it happened: After the catalog came, prospective homeowners who already owned a lot or property pored over the catalog to find a house that fit their needs, family size, and wallet; choices came in three categories: the Honor Built (the priciest), Standard Built (mid-costs that Sears advertised were best meant for Southern states), and Simplex Sectional (simpler designs). Each category offered an array of choices that included a number of bedrooms, differently sized kitchen choices, bathrooms (though they weren't mandatory), and other customizations that homeowners might have to individualize their new home. Though they were not offered in all homes, central heating, electricity, and indoor plumbing were newfangled features that Sears made available to some customers, and the company kept up with technology. Sears offered the help of at least one central designer for those who were simply stymied by the choices.

Sears, it should be noted, did not discriminate by race; anyone could buy a home, which was a bold move at a time when Jim Crow laws were in effect in the South.

Once the cash exchanged hands or financing was complete, there was a short wait between ordering and the day the railroad brought the first of the new materials to the nearest depot: by "materials," that means kits weighing upward of 25 tons and 100,000 individual pieces that Sears guaranteed even the least experienced homeowner could assemble alone in just 90 days—although some neighborhoods were constructed with neighbor-crews, and Sears furnished labor once in a while for a cost.

Despite it sounding like a DIY project to beat all DIY projects, these homes were no hovels: the floors were oak, the roofing consisted of quality shingles, and ceilings were often a now coveted 10 feet from the floor. Look online for Sears Home Plans, and be prepared

HANDY FACT

From *The Handy History Answer Book*, 3rd edition, by David L. Hudson Jr.: "The mail-order business was pioneered by retailer Montgomery Ward & Company, founded in 1872 in Chicago when American merchant Aaron Montgomery Ward (1843–1913) set up shop over a livery stable and printed a one-sheet 'catalog' of bargains.

"In 1886, American Richard W. Sears (1863–1914) entered mail-order business, opening operations in Minneapolis, Minnesota. He moved the business to Chicago the following year and sold it in 1889.

"Montgomery Ward and Sears Roebuck were aided by the U.S. Postal Service's expansion into remote areas: beginning in 1896, mail could be delivered via the RFD, Rural Free Delivery. In 1913 parcel post was added to the postal service's offerings, further benefiting the mail-order houses and their growing lists of customers."

to be green with envy: The Palmyra, a ten-room, three-floor beauty with gorgeous woodwork, a library, a balcony, and a wraparound porch sold in kit form for as little as $1,993. The Magnolia, a veritable mansion complete with columns in front and two and a half bathrooms, sold for under $6,000. Heavy sigh.

Alas, the Depression made its unfortunate mark on Sears Modern Homes, and the last one sold in 1940. If you know what you're looking for, you may be able to spot a modernized Craftsman-style home, but since most of the original records were lost years ago, there's no way of knowing how many Sears Modern Homes survive. Chances are, though, you may have entered or even lived in one!

BRITISH HISTORY: THE ROYAL FAMILY, PART II

AS A RESULT of being in Kenya when her father died and not having any mourning clothes, Queen Elizabeth II reportedly always travels with an outfit appropriate for a funeral.

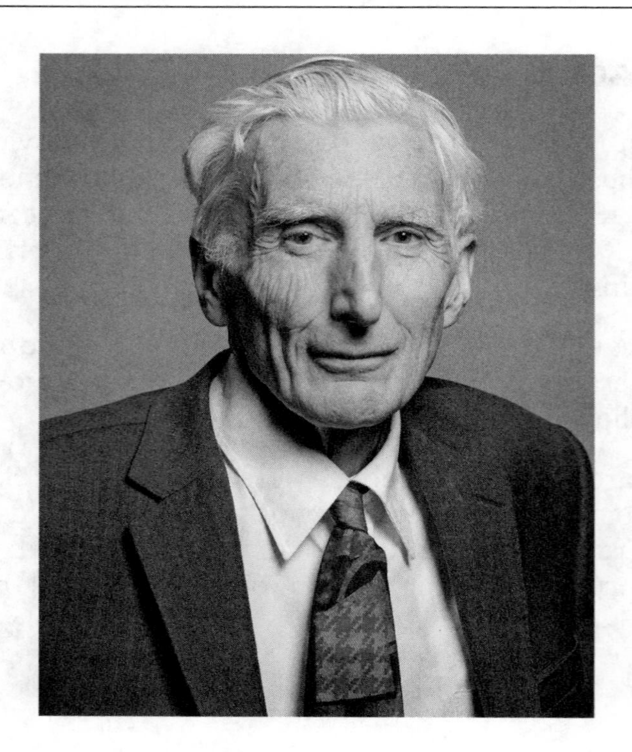

Lord Rees of Ludlow currently holds the post of Astronomer Royal for the British royal family.

IT WAS REVEALED a few years ago that the queen uses small weights to keep her dresses from flapping about in the wind.

BEFORE THE BIRTH of Prince William, royal fathers-to-be were not allowed in the delivery or birth room when the next heir was born. Prince Charles was the first to ignore this protocol. Likewise, Diana was the first to buck the trend of birthing the heir to the throne at home, when William was born in a London hospital.

YOU WILL ALWAYS see royal sons in shorts until they are eight years old. It's been a tradition in upper-crust British families since the sixteenth century.

THERE HAS BEEN an Astronomer Royal on the staff of the royal family since 1675, when King Charles II established the position. There is also someone on staff who tends to the royal stamp collection, and someone is paid to break Queen Elizabeth's shoes in for her. Odder still: during Henry VIII's time, there used to be a Groom of the Stool who acted basically as the royal bum wiper, which makes your job not so bad, eh?

QUEEN VICTORIA IS said to have disliked babies so much that even her beloved Prince Albert was dismayed at the way she

HANDY FACT

According to *The Handy History Answer Book*, 3rd edition, "During her sixty-four-year reign, [Queen] Victoria presided over the rise of industrialization in Great Britain as well as British imperialism abroad. Architecture, art, and literature flourished, in part due to the influence and interests of Prince Albert. It was the prince who, in 1851, sponsored the forward-looking Great Exhibition at London's Crystal Palace. It was the first international exposition—a world fair. The grand Crystal Palace remained a symbol of the Victorian Age until it was destroyed by fire in 1936."

treated her own children. Indeed, had she been a commoner and a modern mother, her behavior might have been considered abusive.

HAD KING GEORGE III (1738–1820) not been a royal, he would have made a fine scientist. He was, in fact, the first royal to study science and accurately forecast the astronomical Transit of Venus, which happened nearly two centuries after he died.

CULTURE: WHO WEARS SHORT SHORTS?

In the not-so-long-ago past, it was a rule at many American schools that boys wore trousers and girls wore skirts or dresses. So, who decided this stuff, anyhow?

The answer isn't easy, and it's an old one.

Of course, you've seen pictures of ancient Greek and Roman orators and lawmakers, and you know they wore togas or longer garments that resembled wraps. For men, women, and children—whether commoner, soldier, or merchant—skirts or skirtlike clothing were business as usual because they were easy to make or fashion and allowed the best amount of freedom of movement. Some garments may have been slightly decorated to indicate status, but the basic shape of the clothing that everyone wore was what we'd consider to be skirt- or dresslike.

There were, however, a few notable exceptions to this: riders and extreme northerners. Scientists who study the history of clothing say that the introduction of rideable animals—particularly horses—seems to be about when trousers appeared because riding wasn't very comfortable without the further protection that pants affords. As for the northerners, well, if you've ever worn a skirt in a blizzard, there's nothing more that needs saying....

Because the occupations that used trousers most—herders, soldiers, long-distance travelers—were mostly men, it evolved that men

King Charles IX of France is pictured here wearing hose in 1566. Such clothing was the height of fashion at the time for men.

wore trousers, while women were relegated to the skirty garments. But by the fifteenth century, men's trousers began to get tighter and shorter until they became hose (see History—Culture: This Part's Got Legs). Even when trousers began to get longer again, you'd look at them funny: they were ballooned, or they were above the knee, or they were "breeches" that most men outside of the theater wouldn't wear today.

And women still had dresses. Long ones that swept the floor and often concealed the ankle. Yards of fabric that was bunched, pleated, petticoated, stretched, laced, bustled, and awkward when wet. Unless one was extremely wealthy, most women made do with just a few dresses.

Despite that, Amelia Bloomer (1818–1894) made her mark early in the Victorian era, the women-wore-dresses ideal lasted until later in Queen Victoria's time, at which point many women, fed up with the restrictions of their expected fashions, joined the Rational Dress Society to gain access to clothing that allowed greater movement. It probably also helped that women had taken to the newfangled contraption called a bicycle and looked at pants and split skirts, recognizing the obvious comfort and utility.

It wasn't until about the first World War that fashions finally relaxed for women, starting with above-the-ankle skirts and the loosening of corsets. The brassiere as we know it appeared in 1913. Flappers took hemlines higher and did away with their mothers' corsets, although they didn't throw them far because corsets gave way to girdles within a few short years of freedom. Hemlines continued to yo-yo throughout history, but there we are again with a skirt.

By 1939, the idea of women in pants in public was starting to take hold. Surely, women had been wearing work pants for years by this time, but slacks or women's trousers were considered daring and even a little masculine. When Katharine Hepburn started wearing them regularly, some gasped, but it didn't take long for most women to add pants to their closets. And that's good: by World War II, when Rosie the Riveters needed pants for wartime work, the garments were already there.

There's a theory that makes the rounds every once in a while that says the height of hemlines follows the stock market. George Taylor (1901–1972) of the Wharton School of Business created this theory in 1925, calling it the hemline index. It says that when hemlines are higher, the economy is bubbling. The shorter the hemline, the worse the economy. Other theories along the same lines use men's underwear, women's visits to their stylists, dry cleaner's bottom lines, and our fast-food-buying habits to predict how we're financially faring.

HISTORYMAKERS: STEP RIGHT UP!

Back before humans were more enlightened, if you were born with a birth defect or suffered from some sort of obvious malady (or had created one for yourself), the likeliest option you had for survival was to sign up to tour with a circus or sideshow; circus managers and owners such as P. T. Barnum (1810–1891), the Ringling Brothers, or one of the Bailey brothers offered a roof, meals, and safety, which was sometimes all a person could get at that time since most employers wouldn't hire a "freak."

Today, the idea of being on a stage and having people gawk at your disability seems abhorrent to most people, but the truth is that many sideshow workers earned decent wages and, in some cases, they actually went on to become famous.

ONE OF THE first to make his mark was Jeffrey Hudson (1619–1682), who was born into poverty but became a favorite in the court of King Charles I (1600–1649). At age seven, Hudson was literally gifted to Queen Henrietta Maria (1609–1669) at a birthday party, and he became her favorite "pet": he enjoyed a life of luxury and privilege, was given an education and taught to hunt, and received a salary starting when he turned eighteen years old. Was Hudson thinking that his position in the court afforded him protection when, in 1644, he challenged a relative of the queen's staff to a duel? Perhaps. At any rate, after Hudson killed the man, the queen had no choice but to banish Hudson from the court. He died in 1682, still in disgrace.

IT'S QUITE UNFORTUNATE that so little is known about Julia Pastrana (1834?–1860). The best guess is that she was born in Mexico in about 1834 with two birth defects that resulted in hirsutism and overgrown gums that distended her jaw. Several versions of her younger life have been told, including that she lived with the governor of the Mexican state of Sinola—but the truth is unknown. What we do know is that she made her appearance in New York in the spring of 1854 as a performer, managed by a man who seems to have been a bit of a scoundrel and a possible pimp and who became her husband. Those who knew Pastrana said that she was gracious, compassionate, intelligent, and sensitive, the latter of which surely allowed her to take offense over being displayed as some sort of savage in a dress. Pastrana died in

March 1860 from complications of childbirth; her baby son suffered from the same birth defects she had and died two days after he was born. To add insult to injury in her sad life, shortly after Pastrana and her son died, her appalling husband allowed their bodies to be embalmed and then bought them back for £800 to be displayed at a freak show.

Hachaliah Bailey (1775–1845) is considered to be the founder of America's earliest circus. He started his circus with just one elephant, a female he named "Old Bet," in 1806.

THOUGH NOT THE tallest people ever (that was Illinois's Robert Wadlow [1918–1940], who was nearly nine feet tall the last time he was measured), Martin Bates (1837–1919) and Anna Swan (1846–1888) toured with sideshows in the late 1870s. Bates was almost eight feet tall; Swan was well over seven feet tall, and their June 1871 wedding made news around the world. He'd been a lieutenant in the Civil War; she was just twenty-four years old. They met while on tour in Europe, fell in love, and had a son and a daughter, neither of whom lived long. Shortly after the death of their first child, they left the sideshow life and retired to a small farm in Ohio, where they lived for the rest of their lives.

BORN INTO SLAVERY in July 1851, Millie and Christine McKoy (1851–1912) were black twins who were conjoined at the lower parts of their spines, sharing one pelvis and one set of legs from the waist down. At the age of one, they were traded by their owner, destined immediately for the "freak show." Then, they were displayed with their mother, but when the girls were just four years old, they were snatched away and taken to Great Britain. Their owner managed to track them down, and they were returned to him in North Carolina after a dispute over who had legal claim to the girls. Once returned, the girls were educated (rare for southern slaves) and were taught to sing and play the piano; even after the Civil War, they chose to stay with their owner because they claimed they loved his wife and could trust her, although they did continue touring for more than twenty years. The twins—often called "Millie-Christine," as though they were one individual—retired to North Carolina in their later years; when they died in the fall of 1912, they were buried in a guarded grave so that their corpses would not be stolen and dissected.

BORN IN TENNESSEE at the end of the Civil War, Josephine Myrtle Corbin's (1868–1928) birth was unremarkable except for one thing: she was born with a birth deformity called dipygus, which means that her body was duplicated, and she had two pelvises,

Shown here around 1880, Myrtle Corbin was born with four legs as a result of a birth defect called dipygus.

four legs, and two complete sets of genitalia. In young Myrtle's case (she took her middle name as her stage name), the outer legs were the stronger of the two, although photos show her using the central pair; even so, a clubfoot on one of the dominant legs made walking difficult. When she was just a month old, it's said that Corbin's father charged a small fee for gawkers, and P. T. Barnum came calling; at the age of thirteen, Corbin joined the Ringling Brothers sideshow. That turned out to be a good call: ultimately, Corbin commanded the vast sum of $450 a week (nearly $12,000 in 2020) and was able to retire at the age of eighteen. One year later, she was married, and at age twenty, Corbin was pregnant with the first of five children. That child had to be aborted due to problems, but Corbin went on to have four more (live) children, graciously allowing the medical community to observe and take note. Like Millie and Christine, Corbin was buried in a guarded grave to avoid body snatchers and grave robbers.

THROUGHOUT HIS CHILDHOOD in Marseilles, Joseph Pujol (1857–1945) was just your normal, average boy. One day, though, when he was still but a lad, he went swimming, and something odd happened: he suddenly realized that he'd spontaneously

HANDY FACT

The Handy Science Answer Book, 4th edition, says: "Siamese twins are identical twins joined at some point of their bodies, most commonly at the hip, chest, abdomen, buttocks, or head. Like other identical twins, they originate from a single fertilized egg; in the case of congenitally joined twins, however, the egg fails to split into two separate cell masses at the proper time. The condition is relatively rare; only about 500 cases have been reported worldwide. Surgery to separate Siamese twins is a complex task and often results in the death of one or both of the twins."

taken a large amount of water into his rectum. What's more astounding was that he learned to do it on command! He then learned that it was possible to suck up air back there and, with practice, was able to release it in the form of different sounds, including music and imitations. What else can a guy who farts like a trombone, a cannon, a thunderstorm, and a singing bird do but take it to the stage? That's what Pujol did, calling himself Le Pétomane, debuting in Marseilles in 1887 and to the Moulin Rouge in 1892 to great acclaim. Alas, an impromptu show, a broken contract, and a lawsuit took the wind out of Pujol's sails, and he went back to being an anonymous baker when World War I broke out. He died at the end of World War II.

CULTURE: I HEREBY RESOLVE TO ...

In the best of all worlds, you would keep your New Year's resolutions. You'd be slim, give up every vice that's ever plagued you, and you'd be wealthy, calm, serene, kind, and wise. More than 4,000 years ago, the ancient Babylonians might have said that same thing. Wonder how that's working out for them....

Back then, New Year's resolutions were more along the line of promises made to gods to pay back debts and return borrowed

Agnes Thatcher Lake Hickok (1826–1907), Wild Bill Hickok's (1837–1876) wife, was the star and owner of Lake's Circus, which had been owned and operated by her first husband, also named Bill. When he was killed in 1869, she picked herself up and continued to parent their daughter and run the circus by herself, thus becoming the first American female to own a circus; it's been said that her managerial abilities and her horsemanship impressed Wild Bill, who married Agnes in 1876.

Julius Caesar of Rome rearranged the calendar when he rose to power, creating a year that began with January 1 in winter rather than with spring in late March, as had previously been the standard.

things, which had to have been easier than losing 40 pounds or quitting smoking. The Babylonians believed that if they honored all promises, the gods would reward them with benevolence; broken promises angered the gods.

Ancient Romans took this one step further, celebrating the new year in the month of January, named for the god Janus, who is depicted by a two-faced man—one looking forward and one looking back. When Julius Caesar became dictator in 49 B.C.E., he rearranged the calendar, putting January first in the lineup, and it likely made sense then to make resolutions that first day of a new year.

By the late seventeenth century, making New Year's resolutions was a common thing in both social circles and religious practices. And so it continues to this day, when nearly half of all Americans make a serious resolution with the best of intentions. Eighty percent of those resolutions are broken by the end of January.

So, resolve to read these things about New Year's Eve:

IN ITALY, TURKEY, and some Latin American countries, the color of your underwear signifies the kind of luck you want. Green undies are for money, red for love, and tighty-whities for peace.

SPANIARDS EAT TWELVE green grapes—one every second leading up to the New Year—for good luck. If you can't stuff the grapes into your mouth by the stroke of midnight, it's said to be bad luck.

IN JAPAN, EACH of the country's Buddhist temples rings a bell 108 times at the stroke of midnight to rid everyone of bad emotions such as anger, sloth, or jealousy.

OLD AND TATTERED dishes are saved all year in Denmark, and at the stroke of midnight on the new year, those plates are smashed against doors—but not your own. You'll want to break the dishes against your friends' doors. The more the mess, the more friends you have.

IN ROMANIA, IT'S believed that animals can communicate with humans on New Year's Day.

IN NAPLES, ITALY, and Johannesburg, South Africa, New Year's Day is the day to get rid of old, unwanted things by tossing them out the door. Mostly, it's reportedly soft or small things, but some revelers have been known to toss sofas and chairs.

HANDY FACT

According to *The Handy Answer Book for Kids (and Parents)*, by Judy Galens and Nancy Pear, "In 45 B.C.E., Roman emperor Julius Caesar instituted what came to be known as the Julian calendar. The Julian calendar was based on a solar year, with a year consisting of 365 days, 6 hours. The year was divided into months that were either 30 or 31 days long (except for February, which has 28 days). Caesar also decreed that the year would begin with January 1; previously, the year had begun on March 25, coinciding with the beginning of spring in the Northern Hemisphere.

"It turned out that the Julian calendar (still in use in some parts of the world), in estimating that a year is 365 days, 6 hours, was off by almost 12 minutes. After several hundred years, those minutes added up, and the Julian calendar was about a week off course from the movements of Earth around the Sun. In 1582, another major calendar reform took place, this time instituted by Pope Gregory XIII. The Gregorian calendar, used in the United States and most other countries of the world today, made further adjustments to align it more closely to astronomical movements." Leap Year, anyone?

IN TALCA, CHILE, the belief is that the dead return to the cemeteries to celebrate the new year, so tradition is to have a party there to join them. The party usually includes feasts, small bonfires, and a sleepover.

HISTORYMAKERS: A BLACK MAN IN THE WHITE HOUSE

In 2006, when Barack Obama (1961–) got the nod to be the Democratic candidate for president in the November election, "blue" voters were excited: here was a chance to make African American history. But did you know that Obama was not the first African American to run for the office of president?

A campaign poster from the 1904 presidential run made by George Edwin Taylor, who was the candidate for the National Negro Liberty Party.

George Edwin Taylor (1857–1925) escaped being a slave by one not-insignificant law that stated that slavery was matrilineal: If the mother was a slave, so was her child. In Taylor's case, his father was a slave and his mother was free, and so was little George when he was born in Little Rock, Arkansas.

Alas, at a young age, Taylor was orphaned and, at age eight, was taken to Illinois, then to La Crosse, Wisconsin, where he was legally remanded to a local black family in nearby West Salem; three years after he arrived in Wisconsin, the Fourteenth Amendment was enacted, granting citizenship to former slaves. Two years later, in 1870, the Fifteenth Amendment granted the right to vote to black men, a fact that young Taylor may not have known but which became important in his life later.

While in Wisconsin, he was given an education, became a journalist, and gained some experience as an activist; he worked at or owned several regional newspapers and became politically active in the state. In 1891, he moved to Iowa, started another newspaper, and changed his political affiliation a number of times; later, he founded the National Colored Men's Protective Association and, having settled on a party to align himself with, became president of the National Negro Democratic League.

Despite the efforts of Taylor and others, however, black men were still being denied their voting rights by the Republican Party. In a last-ditch effort to do something—to at least excite African American voters—the National Negro Liberty Party was formed and, after four men had either turned down the nomination or had been removed from the ticket (and after the nominee before him was arrested and embarrassed the party), Taylor became its candidate for president in 1904, running against Theodore Roosevelt.

Some accounts say that many whites supported Taylor and his reach for the highest office in the land, but white America mostly poked fun at his race. The party had made promises to him that it didn't keep, some states left his name off the ballot entirely, and racism still kept a lot of African Americans from the polls. On election day, Taylor received a tiny, tiny fraction of the national vote.

411

Shirley Chisholm (1924–2005) became the first African American woman to seriously run for president in 1972. There were, in fact, a lot of "firsts" in Chisholm's political career: she was the first black woman elected to the U.S. Congress in 1968. She was the first African American candidate to ask a major party for the nomination for the office of president of the United States and the first woman to try for the Democratic Party's nomination, and she was the first woman to appear in a national presidential debate.

HANDY FACT

According to *The Handy History Answer Book,* 3rd edition, "The first political parties were the Federalists and the Democratic-Republicans. The Federalists favored a strong central government, favored the mercantile and banking interests, and often took a pro-Great Britain position (at least compared to their opponents).

"The Democratic-Republicans favored a less powerful central government, retention of power by the states, favored the interests of farmers along with those of the lower and middle class, and often took a foreign policy stance more in alignment with France.

"Earlier, there were political leaders known as Anti-Federalists, but the Federalists and the Democratic-Republicans were the first political parties after the U.S. Constitution was signed."

GENERAL HISTORY: THIS AND THAT AND RANDOM FACTS

THE VERY FIRST shopping mall was created by Victor Gruen in 1956. Southdale Mall opened in Edina, Minnesota, and was meant to become a hub of the community, complete with theater, recreation, and possible arenas for education.

AT THE TURN of the last century, the life expectancy for the average American was just forty-seven years old. On that note, today, more people die at 11 A.M. than at any other time of day. Furthermore, Las Vegas has the highest suicide rate in the country.

THE VAST MAJORITY of the world's residents don't have a street address. This includes a fair number of U.S. residents, but government officials are working on this problem.

SURNAMES WERE BASICALLY unknown until the time of the Roman Empire. China began using them about 2 B.C.E., and by the Middle Ages, most Europeans had a surname. Still, a surprising number of people in the world don't have last names.

The Southdale Mall, which opened in 1956, is the first enclosed mall to open in the United States. Renamed Southdale Center, it is still open and operating in Edina, Minnesota, today.

IN NEW YORK City, buildings are restricted by height, but if you're planning some construction next to a shorter building, you can purchase the airspace above the shorter building that that building is not using. This gives you "ownership" of that airspace for your use; with the extra added to your building's airspace, you could build a taller structure than "code" would normally allow.

THE FIRST AMERICAN personal income tax was enacted in 1861 to help pay for the Civil War; that tax was rescinded in 1872. Twenty-two years later, a flat-rate federal income tax was enacted by Congress, but it was struck down a year later by the Supreme Court because it was considered unconstitutional. In 1913, the Sixteenth Amendment was passed, allowing the U.S. government to levy income taxes on individuals.

BY MOST ACCOUNTS, when Franklin Roosevelt chose Harry Truman as vice president, FDR barely knew Truman. They had, in fact, met just a small handful of times prior to Roosevelt's fourth inauguration.

MAHATMA GANDHI WAS married at age thirteen and was a father by age sixteen.

IN 1927, SOME 70 percent of the cars in America were owned by people who lived in rural areas or smaller towns.

HANDY FACT

If you think the Dust Bowl years didn't leave a lasting environmental impact, check this out: *The Handy History Answer Book*, 3rd edition, says, "On May 11 [1934], experts estimated that 12 million tons of soil fell on Chicago as [a] dust storm blew in off the Great Plains, and the same storm darkened the skies over Cleveland. On May 12, the dust clouds had reached the eastern seaboard. Between the two storms, 650 million tons of topsoil had blown off the plains."

IN SOME AREAS in America, telephone numbers used to have word prefixes called telephone exchange names or central office names. Each name indicated a certain area of town, sometimes even down to the street, and the first two letters of the word corresponded to the digits on a rotary dial phone; for instance, if your phone number was KEystone 4-0577 (that's the way it would appear in the phone book), someone wishing to call you would dial 534-0577. If you told someone in the know what your phone number was, they'd know roughly where you lived.

CULTURE: SUCK IT UP, BUTTERCUP

So, consider this: everything has a history. Even, as it turns out, something so ubiquitous as the lowly drinking straw. Imagine this: you're in the middle of a hot summer and a cold brew, idly sucking up that wet goodness with a straw that doubles as both a delivery method and plaything. That's roughly what you might've gotten had you been one of the elite in an ancient Sumerian city some 5,000 to 7,000 years ago. Tomb frescoes show that tableau, and scientists quickly recognized what were early drinking straws. Ancient Chinese imbibers did the same with hollow reeds, and in other cultures, drinking straws made of gold, precious stones, and early metals have been discovered in ancient graves.

And that's the way it was for more than 8,000 years.

When sixteenth-century South Americans got a taste for a certain kind of drink, the yorba mate, made regionally from plants, they needed something that would filter the leaves of the plant out of the beverage as it was being enjoyed. Enter the bombilla, a metal straw-and-spoon device with a tiny filter at the bottom end, so you could drink, stir, and filter all at once. Bombillas, by the way, are still available (and still pretty handy).

And then, again, nothing changed for a very long time.

In the late 1800s, beverage consumers were delighted by the manufacture of rye grass straws that were literally made of naturally hollow rye grass. Of course, you could argue that anyone could walk into a field of rye grass and pick their own, but accessibility to the plant was a problem, and the truth was that rye grass straws did require a certain amount of processing by hand. Besides, it was a novel thing to get a free straw with a drink—even though rye grass straws weren't the best solution for sucking up a drink.

And that was enough of a reason for Marvin C. Stone (1842–1899) to act.

It's a fact that rye grass is a sustainable, natural product that is absolutely biodegradable (things late-nineteenth-century folks didn't

The modern paper straw was patented by Marvin C. Stone in 1888.

worry about). But that presented a couple issues—specifically that the straw often got all gummy from the liquid being sucked up into it and that a straw that will disintegrate can disintegrate in a drink, thereby changing the taste of said drink.

Tired of all that, Stone, who had expertise in manufacturing things of a cylindrical nature and had not long before filed a patent for a device to hold fountain pens, took some paper, wrapped it around a pencil, fastened it with a bit of glue, and voila! A straw was born.

Pleased with his device, he made a handful of the straws and left them at his favorite watering hole, where he loved to drink mint juleps. He meant for them to be for his own use, but other bar patrons saw the paper straws and wanted some for themselves, and there you are. Stone patented his invention, and by 1889, his factory was producing paper drinking straws at a rate of two million per day.

In the 1930s, Joseph Friedman (1900–1982) was idly watching his young daughter as she struggled to drink a milkshake, and he saw a problem to be fixed. Friedman was an inventor, and so he tinkered with a paper straw and a regular old screw from his tool kit, wrapping the former around the latter until the pleats would allow a bend in the straw. He patented his idea, too, and created a business to manufacture his product.

And that was the way the world's drinkers sucked up their drinks, but paper straws, bendy or not, weren't without problems of their own. Yes, they were cheap, but they didn't last very long; if the liquid inside wasn't enough to destroy paper, being marinated in liquid would do it. Paper straws became misshapen on the top part, too,

HANDY FACT

According to *The Handy Answer Book for Kids (and Parents)*, by Judy Galens and Nancy Pear, "Soda pop gets its fizz from carbon dioxide, a harmless gas that makes up part of the air we breathe. It is mixed into soda pop to make it light and fun to drink. When a soda bottle or can is sealed, the gas can't escape and stays mixed with the beverage. But when soda is opened or poured into a glass, carbon dioxide—which is much lighter than the liquid—rushes to the surface in the form of bubbles and escapes into the air after bursting. Sometimes the bubbles rise so fast, bringing soda pop with them, that they make a frothy, fizzing foam. Soda pop "goes flat" when it has been left in an open container for a long time and has lost all of its bubbles, or carbonation."

and—ask anyone of a certain age who might remember using them—it was no fun having a lip full of soggy paper bits.

In light of this, it's somewhat of a surprise that it took until the 1960s for straws to be made of plastic—especially when you consider that plastic had been around for a few decades by then. It didn't take long for them to become cheaper to manufacture than their paper cousins, and the new product's popularity grew as consumers realized that plastic straws lasted much longer than paper ones.

And therein lies a problem: plastic straws last much longer than a lot of things; in fact, it could take some 200 years for a plastic straw to decompose. Since Americans alone use some 500 million straws every day, you can see the problem. Landfills are inundated with plastic straws, some of which can end up on curbsides, empty lots, and in the ocean, which can harm wildlife. Many communities and hotel chains have banned the use of plastic straws, and many more are limiting the use or considering the ban.

So, what's the solution? We're back at square one: manufacturers are now making straws of reusable metal and biodegradable, all-natural materials all over again.

BRITISH HISTORY: GO DIRECTLY TO JAIL

If you were a resident of Great Britain prior to the middle of the eighteenth century, you wouldn't have wanted to be a criminal. Punishment ranged from humiliation in the public stocks, which meant being detained by hands, feet, or head (or all three), ridicule, and possibly being the target of animal, vegetable, or stone missiles at best or whipping in a public place at worst. The most incorrigible offenders were executed in some manner befitting the severity of the crime.

After a time, though, these punishments were seen as barbaric, and England began a sort of reformation that saw what were much more barbaric than being a laughingstock, and that included the penal treadmill, which was a large and rather dangerous contraption that operated something like a stair-stepper machine but that prisoners were forced to use for hours and hours on end. That, too, was eventually believed to be inhumane, and the idea of shipping criminals out of England began to take hold.

To be truthful, it wasn't a bad idea—for Great Britain. The country badly needed the manpower in its new colonies to build and help settle. For the convicts, though, it was a pretty raw deal since they were treated rather badly and worked awfully hard. For a time in the later 1600s and for much of the 1700s, British convicts were even sent to the New World, but the American Revolution pretty much ended that little practice.

That was fine for the British; they found somewhere else to send their bad guys and girls when, in 1770, Captain James Cook's (1728–1779) *Endeavor* landed on the east coast of Australia in Botany Bay in what is now New South Wales. On the spot, he claimed the land for the Crown, and it didn't take long for someone to realize that Great Britain had just found its newest punishment.

Eighteen years after Cook made his claim, Captain Arthur Phillip (1738–1814) of the Royal Navy was tasked with creating a new settlement in Port Jackson, which is now a part of Sidney Harbor, consisting almost entirely of convicted people—775 of them, total, aboard six transport ships. At that time, they were joined by a small number of other nonconvicted, free Brits, for a total of perhaps a thousand souls.

If the voyage wasn't hellish enough—estimates were that a quarter of those aboard the ships lost their lives—the convicts' new home was a great unknown. No one seemed to have a clue about the natives in Australia or their friendliness or lack thereof. The pres-

An illustration of the Endeavor *beaching on the coast of Australia in 1770. Captain James Cook claimed the land for England, and it later became a place for that country to dump its prisoners.*

ence or absence of poisonous or dangerous wildlife was a big question, as was the terrain, the weather, and the seasons. The skills of the convicts weren't exactly up for what was, in the beginning, a harsh existence, either: back in Great Britain, it was cheaper and easier to execute the most violent criminals, which meant that most of those sent to Australia were petty criminals, minor thieves, old women, or small children. Few of them had any knowledge of crop-raising or building-raising. Subsequent shipments of convicts only exacerbated the problems already on the ground, and food became scarce for a time.

To the good, though, Phillip recognized immediately that Australia was going to be like no other punishment, ever. For that reason, he subtly elevated criminals and thieves to the same position as freed men, giving them the ability to move about at will and live in their own cabins—something he almost had to do to get the mutual cooperation required to start what turned out to be a town more than a prison, though it's been said that the guards were aware that Australia was still a penal colony, and they were a cruel bunch, quick to use the lash. Even so, through the punishment and the harsh life, for many in the new colony, being forcibly moved ended up being a chance at a fresh start. If a convict had exhibited good behavior, he or she was sometimes allowed the option to return to England, but few actually did.

The success of the community at Port Jackson led the British to bring more convicted prisoners to Australia—eventually, some 160,000 of them in the years between 1788 to 1868. They settled in a series of communities throughout the continent, some of which became prisons within the penal colony (because some people obviously never learn). By that time, in 1850, gold had been discovered there, and dreamers began to intentionally migrate to its shores. Twenty-two years after prison convoys were ended, more than three million people called Australia home; one hundred years after convoys stopped, some twelve million people lived on the continent.

HANDY FACT

According to *The Handy Answer Book for Kids (and Parents)*, by Judy Galens and Nancy Pear, "The smallest continent of the world is Australia. It has almost three million square miles (close to eight million square kilometers) of area. That makes it almost four times the size of the world's biggest island—Greenland—which has an area of around 840,000 square miles (2,175,000 square kilometers)."

Today, Australia is an independent country but is still a member of the British Commonwealth. Sources say that around 20 percent of Australia's residents can trace their ancestry back to at least one former British convict.

CULTURE: WHEREFORE ART THOU, OSCAR?

Given its nearly 100 years of existence, it would be very hard to find anyone in civilized society who wouldn't instantly recognize one of Hollywood's Academy Awards statues. "Oscar" is 13.5 inches tall, weighs 8.5 pounds, is gold-clad, and holds a sword. His face is expressionless, his head is bald, but you gotta love the guy—especially if you're an actor, director, costumer, or any one of the dozens of other categories whose winners' names are engraved on Oscar's base. But despite that everybody knows him, there have been a few times in history—eighty of them, at the time of this writing—where Oscar has gone AWOL.

Oscar was just a kid the first time he supposedly went missing. It was 1937, and he was meant to go to Alice Brady (1892–1939), who won for her role in *In Old Chicago*. Brady couldn't be at the ceremony, though, so a man stepped onstage and accepted her award on her behalf. Not long after the awards ceremony, when Brady contacted the Academy to find out how she could collect her plaque (as awards were given then), officials realized they'd been had and that Brady's award had been stolen.

Or, at least, that's the legend. Truth was that the man was none other than Henry King (1886–1982), the director of *In Old Chicago,* and he gave it to Brady later that evening.

411

During World War II, Oscar statuettes were made of plaster due to wartime shortages.

The next time Oscar went on a little unplanned journey was in 1954, when Margaret O'Brien (1937–), who won an outstanding child actor Oscar, lost her Academy Award to a housekeeper who allegedly pilfered it. Her Oscar was missing for forty years, until it was recovered at a flea market in 1994 and given back to O'Brien.

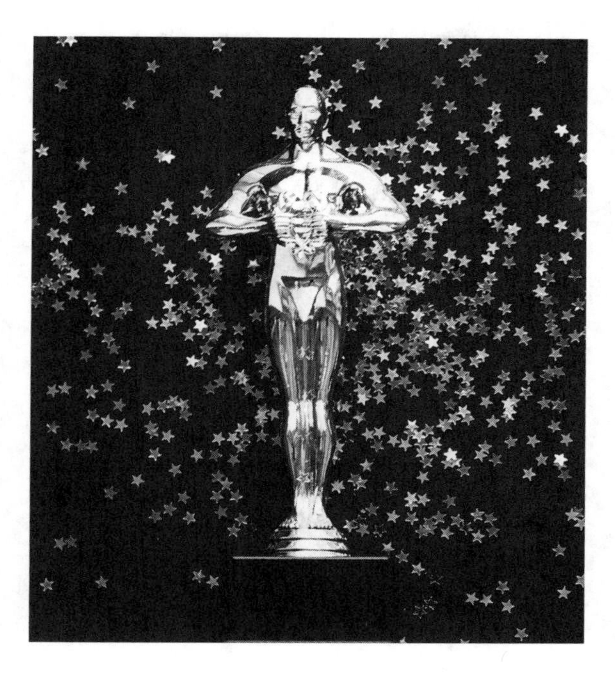

The Academy Award, or "Oscar," is one of the most recognizable and coveted trophies in the world.

In "one of the happiest moments of" her life, actress Hattie McDaniel (1893–1952) was nominated for an Oscar for Best Supporting Actress in *Gone with the Wind*—and won. As she was dying from breast cancer later in her life, she decided that the best place for her plaque was with Howard University in Washington, D.C. Alas, at some point in the late 1960s or early 1970s, McDaniel's award went missing. Nobody seems to know exactly when it disappeared, and nobody knows where it is today.

Actor Gig Young (1913–1978) won an Oscar for his role in the 1969 film *They Shoot Horses, Don't They?* After Young's suicide, his Oscar ended up willed to his agent. While it wasn't exactly missing, it took more than a dozen years for the statue to be returned to Young's daughter.

Gone with the Wind art director Lyle Wheeler's (1905–1990) possessions were seized from a storage locker for nonpayment and were auctioned off. Among the items were the five Oscars Wheeler had won. Because of a loophole in Academy rules, his statues were able to be sold to the highest bidder.

Perhaps the biggest, most well-known Oscar heist came in 2000, when five cases of Oscar gold went missing. They left the Chicago manufacturing plant, where they had been made, on time. They arrived in Los Angeles on time, all fifty-five of them on one giant pallet

Hattie McDaniel was the first African American to win an Oscar.

on March 8, 2000. But at some point in the following twenty-four hours, the Oscars went missing from the trucking company's tracking system.

411

Want your very own authentic Oscar? Be patient: several Academy Award winners have put their own statues up for sale, and some have been sold via estate sales. Be ready to pony up a pile, though: most of them sell for six figures or more. The thing is, you can't buy any Oscar awarded after 1950. That was the year the Academy instituted a rule that all winners since 1950 have signed, stating that they cannot sell their Oscar, except to the Academy for the princely sum of one dollar.

Five days later, when word finally reached Academy officials, they immediately contacted authorities. Even the FBI got involved in the search for the statues and the people who stole them, and a sizable reward was offered. As it turned out, a couple of dock workers had stolen the statues and ditched them—or did they? A few hours after arrests were made, the boxes with statues intact were discovered near a grocery store dumpster. The man who found them received an award for the recovery of most of the statues. The stolen-but-recovered statues were ultimately destroyed because who wants a stolen statue? You shouldn't, that's for sure: Oscar statues now sport serial numbers, making it very hard to own or sell them illegally.

Ten years before that, Whoopi Goldberg's (1955–) statue went missing and was found in a garbage can. One of Katharine

HANDY FACT

The Handy History Answer Book, 3rd edition, says, "On March 22, 1895, the first in-theater showing of a motion picture took place in Paris, when the members of the ... National Society for the Promotion of Industry ... gathered to see a film of workers leaving the Lumière factory at Lyons for their dinner hour. The cinematography of inventors Louis (1864–1948) and Auguste (1862–1954) Lumière, ages thirty-one and thirty-three respectively, was a vast improvement over the kinetoscope, introduced in 1894 by Thomas Edison (1847–1931), whose film could only be viewed by one person at a time...."

Hepburn's (1907–2003) four Academy Awards was stolen in 1992. The thief who stole Olympia Dukakis's (1931–) Oscar tried to sell it back to her (she declined and called the Academy for a replacement). And at the time of this writing, two Weinstein Company Oscars are still missing after having disappeared during the time of Harvey Weinstein's (1952–) sexual assault accusations, arrest, and trial.

AMERICAN HISTORY: BORN HERE

Imagine that you've lived in the United States your whole life. Your father was here his whole life. Your grandparents—heck, your whole entire family—was in this country even before *everybody else* got here. So, what would you have to do to get citizenship?

The answer to that came in 1924, when Congress signed the Indian Citizenship Act.

In the earliest days, before America became a nation, Native tribes practiced self-governing. Decisions, laws and rules, and social nets were made or created within each group's own people.

But Americans thought that Natives should assimilate. In the earliest history of America, that belief often meant that Native Americans were persecuted, subject to what we would call harassment, and they

Native American students at the Carlisle Indian Industrial School, which was founded in 1879 by Richard Pratt as the first boarding school for Indigenous boys. This and other schools were designed to culturally assimilate Native Americans.

existed within a small patchwork of rights. Before the Civil War, citizenship was given to some Native people, mostly those who were half Indian or less. After the Civil War, some Republicans tried to confer U.S. citizenship on tribes that were friendly to the government, but individual states rejected or constricted those wishes.

In 1887, the Dawes Act moved to assimilate Native Americans by doing away with individual tribes and putting a cap on how much of the tribal lands each person would own; leftover land was offered to settlers then. Indian schools were closed, and Native children were sent to "white" schools where they were taught white culture and language, a practice that lasted in some places well into the twentieth century.

One year after the Dawes Act was passed, most Native American women who were married to white citizens were given citizenship in the United States, but they were still not allowed to vote; that wouldn't come until years after white women were granted that right. That year—1868—was also the same year that the Fourteenth Amendment was ratified, stating that citizenship was given to "all persons born or naturalized in the United States," but the law was somehow interpreted by the Supreme Court to disinclude Native Americans. To add to that, the Fifteenth Amendment, ratified in 1870, specifically gave African American men the right to vote.

HANDY FACT

According to *The Handy American History Answer Book*, by David L. Hudson Jr., "In 1830 Congress passed the Indian Removal Act, a measure that President Jackson supported and signed into law that May. It called for the creation of an Indian territory in Oklahoma. Technically, the law called for the voluntary removal of various native tribes from the southeastern part of the United States. In practice, the law led to the forced removal of the Native Americans from their lands. The most famous of these removals was of the Cherokees from Georgia to what later became Oklahoma. This arduous journey became immortalized in history as the 'Trail of Tears.'"

President Andrew Jackson

Native American veterans of World War I were offered U.S. citizenship in 1919. It was obvious by then that the Dawes Act wasn't working, so in 1924, Homer P. Snyder (1863–1937), Republican representative of New York, proposed to grant all Native Americans full status as U.S. citizens.

The Indian Citizenship Act, also known as the Snyder Act, was signed by President Calvin Coolidge (1872–1933) on June 2, 1924, and granted not only citizenship but also made allowances for health care and better classrooms. Even so, many individual states were slow to grant the act's full rights, and some Native Americans in certain states had to wait more than forty years to get their full citizenship.

HISTORYMAKERS: AN ATHLETE AND A HERO

In the last year or so or longer, you've undoubtedly learned that superheroes walk among us. Some of them take their accolades and appreciation given and go home. Others fade away into history....

When he was just a lad of fifteen, Shavarsh Karapetyan (1953–) experienced something that would mark any normal boy for life: a group of bullies grabbed him, beat him up, tied his hands, fastened a heavy rock around his neck, and threw him into a deep lake.

Somehow, he managed to break the bonds on both hands and neck and escape the weight. Somehow, he swam to the surface of the water and saved himself. This, as he said later, and the encouragement he received from friends and family, spurred Karapetyan to take swimming lessons, and he discovered that he was quite good at it.

Two years later, he was good enough to be a champion swimmer, but he was not allowed, for some reason, to join a national Armenian team. Undaunted, he took up finswimming, which is a water sport that uses a snorkel and fins, and within half a year, he began winning major races. Ultimately, Karapetyan became a world champion finswimmer seventeen times, a European champion thirteen times, and a Soviet champion seven times. He broke world records. He was a phenomenon.

And if this was all that he ever did, it would still be pretty remarkable....

But in 1974, while aboard a runaway bus, Karapetyan prevented the bus from plunging into a gorge, saving the lives of everyone on the vehicle, including himself.

Among his many honors for heroically saving dozens of people's lives, Shavarsh Karapetyan had an asteroid belt named after him (3027 Shavarsh) in 1986, and he also carried the torch at the 2014 Winter Olympics in Sochi, Russia.

But there's more.

On September 16, 1976, Karapetyan and his brother, Kamo, were running for a workout near the Yerevan Lake in Armenia when they heard a trolleybus as it went off the rails, broke through a wall, and plunged into the lake with a load of passengers, some several dozen feet offshore. Karapetyan immediately dove into about thirty feet of freezing water, broke the windows of the submerged bus with his feet, and rescued as many people as he could: more than twenty, although just twenty survived.

His efforts left him injured and hospitalized for more than a month, and it ended his finswimming career. For two years, nobody knew what he'd done; photographs and eyewitness accounts were kept locked away by Russian sources, but by 1982, the story had been passed from person to person until it had finally gotten to the newspapers. Suddenly, Karapetyan was a famous man and was awarded two medals for his bravery, though he still seemed to maintain his quiet humility.

And then....

On February 19, 1985, Karapetyan was coincidentally near a sports and concert arena in Yerevan when the building went up in flames. Unselfishly, he was one of the first into the burning building to help firefighters rescue the people inside. His efforts once again resulted in injuries—burns and smoke inhalation this time—and Karapetyan was again hospitalized.

Since his last act of heroism, Karapetyan is said to be living a quiet, unassuming life in Moscow, where he owns a shoe company. He'll never be forgotten, though: a main asteroid belt was named after him in 1978.

HANDY FACT

The Handy Answer Book of Physics, 3rd edition, says that Daniel Gabriel Fahrenheit (1686–1736) "did not define 32° as the freezing point of water. Instead, he defined 0° as the freezing point of a water and salt mixture. Since salt lowers the freezing point of water, the freezing point for this mixture was lower than it would have been for plain water. Upon defining the degree intervals between the freezing and boiling points of the water and salt mixture ... he found that water itself freezes at 32° F.

AMERICAN HISTORY: SIT. DOWN. STAY IN THE OVAL OFFICE

At various times throughout history, it's been said that a cur occupies the Oval Office. Some have said that a wildcat or a mule lives there. For more than a century, the occupant has been represented by either a donkey or an elephant. But seriously, even political animals love their furry friends....

Though, technically speaking, he didn't live in the White House, our first president loved his dogs. George Washington (1732–1799) always had dogs around; in fact, he's credited with creating the American foxhound breed of dog. In addition, he had turkeys, cattle, and, of course, mules: not only was he a Founding Father, but his love for the mule and his efforts at breeding them gave him the nickname of the Father of the American Mule.

John Adams's (1735–1826) horses were the first residents of the newly built White House stables in 1800 (Adams only lived there dur-

President Gerald R. Ford is shown working while getting a little stress relief from his dog, Liberty, a golden retriever. Many U.S. presidents have had pets, especially dogs, in the White House.

ing the last four months of his term). Adams's successor, Thomas Jefferson (1743–1826), owned two bears that were a gift from explorer Zebulon Pike (1779–1813); when they got too big to have at the White House, he sent them to live in a Washington, D.C., museum. That's probably a good thing, too, since Jefferson's successor, James Madison (1751–1836), had a flock of sheep that grazed on the White House lawn, and sheep don't live long around bears.

James Monroe (1758–1831) owned a single dog—a Siberian husky. Not to be outdone, his successor, John Quincy Adams (1767–1848), owned an alligator that lived in the White House, and Andrew Jackson (1767–1845) had a parrot with a very foul mouth. Martin Van Buren (1782–1862) had a pair of tiger cubs that were a gift from the Sultan of Oman, but he didn't keep them long because Congress wouldn't allow it. And our fourth president, William Henry Harrison (1773–1841), kept living lawn mowers on the White House grounds during his brief month as president: his pet cow, Sukey, and his goat, Whiskers.

John Tyler (1790–1862) and James Polk (1795–1849) both loved their horses very much, as did Zachary Taylor (1784–1850). Remember, back then, a horse was more than just a ride: it was a means of transportation and was much faster than walking, so it's natural that a man might get attached to his steed.

America didn't get its first petless president until 1850 when Millard Fillmore (1800–1874) moved into the White House. The fact that he didn't have a furry friend living with him doesn't mean that he didn't love animals; in fact, Fillmore founded the Buffalo, New York, chapter of the American Society for the Prevention of Cruelty to Animals after he left office. (The only other petless presidents were Chester Arthur [1829–1886] and Donald Trump [1946–]).

And then, America went to the dogs again: Franklin Pierce (1804–1869) had two tiny dogs that he doted on while he was in office, and James Buchanan (1791–1868) had a dog, as did Abraham Lincoln (1809–1865). On that note, Lincoln and his entire family were big "animal people." In addition to their dogs, the Lincolns had turkeys, horses, goats, pigs, rabbits, and cats. Lincoln, it's said, was the first president to perform the official pardon of the White House turkey at Thanksgiving—although doing so wasn't an annual tradition for several more presidencies. Turkey, it seems, was a popular dish at the White House, although every one since the tenure of George H. W. Bush (1924–2018) has made it off the grounds alive.

Andrew Johnson (1808–1875) befriended mice. Ulysses Grant (1822–1885) had a dog, birds, ponies, and horses. One of the first Siamese cats in the United States belonged to Rutherford Hayes's (1822–1893) wife, Lucy (1831–1889).

James Garfield (1831–1881), despite his comic book namesake, had a dog.

If you'd visited the White House of Grover Cleveland (1837–1908), you'd be forgiven for mistaking it for the zoo. Cleveland had many dogs, rabbits, horses, and chickens. You have to wonder if Benjamin Harrison (1833–1901) saw it and liked it: his Oval Office tenure included dogs and a goat. It must've been quite a shock for White House staff when William McKinley (1843–1901) moved in with just a parrot.

It's hard to top the menagerie that Teddy Roosevelt's (1858–1919) family brought to the Executive Mansion: there were dogs and a snake, a parrot, ponies, more dogs, guinea pigs, a rooster, bears (real and of the Teddy sort), and a raccoon. Did the staff feel whiplash when William Howard Taft (1857–1930) moved in with his quiet cow named Miss Pauline Wayne?

Woodrow Wilson (1856–1924) had a flock of sheep that he kept on the White House grounds, but they weren't just pleasant to look at—although … can you imagine sheep on the mansion lawn these days? Wilson and his wife had the sheep sheared and the wool auctioned off to benefit the Red Cross and the war effort during World War I.

Warren Harding (1865–1923) had a dog, but when Calvin Coolidge (1872–1933) moved into the White House, he brought with him another menagerie of dogs, cats, birds, and a spoiled raccoon named Rebecca. The Coolidges also enjoyed the company of a donkey, an antelope, and several formerly wild animals. Not to be outdone, Herbert Hoover's (1874–1964) kids had alligators in addition to dogs.

Franklin Roosevelt (1882–1945) famously had a Scottish terrier named Fala, and Harry Truman (1884–1972) was given a puppy one day during his White House tenure, but he turned it down and tried to give it away. Imagine the kerfuffle *that* was! Ultimately, though Truman was a dog lover, he "kept" the canine gift by letting the pup live at a Maryland naval base.

Surely, Mamie Eisenhower (1896–1979) could understand Truman's dilemma: her husband, Dwight (1890–1969), had a dog he adored, but Mamie was less than thrilled with the animal.

John Kennedy's (1917–1963) White House was filled with a cat, a Russian mongrel, a salamander, and a pony named Macaroni. Lyndon Johnson (1908–1973) had two beagles, Her and Him, and you can well imagine the howl of protest when LBJ was photographed lifting his dogs by their ears. Richard Nixon (1913–1994) had Checkers, a dog; Gerald Ford (1913–2006) had Liberty, a dog; Jimmy Carter's (1924–) White House had a cat; Ronald Reagan (1911–2004) had Lucky, a dog; George H. W. Bush had Millie, a dog; Bill Clinton (1946–) had Socks, a cat; George W. Bush (1946–) had Barney, a dog; Barack Obama's (1961–) White House had two dogs; and Joe Biden (1942–) has two German shepherds, one of which was adopted from a shelter.

Who knows what kind of political animals will move in next?

CULTURE: THE CLUB YOU CAN'T BELONG TO

Chances are, you grew up with the idea that you can do anything and be anything, but you should know by now that that's not entirely true. One thing you can aspire to is to be a member of one of America's most exclusive clubs, but you should know right now that your membership is extremely highly unlikely.

It should come as no surprise that gentlemen's clubs began with the uppermost crust of British society in the later eighteenth century and that "gentlemen" meant just that: women or men of lower classes needn't have applied. These were not "gentlemen's clubs" as we euphemistically know them now; there were likely libraries in gentlemen's clubs then, probably billiards, and maybe a bit of furtive gambling. The atmosphere of a 1790s club was aristocratic, male-interest-focused, and probably also politically charged.

By the mid-1800s, one could find gentlemen's clubs all over London, were one to search. And, of course, the idea of an exclusive enclave came to America—to San Francisco—and to a group of artists, writers, actors, social leaders, politicians, and movers and shakers. In 1872, they founded the Bohemian Club, named for the intellectual meaning of the word rather than its devil-may-care implication; six years later, the club established a campground called Bohemian Grove, located in Monte Rio, California.

While the structures may have changed somewhat and were undoubtedly updated for modern amenities, the campground is basi-

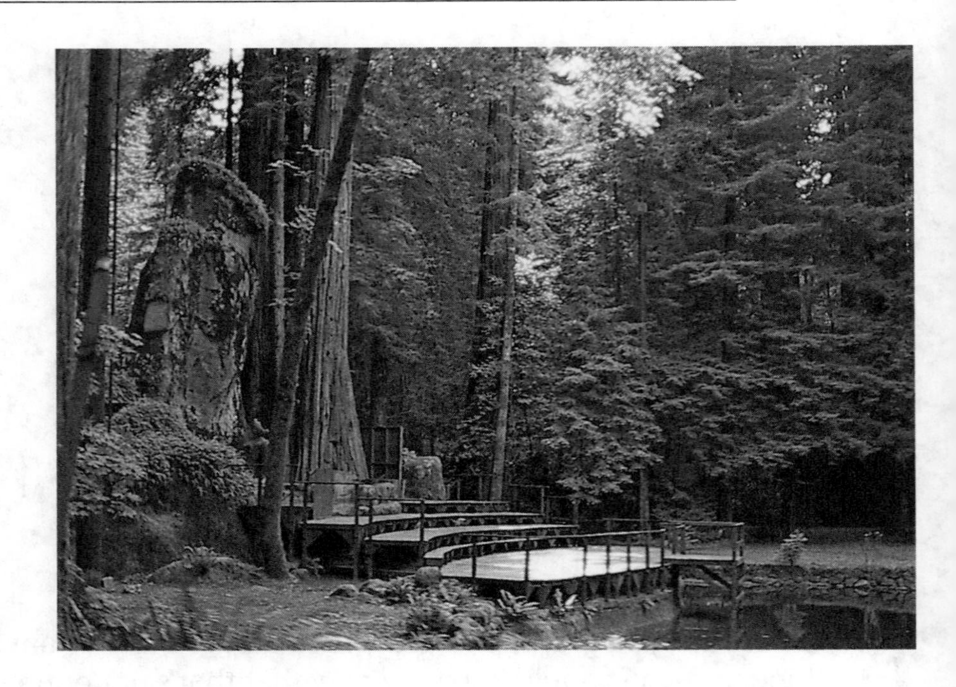

The owl shrine and state at Bohemian Grove.

cally the same as it was in late June 1878, when 2,700 acres of forest and ancient redwood trees became Bohemian Grove. A seventy-foot statue of the Buddha has mostly disintegrated, but the thirty-foot owl at the head of the Grove's lake still stands. Within the vast acreage, there are dozens of smaller camps that are governed by a camp "captain," who also ensures that the camp is maintained—often by local landscapers who are not members. There are several larger buildings, delineated areas, and a main dining room for relaxation, entertainment, and gathering purposes and where rituals are performed, including one with hooded, red-robe-clad men and a fake coffin full of "cares" to burn.

Controversially, the issue of logging has also plagued the Bohemian Grove: the grounds are full of mature redwood trees, and "harvesting" happened for years, as did protests. At the time of this writing, logging on the grounds is once again forbidden by law. Another controversy: most of the club's members are white men; aside from a small handful of female honorary members who were granted very limited membership, women have been and continue to be denied as full members.

Though spy novel lovers may scoff, this camp—which was once routinely used for sixteen days every summer prior to the twenty-first century—isn't meant for conspiratorially juicy reasons, dark ceremonies, or backroom deals, though it's said that a rare meeting relat-

HANDY FACT

You have to think that Ronald Reagan must've greatly enjoyed Bohemian Grove. According to *The Handy American History Answer Book*, "Reagan was called the Great Communicator because he was a powerful speaker who had the ability to charm his viewing and listening audiences."

ing to World War II's Manhattan Project was held on Bohemian Grove grounds in 1942. More likely, as a handful of journalists who've managed to infiltrate the Grove claim, the annual meeting is filled with roasts, rowdiness, and rambunctiousness, resembling *Animal House* more than *Wall Street*. If you can imagine the likes of Ronald Reagan, Richard Nixon, Jack London (1876–1916), Clint Eastwood (1930–), Henry Kissinger (1923–), or Charles Schwab (1937–) pulling middle-school pranks, well, there you go.

As for actually seeing the Bohemian Grove itself, there are five ways to do so: you can try to sneak in, but security is high-tech, they don't play games, and you should just assume that you're going to be arrested. You can become a journalist and offer up your credentials, but Bohemian Grove staff probably won't want cameras or voice recorders around. You can "know somebody" and snag a rare guest spot, but if you're a female or a child, you'll have to vamoose by sundown. You can become staff, but if you're a woman, there is supposedly *literally* a line drawn in the sand to indicate how far you can go on the grounds, and you, too, would have to leave at sundown.

Or you can join the club, but good luck with that.

HISTORYMAKERS: ANNNNND, THEY'RE OFF!

When it comes to sports, there really aren't a lot of them older than horse racing. It's believed that racing horses dates back some 6,500 years, to the Central Asian people who first domesticated the animals. Nobody knows for sure, but they might have gambled on the outcome of the ponies; if they did, they'd be in good company. In many parts of the world, horse racing is one of the few sports on which it's legal to gamble. It's popular in many countries and cultures, with millions of spectators and billions of dollars in wagers and wins each year. But there's more to horse racing than just watching ponies run in a circle around the track....

THE VERY FIRST winner of the Kentucky Derby was Aristides in 1875; the jockey aboard Aristides was Oliver Lewis (1856–1924), who was African American. Once again riding Aristides, Lewis won the Belmont Stakes that same year.

EARLY IN THE sport in America, African American jockeys were considered to be the best in the world, in part, it's said, because the jockeys were groomers and exercisers both during slavery and in the time of Reconstruction, and they knew the horses better than did their owners. In the first twenty-eight years of official racing, fifteen of the Kentucky Derby winners were African American; one of them, Isaac Burns Murphy (1861–1896), won three times, in 1884, 1890, and 1891. That feat ultimately gained him admission to the Hall of Fame at the National Museum of Racing and Hall of Fame. Murphy once calculated that he'd won around 44 percent of his races—and if that's correct, it's a record that hasn't been equaled and may never be bested.

THE TERM "TRIPLE CROWN" wasn't used officially until the *Daily Racing Form* popularized it in 1930.

NO MATTER WHEN he (or she) was born, a racehorse's birthday is January 1. This is done to group the animals by age, but it's an imperfect method; it means that a horse born at the end of January in any given year could potentially race against a horse born at the end of November of that same year.

SO FAR, THERE has never been a racehorse over age eighteen to win a professional race.

YOU WILL ONLY ever see three-year-old horses in a Triple Crown race. If they don't win any of the three races (the Kentucky Derby, the Belmont Stakes, or the Preakness Stakes), they have no second chances or do-overs. There's just one shot at winning those races.

WINNING A TRIPLE Crown ain't no small thing. At the time of this writing, just thirteen horses in the modern history of racing have accomplished the feat. They are: Sir Barton (1919); Gallant Fox (1930); Omaha (1935); War Admiral (1937); Whirlaway (1941); Count Fleet (1943); Assault (1946); Citation (1948); Secretariat

(1973); Seattle Slew (1977); Affirmed (1978); American Pharaoh (2015); and Justify (2018).

THOUGH THERE IS no weight limit per se, thoroughbred racehorses generally carry, on average, around 126 pounds worth of total weight during a Kentucky Derby; for the Belmont, it's recommended that fillies carry no more than 121 pounds. Jockeys generally weigh between 108 and 118 pounds, which takes effort for most jockeys to maintain and is controversial; still, this is up considerably since the 1920s, when jockeys weighed as little as 95 pounds. No matter what he's carrying weight-wise, today's thoroughbreds still manage to run at about 40 miles per hour.

JUST AS THERE is no official weight restriction, there is no official height restriction for American jockeys. Even so, most of them are between 4'10" and 5'6".

IN 1967, KATHY Kusner of the U.S. Equestrian Team applied for a jockey license in Maryland but was turned down; a judge ruled in 1968 that Maryland had to certify her, but she was injured before she could ride. Later that year, Penny Ann Early was on the roster to ride in Churchill Downs, but the male jockeys boycotted her appearance. They likewise promised to boycott if Barbara Jo Rubin followed through on her ride at Tropical Park in early 1969. The first female to finally ride in an American pari-mutuel race was Diane Crump (1948–), who rode Bridle n' Bit at the Hialeah Race Track in Florida on February 7, 1969, finishing tenth out of twelve. A little over a year later, Crump became the first woman to compete in the Kentucky Derby, riding Fathom, and finished fifteenth out of seventeen. Crump was twenty-one years old.

JOCKEY FRANK HAYES (1888–1923) had not won a single race aboard any horses he'd ridden in all his years as a jockey, and he

A c. 1901 photo of Churchill Downs, home of the Kentucky Derby.

HANDY FACT

Where did the saying "Slow and steady wins the race" originate? *The Handy History Answer Book*, 3rd edition, says it came from *Aesop's Fables*, which were "short, moralistic tales ... handed down through the oral tradition...." Those stories "include the well-known story of the tortoise and the hare," from which the saying comes.

never thought he'd win riding Sweet Kiss, a pony with 20–1 odds at Belmont Park. It was June 4, 1923, and the pair were up against fan-favorite Gimme when Hayes stepped his steed into the starting gate; he was thirty-five years old and had spent years as a stable hand, had ridden as a substitute jockey a time or three, and was outright scheduled to be on Sweet Kiss that afternoon. The horse did its job: Sweet Kiss surprised everyone by beating every other pony on the track, including Gimme. The jockey, alas, didn't know that: at some point during the race, Hayes, who'd been in good spirits in the hour before the race, had a massive heart attack and died astride his horse; no one knew it until he fell from his saddle at the end of the race. Hayes became the first dead man ever to win one leg of the Triple Crown; officials at Belmont gave him the win—the only win Frank Hayes would ever have astride a horse.

CULTURE: MATCH THOSE NUMBERS

Life will be great when you finally win the lottery, won't it? Somebody has to win eventually, so you play because, well, after all, it might as well be you, right? Or maybe not....

While having a national lottery is a relatively modern concept, the idea of a lottery itself is not very new at all: as early as 205 B.C.E., gamblers in China's Han dynasty enjoyed playing a version of Keno, possibly in order to fund buildings and other projects. Lotteries were used in ancient Greece as a method of filling government positions, and the Romans used lotteries for punishment. During the Renaissance, Italians edged the lottery into a more recognizable (for us) form, when certain municipalities used it to raise needed funds while awarding gifts and cash to players. By the latter 1500s, Queen Elizabeth I had turned the lottery into a way of filling the public-fund coffers and, in fact, used lottery money to fund the New World colony of Jamestown.

At the time of this writing, the largest amount ever paid to a single Mega Millions winner was more than $1.5 billion. The anonymous South Carolina winner opted to take a lump-sum payment of nearly $900 million instead.

In part because of that, having a lottery was in the cards for early Americans, and lots of lotteries were run by our forefathers, including Ben Franklin, John Hancock, and George Washington. They, too, used the money to fund projects, including colleges and other institutions that the new country needed but, alas, people couldn't get past the fact that, at its most basic, lotteries are gambling, and scandals of some sort or another seemingly always surfaced. Add that to the morality (or lack thereof), and a huge outcry that followed temporarily banished most lotteries. By 1890, every state but Delaware and Louisiana had banned lotteries altogether.

Don't think that lotteries disappeared then, though. They didn't, but they also didn't exactly advertise that fact.

Puerto Rico was the first to bring a U.S.-run lottery back, which happened in 1934. Thirty years later, New Hampshire offered the lottery for its residents and visitors. Scratch-off tickets were offered in the 1970s, and "ball" lotteries followed a short time later in most states.

Absolutely, lotteries and such kinds of state-sanctioned games of chance have been a great thing for municipalities. Lotteries are moneymakers, for sure, and many fund public education within the state. Should you want to take your chances with a lottery, you can find hundreds of books and websites that offer tips and ways to beat the system and get rich. But let's do some real math....

While the average "ball" ticket is a buck or two, and you can spend upward of $10 for a scratch-off, the vast majority of lottery tickets either yield a payout of exactly what was spent to purchase them or nothing at all. That means that while Americans spend more than $70 billion on lottery tickets per year, or about $300 per person, most of the time, there's a return of $0. You can have all the "systems" you want, but the truth is that the lottery is a game of chance, and the odds are not in your favor. If you win, you won't get the entire jackpot amount; you'll have to split it in some way or another down the line, and then there are the taxes involved, both for the state (if applicable) and for the fed-

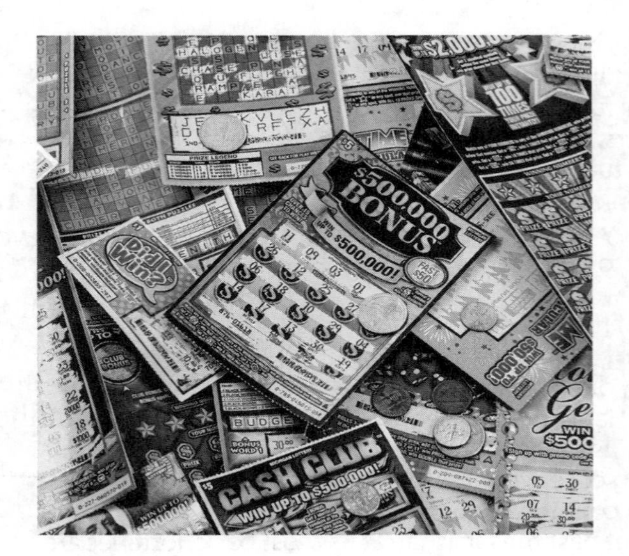

Lottery games like these instant-win scratch cards from Michigan are often used by local governments to fund such public projects as schools.

eral coffers—not to mention the nightmare for your heirs, should you die before getting all of an annuity if that's your choice.

But okay, let's say you win the whole enchilada. It sounds like a dream, but interviews with past winners show that money, indeed, does not buy happiness. Many big lottery winners are sad to find that family, friends, and acquaintances treat them like walking wallets. There's jealousy, greed, and sometimes violence right outside the door. Money doesn't protect anyone from divorce, death, scams, frauds, thieves, blackmailers, or heartbreak. A surprising number of big-ticket winners find themselves bankrupt quicker than they ever thought possible.

It's fun to dream, but remember this: state lotteries are exempt from truth-in-advertising laws. It's true, somebody's got to win, but the real statistics are that it's usually many *millions* to one that you won't. Experts say that investing your money is a wiser bet.

HANDY FACT

The Handy History Answer Book, 3rd edition, says that poker, "in which a player bets that the value of his or her hand is higher than those held by the other players, was invented in the 1820s by sailors in New Orleans. The sailors combined the ancient Persian game *as nas* with the French game *poque*, which itself was a derivation of the Italian game *primiera* and a cousin of the English game brag."

HISTORYMAKERS: CRUNCH, CRUNCH, CRUNCH

For many people, there are three main signs that fall has arrived: changing leaves, pumpkin spice everything, and an abundance of freshly picked apples.

As for the latter, while the Old Testament says they were in the Garden of Eden at the beginning of creation, scientists say that apples likely sprang from mountainous regions of Central Asia, particularly Kazakhstan, millions of years ago and spread in some areas of Asia and Europe. Undoubtedly, humans ate wild apples, but they ultimately determined that diversity was the key to palatability: humans as early as 3000 B.C.E. were believed to have cultivated the plants and selected them based on taste, consistency, and firmness. The origin of grafting to create a new type of plant is unknown.

Different flavors of apples grew in different climates and areas of Europe and crabapples sprouted in the New World, but as people migrated to America, they brought with them the seeds of their own favorite strains. Jamestown residents brought apple cuttings and plants with them to make cider. Individuals planted a seed or three here or there on their own little plot of land, but America's first official orchard was started in 1625 in Boston by Reverend William Blaxton (1595–1675). Seeds were traded with the Natives or spread on trade routes within the colonies and, of course, John Chapman (1774–1845), aka Johnny Appleseed, did his share of sharing seeds of the fruits. Thomas Jefferson was said to be the forefather, so to speak, of the Fuji apple; George Washington and Benjamin Franklin both were big apple fans. By the mid-1800s, hundreds of different kinds of apple cuttings or seedlings were available for purchase, sent to anywhere in the country. As better horticulture methods were found, that number increased.

Yes, apple seeds contain cyanide and are poisonous. But considering that there are about 5 apple seeds in the average apple (but sometimes more), you would have to eat the seeds of nearly 40 apples to get anywhere near a fatal amount of cyanide.

Today, no one knows exactly how many varieties of apples existed in North America a century ago, but one estimate is in the low five figures. Part of the problem with identifying them now is with the overlap of names that were misspelled or colloquially miscalled when a strain was moved prior to good record keeping. Other issues came with

Unlike some American legends, Johnny Appleseed's story is based on a real man named John Chapman, a pioneer, missionary, and nurseryman who spread a message of conservation along with his apples throughout much of New England, the Midwest, and Ontario.

changing tastes and strains that were intended for exact uses, such as cider-making, baking, or eating. Experts fear that there are strains they don't know about, apples that were lovingly brought to America by ethnic groups and planted, but that may be about to become extinct. Only recently, apple fans of the Lost Apple Project recovered a few precious trees in Oregon that were thought to be lost.

HANDY FACT

The Handy History Answer Book, 3rd edition, says that Steve Wozniak (1950–) and Steve Jobs (1955–2011) "spent six months working out of a garage, developing the crude prototype for Apple I," a computer that "was bought by some 600 hobbyists—who had to know how to wire, program, and set up the machine. Its successor, Apple II, was introduced in 1977 as the first fully assembled, programmable microcomputer, but it still required customers to use their televisions as screens and to use audiocassettes for data storage. It retailed for just less than $1,300."

Undoubtedly, there's a lot left to be discovered about America's love affair with apples, but in the meantime, you can have apples in many different forms, including yogurt, applesauce, pastries and cakes, apple butter, dried apples, caramel apples, and, of course, crunchy-fresh. And, if you're feeling particularly adventurous, there are some 7,500 different kinds of apples grown around the world and a hundred different kinds of apples that are available commercially on a regular basis.

CULTURE: PRETTY FOR THE POTTY

If you could somehow travel back seven decades, you might arrive at your destination needing a rest stop. Once inside the little room, though, you'd get a surprise.

In the 1950s, when folks were flush with cash and upgrading the rooms in one's home was suddenly possible with a postwar economy, it became common, even de rigueur, to have colored bathroom fixtures.

Blue and pink were the most popular bathroom colors, but green and yellow were also quite a thing then, and the whole decorating idea was literally a floor-to-ceiling endeavor. Plumbers sold colored toilets in robin's-egg blue or petal pink. You could have a sink (with or without matching countertop) and a tub plus tile in buttercup or soft green. Your flooring—which might be carpet or rugs—could match the fixtures. Walls, too, and if you were *really* into this fad, you could buy towels and washcloths, a toilet seat cover, a bathroom scale, wastebasket, drinking glass, bars of soap, and light switches in the color you'd chosen for your bathroom.

The crowning piece was the color-coded toilet paper, which was sold well into the 1970s. It came in blue, green, pink, and yellow, although beige and purple showed up occasionally; if you lived large, you might be able to find TP with a floral design, to shake things up a little. Some of the fancier stuff came lightly scented, while some toilet paper also had matching tissues if you dared (and, for that matter, paper towels for the kitchen).

So, what happened, and why can't you find your blue toilet paper at the grocery store or from a local retail outlet anymore?

Toilet paper was first used as a method of hygiene by the Chinese in the sixth century. By the late 1300s, Chinese manufacturers began mass-producing the product; its use was not fully popularized in Europe until later in the 1700s.

It seemed that three things occurred nearly simultaneously: health-related reports were released in the 1980s, and doctors began to say that the dyes used in toilet paper weren't good for delicate skin or sensitive areas. Environmentalists also began to question the use of artificial dyes on discards that had potential to affect the water system; never mind that white toilet paper is bleached paper.

But perhaps the biggest reason for the demise of colored toilet paper and facial tissue came from evolving American tastes. Colored bathroom fixtures and the fad of matching everything in the smallest room in the house fell quickly out of favor. Homeowners in the 1980s wanted sleeker looks with a basic, white tub and sink, so color was *out*, and it took those blue, green, pink, and yellow rolls with it.

Even so, in certain areas, blue toilet paper was still available well into the twenty-first century.

Today, you can find boldly tinted TP and tissue if you know where to look for it, and because color comes with a price if you're willing to pay for it. And since fads tend to circle back again, keep a lookout: you never know when green toilet paper may again be seen in a bathroom near you.

HANDY FACT

Ancient Romans used the remains of a dead sponge (a primitive type of animal) on a stick for purposes of hygiene. According to *The Handy Science Answer Book*, 5th edition, compiled by the Carnegie Library of Pittsburgh, a live sponge "that is 4 inches (10 centimeters) tall and 0.4 inches (1 centimeter) in diameter pumps about 23 quarts (22.5 liters) of water through its body in one day. To obtain enough food to grow by 3 ounces (100 grams), a sponge must filter about 275 gallons (1,000 kilograms) of seawater."

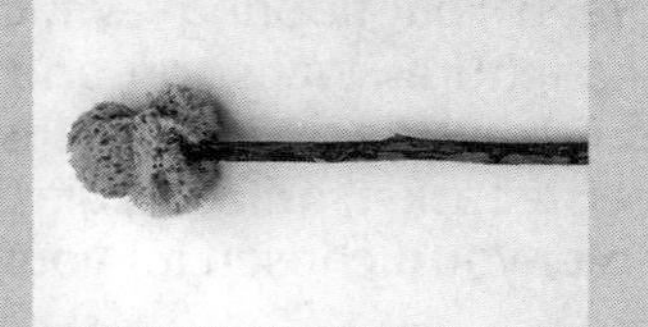

The technical word for the sponge on a stick used in ancient Rome for cleaning oneself after defecating is "xylospongium" or "tersorium." Next time you get in an argument with someone, you might suggest they "go suck on a xylospongium!"

HISTORYMAKERS: THE LEGEND OF WATKUWEIS

It's hard to imagine now, but for more than three dozen years after Meriwether Lewis (1774–1809) and William Clark (1770–1838) returned from their Corps of Discovery mission, their efforts were largely ignored by their fellow Americans, as if the whole long trip was a big nothing. What many people don't know is that they almost didn't return at all.

First, some backstory....

Just days after Thomas Jefferson purchased 828,000 square miles from the French for $15 million, (or about $18 per square mile), he sent Lewis and Clark on a journey that was multipronged. Most famously, the scientist in Jefferson wanted an idea of the natural re-

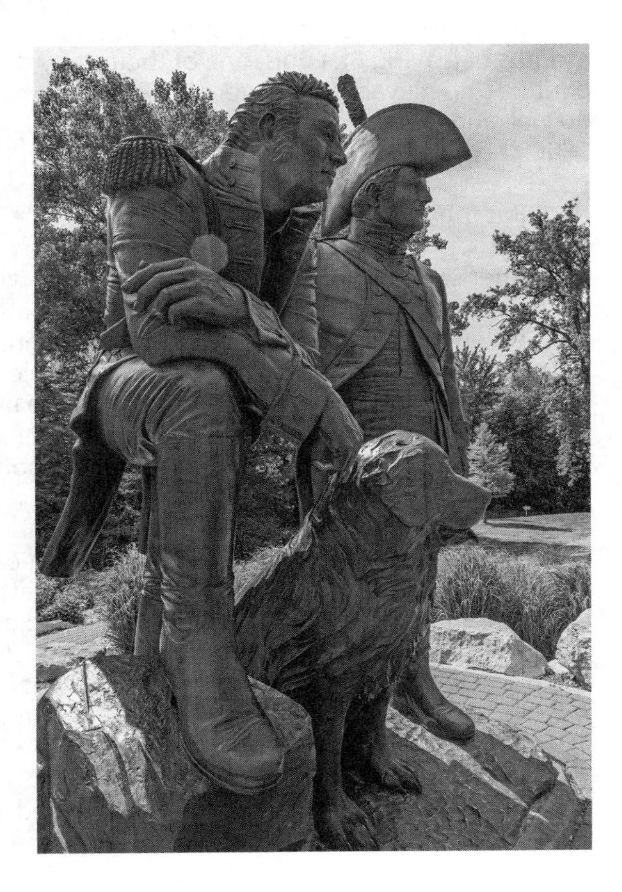

The Lewis & Clark Monument is found at Frontier Park in St. Charles, Missouri.

sources that the land held both scientifically and in its potential for economic gain. He also wanted a map of what he'd just purchased; specifically, he wanted a map to show routes for travel and trade, but that's not all. Jefferson also wanted to present a strong American presence in that territory to show potentially encroaching British and French troops who really owned the area. He also wanted to declare American sovereignty to the Native American groups within the Louisiana Purchase.

Lewis and part of the crew left Pittsburgh on August 31, 1803, and picked up Meriwether; his slave, York; and nine other men in Kentucky; the group left there and arrived in St. Louis, Missouri, in early December that year. They wintered there, leaving St. Louis in May of 1804 with Jefferson's wish to take their exploration past what had officially been purchased and on to the Pacific Ocean before other Europeans could claim the northwest areas, in what is now the Pacific Northwest.

They took plenty of knives, arms and ammunition, gifts for the Natives to offer in peace, and cartography equipment for mapmaking.

Alas, provisions were no protection from problems, and within months of leaving St. Louis, one of their party died of a ruptured appendix. Unbelievably, however, the rest of the expedition appeared to go well and, aside from the deaths of two Blackfoot warriors in July of 1806, there were no other deaths reported.

But there was that little "almost"....

After crossing the Bitterroot Mountains in what is now part of Idaho and Montana, the convoy apparently fell into a bit of trouble. It was September of 1805; winter had already set in on the mountains, and the party was not as prepared as they should have been. Snow covered the mountains, which were larger than the men of the expedition thought they'd be, and frostbite was a constant fear. After eleven days in the Bitterroots, the group was discouraged, bone-tired, and starving.

Oral legend has it that the Nez Perce had been watching the Corps of Discovery members, and they weren't entirely happy with what they saw. It would have been a simple matter for the tribe to kill the party and take their weapons, thus gaining a not-inconsequential pile of the best firearms around, ridding themselves of the white men, and also asserting dominance over other tribes in the area. A council was assembled to discuss the idea of a raid, complete with the murder of everyone in the party.

And that's when an elderly woman named Watkuweis stepped forward and told the tribe to leave the white men be.

Legend says that Watkuweis had had a harsh life. As a young child, she abruptly and quietly disappeared from her tribal land one day, having been captured, it's said, by a war party based in nearby North Dakota. From there, Watkuweis—who, legend says, was the first Nez Perce to see a white person—was reportedly taken to Canada, traded as a slave again and again, and mistreated. Once in adulthood, her story diverges: some storytellers say Watkuweis was married happily to a white man, while other accounts say that he was her captor. Either way, she decided to leave her white husband. To escape him, she relied on white neighbors and settlers for help; they supposedly sheltered her and assisted her escape, giving her a horse and enough provisions to leave. She made her way back from the Great Lakes area (a trip of over a thousand miles) to the Nez Perce, and she told them where she'd been and about those who showed her compassion.

Although the Nez Perce listened to her story and apparently digested it, when Lewis and Clark and their party arrived, bedraggled, on tribal land, the Nez Perce were said to have been sorely tempted by the things they would get by killing the white men—the least of which were guns and the respect that came with firearms. Watkuweis, they say, got wind of the discussion and hurried to the chief's camp. She urged him to leave the Corps of Discovery men alone.

"Do them no harm," she supposedly said.

And so it was.

HANDY FACT

The Handy History Answer Book, 3rd edition, tells us that "after the expedition, [Meriwether] Lewis was made governor of Louisiana Territory, a post he served from 1807 to 1809. [William] Clark resigned from the army in 1807 and became brigadier general of the militia and superintendent of Indian affairs for Louisiana Territory. In 1813 he became governor of the Missouri Territory (the Louisiana Territory less the state of Louisiana, which was organized as a state and admitted into the Union in 1812), a post he held until 1821."

CULTURE: I LOVE THE KNIGHT LIFE

Imagine being the guy (almost always a guy, though it was sometimes a woman) who saves the world—or, at least, everybody thinks he does. Imagine the clink of sword, the pageantry of competition; yes, the life of a medieval knight must have been an exciting one. Think of it: powerful rousting, jousting, and adverturing.

Or not.

THE KNIGHT IN SHINING ARMOR

Okay, let's start with this: the practice of wearing a full suit of armor wasn't the first kind of protection that knights wore.

Starting in at least 3 B.C.E. through the Iron Age and into the earliest Middle Ages, knights wore suits of chain mail (which may be redundant; calling it "mail" is probably best), including coats of various lengths, mittens, shoes, leggings, and mail to cover every body part but the face. Special underlayers were made for comfort and coolness, or everyday clothing might be worn beneath the links; many warriors also wore an overcoat as well. Mail, which weighed roughly 30 pounds for a full suit, could potentially stop a sword but was no help in deflecting arrows or a heavy cudgel and was therefore of rather limited protection.

Chain mail was an effective way to protect the body from injury and also had the advantage of being flexible.

In about the year 14, full armor became available. You can be sure that showing up to any battle in a full suit of the stuff was intimidating, but there was one obvious problem: all that metal.

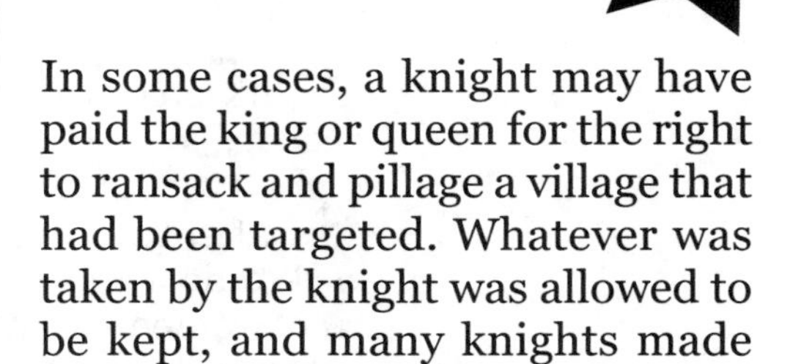

In some cases, a knight may have paid the king or queen for the right to ransack and pillage a village that had been targeted. Whatever was taken by the knight was allowed to be kept, and many knights made their fortunes in this manner.

Wearing a suit of armor wasn't all it was cracked up to be, as it turns out. First of all, there's the weight of the whole contraption: a full suit of armor weighed from forty pounds up to more than a hundred pounds, not counting the helmet and all the weapons and other things a good knight had to carry. If the armor was worn over chain mail, that just added to the heft. This weight meant not only sore muscles at the end of a battle, but it also meant that a knight couldn't run very fast or very far. Even just moving around a little bit was difficult and exhausting, even though most suits had rivets and joints that allowed some extent of freedom of movement.

Secondly, a suit of armor was hot both inside and outside. Wearing it with the sun beating down on it was a good recipe for heat exhaustion and dehydration; taking it off and touching it could mean a nasty burn if the battle took place in a hot desert. You have to think that taking part of a suit of armor off must've felt good, but doing so also exposed one to danger.

Lastly, a suit of armor wasn't the protection you might expect it to be. The plates that made up the body of the armor left areas of exposure open for arrows. The armor wasn't thick enough to withstand the force of a charging horse with a lance. The helmets made it hard to see if danger was heading toward a knight, and some of the vents that were forged large enough to breathe were also large enough to be pierced by an arrow. Worst of all, even if the armor wasn't pierced by an enemy sword or cudgel, the force of weapon on metal could hurt a man wearing armor enough to kill him.

Adding insult to (literal?) injury: a knight was often responsible for purchasing his own battle-ready armor. Other options were that he could steal his armor, take it after he'd vanquished another knight, or take his chances with the cheap stuff that merchants and burghers wore. Despite the fact that the life of a knight was what every boy aspired to, any way you looked at it, the cost of a life-saving suit of armor wasn't for misers: to get a decent suit cost roughly the equivalent of three years' labor for an average medieval craftsman.

HOW ONE BECAME A KNIGHT

In the early days of chivalry, knights were merely noblemen who went to battle. Being a knight then was a simple matter of merely taking up arms and riding off.

It's later, in about the twelfth century, that things began to change....

After that, when a firstborn boy child came into the world into a well-to-do medieval household (or a girl child; history records a handful of women who were beknighted), his destiny was likely decided before he learned to walk.

At around age seven, the child was taken from his home and went to live with the knight and to learn the ways of knighthood. This meant being a page first, which looks a lot now like play: a page would pretend-practice with faux weapons, he'd learn to ride well, and he might be formally educated. He would have minor chores and tasks to do for the knight or his family but not a lot of responsibility.

Once a page became more mature—generally at around age thirteen to fifteen, he graduated to squire. This was serious stuff: a squire was a sort of right-hand man to a knight. He was often responsible for keeping a knight's armor in good shape, without rust, and clean. He may have cared for the knight's horse or at least had a hand in it; he might have worked to care for the knight as well by

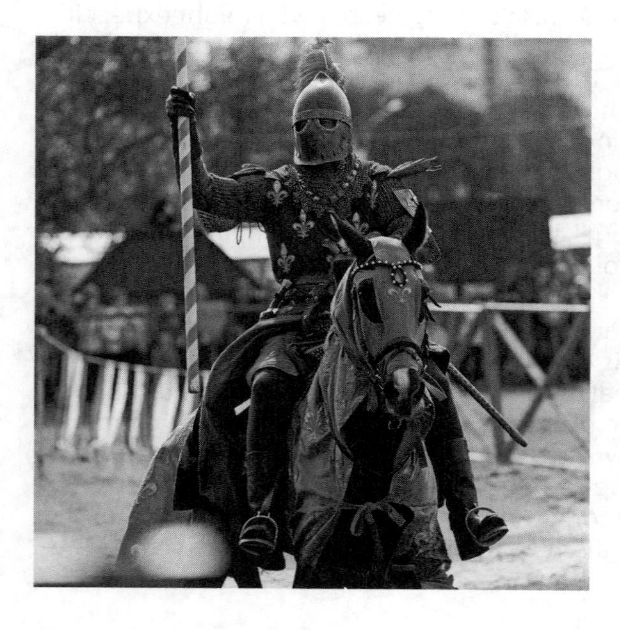

There was a definite process of training and apprenticeship to become a knight. One worked his way up from page to squire and then knight.

bringing him food and drink when needed. As a squire, he was encouraged to practice with real weapons; if there was a battle, he might go along to assist the knight.

Once a young man had mastered all the skills his mentor taught him to the mentor's satisfaction, then he became a knight—although usually not with the fanfare we expect with today's knighting ceremonies. Yes, there was some pomp involved, a swearing-in ceremony of loyalty, and what were certainly meaningful gifts bestowed upon a new knight, but for the most part, when a man became a knight, he *just did*. He no doubt then started casting about for his very own page and squire.

As for wealth, though, it's complicated. Being a knight didn't indicate wealth—although, to be sure, a knight was better off than were the peasants. In early times, knighthood was conferred upon men of any means, but later medieval knights came with a background of some privilege. Most, however, were not landowners; even so, being a knight offered the status of minor nobility, in many cases. With castles being costly to run, it's unlikely, therefore, that a knight had his very own fortress, complete with moat or drawbridge.

KEEPING THE PEACE

Being a medieval knight was a great way to blow off steam, apparently.

While it's true that knights were supposed to be chivalrous, that didn't always happen, according to many sources. Shaking down the local peasants was permissible in times of war, and that meant robbing them (at best) and maiming them so they could no longer serve their lords (at worst). This is a surprise since codes of chivalry were instituted to ensure that knights protected a lord's peasants, but that obviously didn't work so well, and plundering, looting, and destruction happened so often that priests sometimes called on lords to reign in their knights. It should be mentioned that codes of chivalry banned any thoughts of assaulting priests or women, so there's that.

On that note, it should be pointed out that because of the power that the Catholic Church had in medieval times, the Church had quite a hand in keeping knights in line.

That may have been a challenge, though: Medieval knights often had plenty of time on their hands to cause mischief. For those of means (or with access to a generous lord of means), there were peasant-servants, who catered to their whims; food and drink were plentiful, heavy on the latter, and parties were not just occasional things.

HANDY FACT

Says *The Handy History Answer Book*, 3rd edition, "Feudal knighthood ended by the sixteenth century. Today, the only vestige of this tradition is found in Britain, where knighthood is an honorific designation conferred by the king or queen on a noble or commoner for extraordinary achievement." Of course, you can likely see a great jousting tournament at your local Renaissance Faire.

Today, knighthoods are honors handed out to recognize people for their charity and good citizenship. For example, the singer and musician Sir Elton John was knighted by Queen Elizabeth II in 1998.

Games, tournaments, and practice filled up their days, as did basic peacekeeping among the peasants and ensuring that the lord of the manor was a happy man.

AMERICAN HISTORY: MY DINNER WITH FIDO, REDUX

Going to a restaurant or getting takeout is always a treat for two reasons: one, because you get to enjoy a great meal that you didn't have to make; and two, because there's almost always enough for tomorrow's lunch, too. So, why do we sometimes say those leftovers belong to Fido?

If you've gotten this far into this section, it shouldn't surprise you that Romans in the sixth century B.C.E. encouraged guests at ancient banquets to bring linens to the party. Initially, the intention was to use the linens to wipe faces and fingers, but later, the intention was that they could take home some of the extra food.

Surely, through the ages, people were either encouraged to take home leftovers or they just snuck them out when nobody was looking, but the distinction must be made between taking leftovers from a private home and taking them home from a paid eatery. The former must have had some degree of shame to it unless the hostess was adamant. On the other hand, while restaurants and inns were available for dine-in as early as the latter 1600s, dining habits were different then (see "Stick a Fork in It" in British history), and there was a real possibility that the establishment's cook had plans for whatever didn't get eaten at a communal meal. As meals became personally ordered and individually served, the rules had to have loosened. Still, even though the argument could be made (as it is now) that the diner paid for everything he was served, at some point, taking home the food that remained at the table was something of an embarrassment, and leftovers were taken with some degree of furtiveness.

All this changed during World War II. Food shortages were a real thing then, and just having a meal out was a special treat—both things that eating establishments recognized. Restauranteurs also knew that Americans, because of the war effort, were encouraged to feed table scraps to pets (and, for that matter, to livestock). Whether it was to help customers avoid embarrassment or to actually feed leftover food to pets, the "doggie bag" or "doggie pack" came to be.

Records conflict as to which establishment was the first to offer doggie bags. Some say the practice started in San Francisco and Seattle. It may have been New York or Ohio or Alabama. But by the 1950s, restaurants across the country would, upon request, happily box or bag up bones and gristle for Man's Best Friend in a handy, wax-paper-lined container that often sported a cartoon dog or words indicating that the customer wasn't taking home leftovers—oh, no.

The very *idea* that the food was meant for human consumption made etiquette mavens clutch their pearls. Taking home food

While animals have been eating table scraps for as long as they've been sharing human mealtimes, the first commercially made pet food appeared in 1860 when American businessman and entrepreneur James Spratt (?–1880), after having noticed dogs eating leftover biscuits from the crew of a returning ship, formulated dog food made of grains, vegetables, roots, and beef-based liquids.

CAUTION.—It is most essential that when purchasing you see that every Cake is stamped SPRATT'S PATENT, or unprincipled dealers, for the sake of a trifle more profit, which the makers allow them, may serve you with a spurious and highly dangerous imitation.

SPRATT'S PATENT MEAT FIBRINE DOG CAKES.

From the reputation these Meat Fibrine Cakes have now gained, they require scarcely any explanation to recommend them to the use of every one who keeps a dog; suffice it to say they are free from salt, and contain "dates," the exclusive use of which, in combination with meat and meal to compose a biscuit, is secured to us by Letters Patent, and without which no biscuit so composed can possibly be a successful food for dogs.

Price 22s. per cwt., carriage paid; larger quantities, 20s. per cwt., carriage paid.

"Royal Kennels, Sandringham, Dec. 20th, 1873.

"To the Manager of Spratt's Patent.

"Dear Sir,—In reply to your enquiry, I beg to say I have used your biscuits for the last two years, and never had the dogs in better health. I consider them invaluable for feeding dogs, as they insure the food being perfectly cooked, which is of great importance "Yours faithfully, C. H. JACKSON."

"36, North Great George Street, Dublin, June 9th, 1874.

"Gentlemen,—Please to forward to my private residence, as above, 4 cwt. of Dog Biscuits as before; let them be precisely the same as those supplied on all former occasions I have much pleasure in bearing personal testimony to their suitability and general efficiency for greyhounds, and in adding that my greyhound, Royal Mary, winner at Altcar of last year's Waterloo Plate, was almost entirely trained for all her last year's engagements upon them. "Yours obediently, WILLIAM J. DUNBAR, M.A."

"Rhiwlas, Bala, 21st June, 1873.

"Sir,—I have now tried your Dog Cakes for some six months or so in my kennels, and am happy to be able to give a conscientious testimonial in their favour. I have also found them valuable for feeding horses on a long journey, when strength and stamina are important objects. It was the opinion of my brother judges and myself that dogs never appeared at the close of a week's confinement in better health and condition than the specimens exhibited at the Crystal Palace Show, and I understand that your Cakes are exclusively used by the manager. "R. J. LLOYD PRICE."

An 1876 ad for Meat Fibrine Dog Cakes from the Spratt company.

HANDY FACT

The Handy Science Answer Book, 5th edition, says, "Aluminum foil is produced by rolling sheet ingots cast from molten aluminum. The sheets are then rerolled to the desired thickness using heavy rollers. Alternatively, aluminum foil is made using the continuous casting method and then cold rolling. Aluminum foil was first produced in 1903 in France and in 1913 in the United States. The first commercial use of aluminum foil was to package Life Savers."

in a bag was gauche. It was a complete social faux pas! Emily Post (1872–1960) spoke out strongly against it.

And yet, not even the great Mrs. Post could deny the thought of leftover General Tso's or half a T-bone for lunch, and restaurateurs knew it. Some even separated the edibles from the bones and gristle and sent the works home with happy diners, who may—or may not—have been feeding them to the pup.

Today, though many foreign visitors to the United States think the take-home idea is a bit on the disgusting side and few restaurants in foreign countries offer a doggie bag, we don't think twice about having our leftovers wrapped tight in foil so we can take home tomorrow's lunch.

HISTORYMAKERS: WORKING FOR PEANUTS

Back when Jim Crow laws were in effect and basic racism largely kept black Americans and white Americans apart, the rules didn't stop at housing, jobs, water fountains, or travel. It extended to the comics, too.

Racism on the funny pages has been around as long as the comics themselves. Many nineteenth-century strips had recurring characters that were stereotypically black spear-carrying cannibals with nose-bones and grass skirts; later black comic strip characters were lazy, unintelligent, and had bulging eyes and pendulous lips. Even the more popular ones, like *Mandrake the Magician* from cartoonist Lee Falk (1911–1999), originally included a savage-acting sidekick, though he was ultimately updated.

Not long after World War II ended, *All-Negro Comics* published and featured a black scientist. While this wasn't a comic *strip* but a comic *book*, this opened the door for more artists to offer black characters and even the genre's first superhero, Black Panther, in July of 1966.

And yet, racism in comics persisted.

The nation reeled in the days following the assassination of Martin Luther King Jr., and most Americans couldn't stop thinking about it and what it meant for the nation. Harriet Glickman (1926–2020), a schoolteacher in Los Angeles, took her thoughts and acted: she wrote a letter, dated April 15, 1968, to artist-cartoonist Charles M. Schulz (1922–2000) of *Peanuts* fame and asked him to create a black friend for Charlie Brown.

Unbeknownst to Glickman, Schulz had been thinking about it already.

By this time, *Peanuts* had become one of America's best-loved comic strips; it had been the basis of animated cartoon specials and a Broadway musical, and the strip was read by millions of fans around

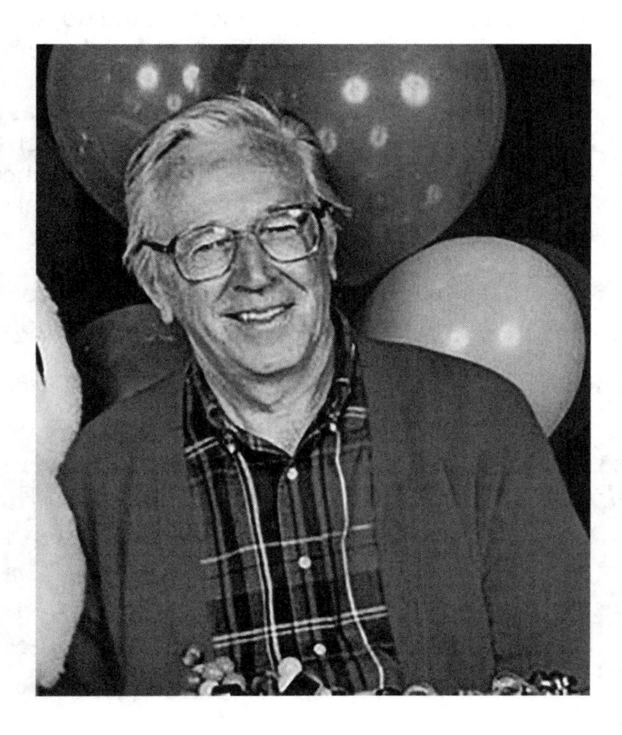

Cartoonist Charles Schulz drew some flak from Southern newspaper editors when he introduced a black boy named Franklin to his famous comic strip, Peanuts.

When *Peanuts* debuted in the fall of 1950, it was a flop. Only seven newspapers took the strip on, and at the end of its first year, the comic came in dead last in one reader's survey. It wasn't until a book of reprinted strips was published that *Peanuts* gained a large and fiercely loyal fan base.

the world. Schulz's characters were sweet and wise and weren't strangers to gentle (or not-so-gentle) pokes at current events, but as much as the notion appealed to him, Schulz fretted that adding a black character to his comic strip during those very tumultuous times might seem "condescending" to black readers. He was quite worried that he, the white son of a white barber from Minnesota, wouldn't be able to understand enough of what it was like to be a black child growing up.

What many people don't know is that for a lot of cartoonists, there's significant "lag time" between the time an idea for a cartoon is formed and the day it publishes—at least a week and sometimes several weeks. Because of this, we can surmise that Schulz didn't think too long before he took Glickman's idea to heart (it's said that they forged a friendship) and acted on it.

On July 31, 1968, *Peanuts* readers found Charlie Brown at the beach, where his "silly sister" had thrown their beach ball in the water, and it floated over to a black boy named Franklin, who was Charlie Brown's age. In subsequent columns, the boys became friends, though the storyline said that they didn't attend the same school (he was classmates with Peppermint Patty and Marcie). Franklin was a good student, polite and conscientious; his father was a soldier in Vietnam, and Franklin loved to talk about his grandfather. In later years, in animated specials, he was shown to be quite a dancer, but beyond occasional nuggets and sporadic appearances, Franklin, as a character, was never quite as rounded out as the other characters in the strip.

In the months after Franklin made his debut in *Peanuts*, Schulz admitted that he caught some flak from a handful of newspaper edi-

HANDY FACT

Fans would agree: Charles Schulz was a top-notch artist. From *The Handy History Answer Book*, 4th edition: "We do not actually know [how long humans have been producing art] because many forms of art, such as wood carving, decorative clothing, and face painting, have not survived. The oldest surviving art is cave paintings. Cave paintings done by *Homo sapiens* may be 40,000 years old, while cave paintings by Neanderthals are possibly 64,000 years old."

tors in the South and from Southern readers who strongly believed in segregation, even in the comics. He stood up to them, telling one editor that he had two choices: "Either you print it just the way I draw it or I quit. How's that?"

THE LAST THING ...

THE LAST AMERICAN World War I veteran was Frank Buckles (1901–2011), who was sent to Europe in 1917 on the RMS *Carpathia* and served in England and France as an ambulance and motorcycle driver. During World War II, though Buckles was a civilian, he was captured in the Philippines and was held in a Japanese POW camp. Buckles died on February 27, 2011, at age 110. Overall, Florence Green (1901–2012) was the absolute last World War I veteran. She served in the Allied Armed Forces. Green was a British citizen.

THE LAST TSAR of Russia was Nicholas II or Nicholas II Alexandrovich Romanov (1868–1918), who ruled from November 1, 1894, until his forced abdication on March 15, 1917.

THE LAST SONG by Lawrence Welk to hit the Top 40 charts was "Calcutta," which reached #1 in February of 1961. It was his only

Lawrence Welk and His Orchestra had a dozen top-100 hits on the Billboard charts, ending in February 1961 with "Calcutta."

#1 hit and was also the last of a long string of Welk songs to hit the charts.

THE LAST TIME Hollywood made a silent film was in 1935. The movie, *Legong: Dance of the Virgins,* was not originally shown in the United States despite its origin here because it was feared that its on-camera female nudity might cause audiences to riot.

THE LAST KNOWN U.S. slave ship was the *Clotilda,* a two-masted schooner that arrived in Mobile Bay in 1860. Between 100 and 160 Africans were aboard the *Clotilda* when it landed.

THE LAST CONTINENT to be discovered was Antarctica in 1820, although no human had actually set foot on the continent until 75 years later. Before Antarctica was discovered, there was New Zealand, which was not discovered until the late thirteenth to early fourteenth century.

THE LAST LIVING Civil War veteran was Albert Henry Woolson (1850–1956), who fought with the Union Army. There may have been others, but Woolson was the last undisputed veteran.

IN CHRISTIAN THEOLOGY, the Last Four Things are death, judgment, heaven, and hell. These four things are recommended for meditation.

"NEVER THINK YOU'VE seen the last of anything."—Eudora Welty

UNTIL 1975, U.S. armed forces included cigarettes in every soldiers' daily food rations.

THE LAST AMERICAN to receive a Civil War pension check was Irene Triplett. She was born on January 9, 1930, to a thirty-something mother and an eighty-four-year-old father who was a Civil War veteran. He died in 1938 at age ninety-two; Triplett received her $73.13 monthly check until she died in May of 2020.

AND FINALLY, TO your last day: From birth to age forty-five, accidents are the leading cause of death. After forty-five, it's heart disease.

SCIENCE

GENERAL SCIENCE: THE FIRSTS

The best way to kick off this part of the book is with a few notable firsts....

THE FIRST ANIMALS in space were fruit flies sent by American military scientists a whole forty-two miles into space on a Nazi-designed V–2 rocket on February 20, 1947.

THE FIRST TIME anyone thought to write down how to do surgery on humans happened nearly four thousand years ago in ancient Egypt. The Edwin Smith Papyrus (named after the man who bought and tried to translate it) shows that even then, humans were pretty savvy about how our bodies worked.

ELIZABETH BLACKWELL (1821–1910) graduated from Geneva Medical College in New York in 1849, thus becoming the first woman in the United States to earn a degree in medicine. Even

Marie Curie won the Nobel Prize twice—once in physics in 1903 for her work on spontaneous radiation and again in 1911 (in chemistry) for her continued work in radioactivity.

HANDY FACT

The Handy Science Answer Book, 5th edition, says, "Evolution of the *Homo* lineage of modern humans (*Homo sapiens*) has been proposed to originate from a hunter of nearly 5 feet (1.5 meters) tall, *Homo habilis*, who is widely presumed to have evolved from an australopithecine ancestor. Near the beginning of the Pleistocene epoch (two million years ago), *Homo habilis* is thought to have transformed into *Homo erectus* (Java Man), who used fire and possessed culture. Middle Pleistocene populations of *Homo erectus* are said to show steady evolution toward the anatomy of *Homo sapiens* (Neanderthals, Cro-Magnons, and modern humans) 120,000-40,000 years ago. Premodern *Homo sapiens* built huts and made clothing."

so, and despite that this feels like a landmark event that might launch a thousand careers, Geneva Medical College refused for a time to admit any more women to its program.

ECOLOGY ISN'T ANYTHING new: in 1856, Eunice Newton Foote (1819–1888) presented her findings on the greenhouse effect to fellow attendees at a meeting for the American Association for the Advancement of Science. Foote was obviously a respected scientist of her time; she was also a forward-thinking women's rights advocate.

WHILE WE'RE ON a roll here, Marie Curie was the first woman to win a Nobel Prize for her work in physics in 1903. She did it again in the field of chemistry in 1911.

A VERY EARLY version of the human genome was published by the Human Genome Project in 2000. Three years later, the genetic makeup of human genomes was 99 percent completed.

ABOUT 3,800 YEARS ago, the Egyptians first developed fractions.

BIOLOGY: TOOT, TOOT, TOOTSIES, HELLO!

From the moment you get out of bed until the moment you tuck yourself in, there's one part of your body that takes a beating: your feet. Read on, and you'll know why you should truly appreciate these oft-overlooked appendages.

Pity the poor foot. Not only do you stuff it in shoes for most of the day, but in an average 24 hours, you might force tons of cumulative weight on it—possibly the equivalent of a loaded dump truck. If you're a high heel wearer, that's increased by up to 75 percent, but that weight would be concentrated on the part of the foot just behind the toes. Ouch! In your lifetime, you'll take those tired tootsies more than 100,000 miles, one step at a time, but that's actually good: plain old walking is one of the best ways to exercise your feet. It's even better if you leave your feet naked.

Really, when you get right down (literally) to it, the foot is a magnificent machine.

One out of every four of your body's bones are found in the feet: that's 26 bones in each foot, plus 107 ligaments, 33 joints, 19 muscles, and 10 tendons apiece. Your big toes have two phalanges (bones) each, while all the other toes have three phalanges, mirroring what you have on your hands; in fact, it's believed that our big toes are thumblike throwbacks to our far-back ancestral days of living in trees. Some people still have that thumblike dexterity in their big toes; in the case of an accident, it's common for surgeons to replace thumbs with toes.

Those toenails, by the way? They grow at half the rate your fingernails do, so if you lose one completely, you can estimate that it'll take a year for a completely new toenail to grow. And there's a reason elderly people have thicker toenails: it's because they've had a lifetime to accumulate injuries, bumps, and slow-sloughing cells. Like almost everything else that ails Grandma, you can blame those thick nails on age.

If you stuff your feet into socks and shoes all day, you know they get hot. It's okay: feet have automatic cooling systems in the form of some 250,000 sweat glands on the pair, which can offload up to a cup of sweat per foot per day. The bacteria held near your feet by your socks dine on the sweat, which is why you may have stinky feet. In addition to those lovely sweat glands, you've also got thousands of nerves in your feet, making them one of the most sensitive parts of your body, per square inch.

So, let's say you need to buy new shoes.

If you like to shop in-store, you'll want to wait until about six to ten hours after you get up. As your day passes, feet tend to swell, so midday or early evening is the best time for a shoe try-on. It's also likely that one of your feet is bigger than the other; in most people, it's the right foot, but it's a good idea to ask if you don't know. Also, don't just reach for the size shoe you used to wear, or you might be in for a surprise: statistically speaking, both men and women in the United States and Great Britain have seen an average increase in foot size and width over the last few decades.

The thing is, you should always be mindful of your feet: up to eight out of ten of us have foot problems at some point in our lives, including ingrown toenails, plantar fasciitis, Morton's toe (in which the second toe is bigger than the big toe), bunions, fallen arches, and plain old tiredness. And if all else fails, sit down and get off your feet.

HANDY FACT

The Handy Science Answer Book, 5th edition, says, "Twelve astronauts have walked on the Moon. During the six Apollo missions to the Moon, each Apollo flight had a crew of three. One crewmember remained in orbit in the command service module (CSM), while the other two actually landed on the Moon. Apollo 11 landed on the Moon on July 20, 1969. The first man to walk on the Moon was Neil A. Armstrong (1930–2012), followed by Edwin E. (Buzz) Aldrin Jr. (1930–)."

SPACE: THE BEST DRESSED MAN (ON THE MOON)

When NASA was ready to send a man into space with the goal of landing on the moon, they couldn't send their astronauts out with just any old outfit—but the problem was that since nobody had ever been on the moon, the construction of the space suits was quite a bit of guesswork.

NASA scientists weren't left to figure it out themselves, though. Owens Corning experts contributed their knowledge for the fabric, as did others who understood the nature of what was needed: specifically, material that wouldn't incinerate at any time during blast-off, moonwalks, or re-entry. Ultimately, a material called "beta cloth," a Teflon-coated fiberglass mixture, was used; Playtex, the women's undergarment manufacturer, further allowed NASA to hire its industrial employees and to use their knowledge and expertise. Playtex's parent company helped with another important material: rubber.

They just needed someone to bring everything together. Enter Sonny Reihm.

Reihm had joined Playtex in 1960, and by 1969, he'd slid over to work on Apollo as project manager. Not quite thirty years old, it was his job to make sure that the space suits were perfectly suited for lunar activity—he didn't even want to *think* about what would happen if Buzz Aldrin or Neil Armstrong tripped over something on the moon and ripped a hole in the knee of what was essentially a one-man space capsule—so he assigned his best engineers to design the suits, and he put his best seamstresses to work sewing them.

But first, the astronauts had to travel to the International Latex Corporation (ILC) headquarters in Dover, Delaware; Playtex was then a part of ILC Dover. There, they had to be measured from head to toe and round and round because each suit had to be custom fitted, with no waste or shortage and no obvious discomfort allowed (there was enough discomfort in the fact that bathroom breaks would be difficult). Each suit required room enough to ensure that the astronauts could do what they set out to do on the lunar surface. And they had to be fireproof.

With the measurements in hand, Playtex engineers and other subcontractors worked to perfect the suits' designs to ensure safety,

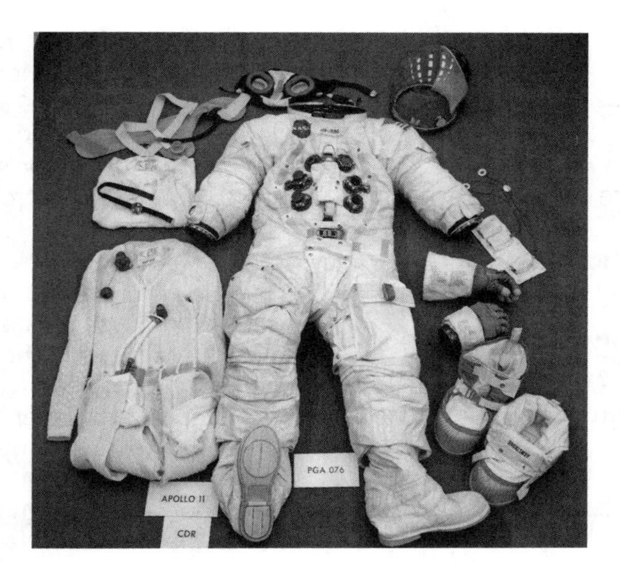

This is a photo taken before the Apollo 11 mission of the suit that Neil Armstrong would wear in 1969.

comfort, and mobility. They added materials to allow for movement in the elbows and knees, and they removed excess fabric where it wasn't needed. At each step, the Apollo 11 crew had to return to Dover to ensure that the suits fit as perfectly as possible.

Once the engineer's jobs were done and the plans for the suits passed muster, Playtex seamstresses—always women because they were said to be more "agile"—worked extra-long hours each day, painstakingly cutting the fabric out and stitching suits that had to be made better than anything that ever sashayed down a runway. More than twenty layers of fabric, insulation, nylon, rubber, and high-tech cloth, as well as flexible, steel cables, went into each suit—sewn layer upon layer, sometimes by hand, stitch by stitch—and not even the smallest mistake was tolerated: NASA condoned only the tiniest fraction of difference in stitch allowance, and one wrong snip or needle-poke meant that the entire suit was scrapped and that particular suit had to be created again from the beginning. While the seamstresses toiled, other women worked next to them to seal and waterproof the suits with liquid rubber.

Seamstresses often didn't know what part of the suit they were constructing, but they knew that they were working on a space suit, and that a man's life depended on the precision of their work. No pressure, right?

In the end, their work led to suits that weighed 280 pounds (127 kilograms) each on Earth (they weigh nothing in space, of course) and cost somewhere around $100,000, or around $800,000 in today's

If you could drive to the moon, it would take you awhile. The moon is nearly 252,000 miles from Earth, at its closest. Assuming you drive 8 hours a day at 65 miles per hour, it would take you between six months and a year, depending on potty breaks.

money. Each of the astronauts' suits was meant to shield the men from temperatures ranging from about −250 degrees F to +250 degrees F, as well as from anything flying around near the moon's surface, and they had to protect the astronaut whether he left the module or not.

Neil Armstrong reportedly said that the suit he wore was comfortable. He looked mighty snazzy in it when he wore it on July 20, 1969; Buzz Aldrin wore one, too, when they stepped together onto the Moon. From his seat at Mission Control in Houston, Sonny Reihm watched, reportedly with his heart in his throat, as Armstrong and Aldrin romped across the Moon's surface.

The suits held fast.

HANDY FACT

Did you know that, like Earth, the Moon also has ground vibrations?

From *The Handy Science Answer Book*, 5th edition: "Similar to earthquakes, moonquakes are a result of the contstant shifting of molten or partly molten material in the interior of the moon. These moonquakes are usually very weak. Other moonquakes may be caused by the impact of meteorites on the moon's surface. Still others occur at regular intervals during a lunar cycle, suggesting that gravitational forces from Earth have an effect on the moon similar to ocean tides."

ANIMALS: THIS IS A JOB ... FOR SUPERCRITTER!

We humans like to think we're superior to other creatures on the planet. And that may be so—or it may not....

A REGULAR HOUSEFLY can handle walking around on a ceiling for hours if there's a reason to do so. If you hung upside down for that long, you would probably die because your lungs can't fully handle the weight of your other organs, and they eventually wouldn't be able to reinflate if you were upside down. Being in that position for too long means you'd ultimately probably asphyxiate.

COMFORTABLY, THE AVERAGE human can walk at a rate of about three or four miles per hour and can jump from a standing position some six or seven feet horizontally—maybe once or twice or a few times but not repeatedly. Compare that to a red kangaroo, who can leap-travel up to 40 miles an hour over short distances. One good leap will carry a red kangaroo 25 feet and they can do that all day long.

THE LOWLY ANT can carry an object that weighs up to 50 times more than the ant weighs. If you are a 200-pound human, that means you would be able to carry something that weighs 10,000 pounds—like, say, a fully grown hippopotamus.

TAKE A DEEP breath, open your mouth, and scream, and you would, on average, be able to reach 100 dB (the record is 129 dB). Compare that to a sperm whale's clicks, which have been measured at 230 dB, and you're practically whispering.

IF SHE TIMES it right, a woman can have two single-birth babies in one calendar year. In that same time, a rabbit could have nearly 85 offspring, and a coral will send millions of eggs out into the sea.

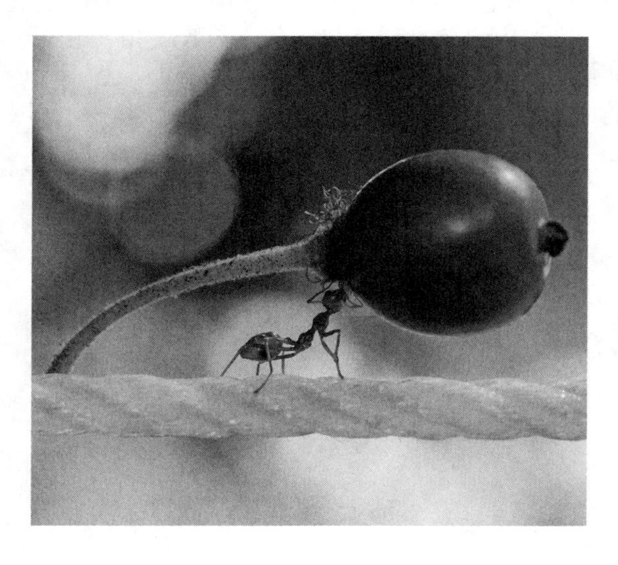

When you consider their ability to carry objects 50 times their weight, ants are truly superhuman in strength.

THE AVERAGE ADULT human consumes four to five pounds of food each day; for women who eat right, that's 1,600 to 2,400 calories per day, and for a man, that's up to 3,000 calories. But once again, we're outdone by a sperm whale, who eats more than three tons of krill each day, or well over a million calories.

AS COMPARED TO your dog's, your nose is pretty pathetic. You have some 6 million scent receptors there—but Poochie has some 300 million scent receptors in his nose. It's been said that if you're making vegetable soup on the stove, your dog can smell each individual carrot in the pot.

WHEN DIVING, A peregrine falcon can achieve speeds of over 240 miles per hour. A cheetah can run at speeds of up to 75 miles per hour in short bursts. Usain Bolt can run at a speed of 28 miles per hour. You likely average a little over nine miles per hour.

JUST IN CASE you're feeling bad, though, here's somewhere you do excel: your memory. Research suggests that most dogs forget minor, random things within two minutes. Rats generally forget overnight (so, within a few hours). A human, however, can recall events that happened seventy, eighty, even ninety years ago.

HANDY FACT

According to *The Handy Science Answer Book*, 5th edition, "Some cats seem to have extraordinary 'memories' for finding places. Taken away from their homes, they seem able to remember where they live. The key to this 'homing' ability could be a built-in celestial navigation, similar to that used by birds, or the cats' navigational ability could be attributed to the cats' sensitivity to Earth's magnetic fields. When magnets are attached to cats, their normal navigational skills are disrupted."

BIOLOGY: IT'S WHAT'S INSIDE THAT COUNTS

You might wonder how it happens that two kids growing up at the same time in the same family can be so different. Some reasons are obvious: they're different people, after all. Still, the explanation may have some hidden parts to it.

Some scientists say that birth order matters. If you were a first-born, you're said to be the driven-leader, type-A type. Second-borns (and subsequent kids) are supposed negotiators. The last sibling tends to be more laid back and easygoing, more peacekeeping, or so it's been said. Of course, other scientists say—and research has borne out—that the birth-order theory is a lot of hooey, that there should be no overgeneralizing, and that myriad other things determine what you're like as an adult. Only one thing seems to be true-ish: research shows that firstborns might have a slightly higher IQ than their younger sibs.

Maybe. *Possibly.*

In this mix, research reminds us that a firstborn was basically an only child until the next kid came along, and that impacted his or her

Scientists often use twins as objects of their studies into how biology affects behavior, intelligence, emotions, and other factors versus environment.

personality. Second-born children, and those who might come afterward, were never the only child, even for a minute. Surely, that had some impact on the way a child becomes an adult.

This argument is overall a small part of the age-old debate of nature vs. nurture, one that's basically been debated for some 500 years: do inner factors like your genetics, your DNA, instinct, and neurons determine your life path, personality, and reactions—or does it all come from outside: how you're raised, educated, and taught? Or is it a little of both? And what if the siblings are identical twins, who have the same genotype and similar (but not the same) DNA? More complications: if identical twin brothers marry identical twin sisters, their children are genetic siblings. How does that play into nature vs. nurture?

Psychologically speaking, if a behavior presents itself very early in life, it's assumed to be one of nature. As a child grows, behaviors and preferences are assumed to come from the nurture side of the equation. Most researchers and psychologists say that the vast majority of what makes us *us* in this nature-versus-nurture argument lies somewhere in between on the spectrum and that there are many things that can be inherited behavior-wise. Scientists believe, for instance, that it's possible that we may inherit phobias. Fear of spiders or snakes seems to be ingrained in many people, and it may go back centuries to our ancient ancestors. Is a lack of fear somewhere in our DNA, too?

When all is said and done, it's like this: even if you and your sibs grew up in the same house with the same parents, the same pets, the

HANDY FACT

The Handy Science Answer Book, 5th edition, says, "More than one thousand genetic tests are available and in use to identify changes in chromosomes, genes, or proteins. The three methods used for genetic testing are:

1. Molecular genetic tests, which study single genes or short lengths of DNA to identify variations or mutations that lead to a genetic disorder.
2. Chromosomal genetic tests, which analyze whole chromosomes or long lengths of DNA to see if large genetic changes exist, such as an extra copy of a chromosome, that cause a genetic condition.
3. Biochemical genetic tests, which study the amount or activity level of proteins. Abnormalities in either can indicate changes to the DNA that result in a genetic disorder."

same economic situation, had the same schools and teachers, and were raised with the same morals and rules, the chances are that you'd come out differently because DNA is never totally identical, and identical experiences are imprinted on an individual differently than they are on someone else.

Or not.

The jury's still out....

CHEMISTRY: WEIRD MEDICAL CURES FROM TIMES GONE BY

How are you feeling today? Fine? Good, but maybe not so much if you lived in ancient times when the cure was just as bad as the disease....

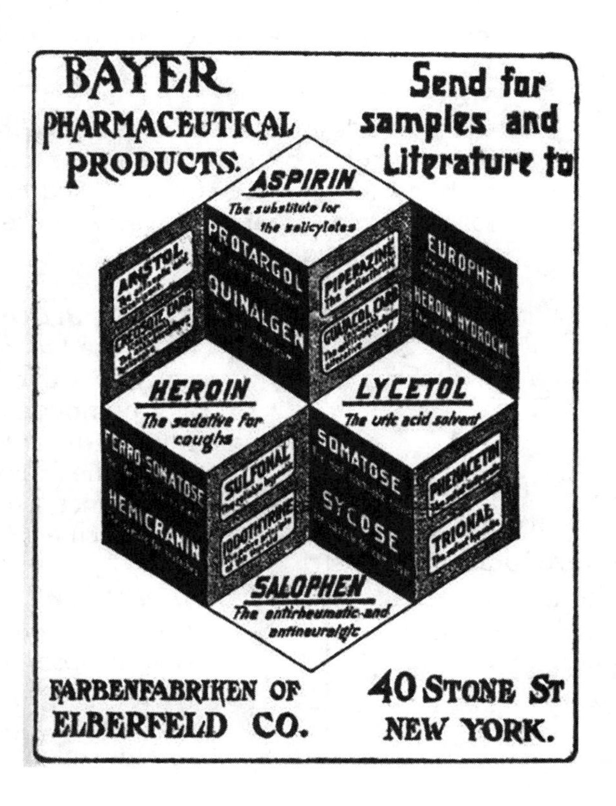

Heroin was once an ingredient in a number of over-the-counter medicines from such companies as Bayer.

ONCE UPON A time not so long ago (like, less than two centuries ago), the parent of a fussy, whiny child could go to the corner store and purchase an elixir that would calm the nerves of their little one. The products often sported comforting names like "Mrs." So-and-So's tonic or "Aunt" Such-and-Such's remedy. Never mind that the product contained cocaine, heroin, opium, laudanum, morphine, or a combination thereof.

NEVER FEAR: ADULTS weren't left out. At about this same time, heroin and cocaine were also used for cough syrup and tooth problems for grown-ups.

MERCURY WAS ONCE believed to confer everlasting life on the person who ingested it. Ironic since mercury poisoning often shortened that person's life.

ANCIENT ROMANS BELIEVED that the best way to cure herpes was to cauterize the sore with a superheated iron. That had absolutely no effect on the virus, but it did leave a nasty, painful injury where nobody wants a nasty, painful injury. In medieval times, hemorrhoids were treated the same way.

THE ROMANS BELIEVED that human urine left teeth whiter. There's no record of the person who thought of it first.

THERE'S A REASON we call the seller of a dubious medical cure a "snake-oil salesman." In the 1800s, traveling medicine shows and salesmen sold what consumers were told was literally snake oil in a bottle that was good for what ailed ya. Later, government officials determined that there was no snake oil—no snake parts at all, in fact—inside the bottle.

ANIMAL BLOOD AND/OR dung were often used alone, together, or in conjunction with other substances to cure any number of maladies. Often, the traits of the animal determined which illness their body fluids were used for; for instance, a rooster's blood might be used to cure a man's impotence.

Physicians used to scoff at some ancient cures and treatments, only to learn that they were totally valid. These included the use of leeches and maggots, fecal transplants, and the use of the nasal cavities to reach the brain.

TAPEWORMS WERE USED for a time in the late 1800s and early 1900s for weight loss. A woman (usually a woman) simply ingested a capsule that contained a tapeworm egg so that the creature might hatch and establish

HANDY FACT

From *The Handy Science Answer Book*, 5th edition, "Leeches have been used in the practice of medicine since ancient times. During the 1800s, leeches were widely used for bloodletting because of the mistaken idea that body disorders and fevers were caused by an excess of blood. Leech collecting and culture were practiced on a commercial scale during this time. William Wordsworth's (1770–1850) poem 'The Leech-Gatherer' was based on this use of leeches. In 2004, the Food and Drug Administration (FDA) approved the commercial marketing of leeches (*Hirudo medicinalis*) for medical purposes."

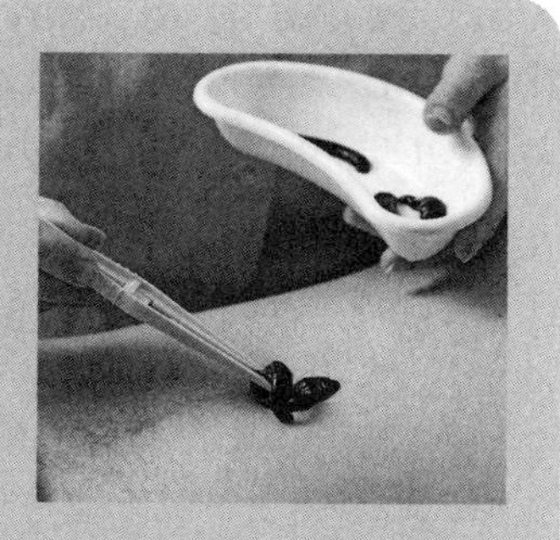

Leeches, it turns out, do have therapeutic benefits, and people are starting to use them again to treat some ailments.

itself in her intestines and gorge itself on whatever she overate. While that's absolutely a dangerous thing, the practice does occasionally make a reappearance.

IN THE 1950s and early 1960s, most high-end shoe stores had fluoroscopes so that children could have their feet measured to ensure the best fit. Good parents brought their kids in every time the kiddos outgrew their shoes, but few people thought to remember that fluoroscopes were X-ray machines, and X-rays meant radiation exposure.

UNTIL THE LATE 1800s, some medicines were made of human body parts, including the liver, blood, bones, and, in the case of ancient mummies, even flesh.

TECHNOLOGY: IT'S NO FLASH IN THE PAN!

As everybody who's ever created a document or spreadsheet on a computer knows, the first rule of doing anything on a computer is: *Back It Up.* Make sure that if the computer crashes, the electricity goes off, Armageddon occurs, you accidentally hit the wrong key, or some joker thinks they're cute and performs an "ESC" (escape key) move, you won't panic because you've saved your work. For that, you have a few choices—one of them being a handy little device that you can slip into a pocket or put on your key chain, safely away from that joker at work.

So, who created that magical bit of plastic, spring, and metal we call the USB flash drive or thumb drive?

The first, easiest answer is: it's complicated. And contentious.

Before the invention of the flash drive, data was generally transferred from computer to computer by use of a floppy disk. "Floppies" were roughly five-inch-square bits of plastic that did what their name suggested: they flopped as you put them into your computer because they were so thin (about 1.6 mm). They were thin on storage, too: most floppies only held about 1.44 megabytes of info. If you were transferring a lot of info, you had to have a lot of floppies. In the 1980s, smaller floppy disks (which, at about 3 to 3.5 inches square and made of harder plastic, weren't floppy at all) were made, but they still didn't hold a whole lot of information.

Neither were very efficient, either, so in the early 1980s, Fujio Masuoka (1943–), who worked for tech giant Toshiba, invented flash memory and introduced it to the world at the International Electronic Developer's conference in San Francisco in 1984. With flash memory, more info could be stored in smaller areas, and flash memory was erasable and reprogrammable. It was said to be very intriguing.

Masuoka went on to develop NAND flash memory, which was a sort of precursor to the flash drive we know today. This was exciting news in the tech world, although at that time, there was no practical

Floppy disks evolved from large (truly floppy), 8-inch disks to 5.25-inch, hard-plastic-case disks and then 3.5-inch "floppies." They were supplanted in the early 2000s by CD-ROM technology.

way to transfer data (at least not for consumers) because the universal serial bus (USB) port wasn't invented until the mid-1990s.

The very first flash drives held just eight megabytes of information. That's about enough to carry the song "In-A-Gadda-Da-Vida" or eight rounds of the "Baby Shark" song or three average-sized photographs from your desktop to your laptop.

In April of 1999, a patent was filed on a "flash disk" by Israeli company M-Systems (which was purchased in 2006 by ScanDisk). Here's where the contention—and more than a few lawsuits—lies: later in 1999, IBM filed an invention disclosure claiming that one of its employees, Shimon Shmueli, had invented the flash drive—an assertion that Shmueli continues to claim; Netac, a Chinese company, has also said that it invented the device; and Pua Khein-Seng from Malaysia is credited with inventing the pen drive. Tech company Trek Technology of Singapore offered the first commercially available "ThumbDrive." By the turn of the century, IBM began selling flash drives in the United States, calling them DiskOnKey.

Understandably, early flash drives were slow and small on space; modern consumers would be frustrated beyond all measure if they had to go back to a USB 1.0 or 1.1. For consumers in the year 2001, however, this was a great little bit of technology, but though exceptionally good, it was also exorbitantly expensive. Flash drives with a mere 128 megabytes were often $30-$50, or well over $43 in today's money. Just as frustrating was that they tended to wear out faster.

Within a decade of its introduction to consumers, flash drives had grown in capacity and shrunk in price. The standard, most widely available in 2003, was a 2GB flash drive, which could hold thousands of documents or hundreds of pictures; they could be had for $20 or so. The first 1 gigabyte flash drive was available about a year later.

IN 2006, CAR manufacturer Kia became the first automakers to put USB ports in their cars.

BY 2009, YOU could purchase a 256GB flash drive, which would hold more than 25 two-disc movies or over 50,000 documents.

IN 2013, 1TB flash drives could be had by consumers; four years later, that capacity had doubled.

IN 2015, FLASH drives that worked with computers *and* cell phones were available.

HANDY FACT

Although, strictly speaking, the computer as we know it couldn't have been invented until there was electricity first, *The Handy Science Answer Book*, 4th edition, says that "the English visionary Charles Babbage (1791–1871) persuaded the British government to finance an 'analytical engine.' This would have been a machine that could undertake any kind of calculation. It would have been driven by steam, but the most important innovation was that the entire program of operation was stored on a punched tape."

Charles Babbage

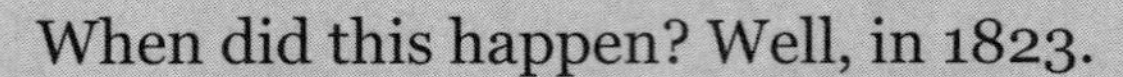

When did this happen? Well, in 1823.

TODAY, IT'S BEEN estimated that well over two billion flash drives are in pockets, purses, lanyards, and briefcases the world over.

That's a lotta data (see also First Computers and Other Tech Forgettables).

PHYSICS: ICE, ICE, BABY

At its very basic, the way to describe ice is that it's solid water—specifically, it's water frozen by low temperatures into a solid state.

That's the easy way to say it. But there's so much more to ice—it cools our drinks, causes accidents, looks so pretty, and saves our lives.

First of all, ice can be found as close to the sun as the planet Mercury and, from what we know, as far away from the sun as the Oort cloud objects some 3.2 light years away; the planet Saturn even has an ice volcano from which ice crystals spew. On Earth, ice can be found in surprising places but is most common in snow form on the highest mountain ranges planetwide, on the northernmost and southernmost latitudes for most of the year, and at lower latitudes during certain seasons (see Science—The Environment: "Fall" and Other Seasonal Things).

Without ice, Earth would suffer both in the water cycle and in the temperature of the atmosphere. It is, for instance, essential for the planet's biosphere that ice exists, partly because ice reflects sunlight back into space, keeping the planet cooler. It's likewise been suggested that floating ice keeps the liquid water it floats on from freezing into one solid, top-down mass, not only protecting the flow of the liquid water but what lives in it, too. Thinner ice on water also allows a bit of light to permeate the surface of liquid water, saving organisms such as algae, which is eaten by water creatures.

And if none of that convinces you yet, check this out: Ice is important because the presence of healthy glaciers impacts the amount of water on the planet and its quality. Nearly 70 percent of Earth's fresh water is found in the world's glaciers, and in some places, glacial melting is already causing problems both in erosion of coasts and in the goodness of water consumed.

Without getting too technical here, ice is, strictly speaking, a mineral because it's a naturally occurring, inorganic solid with an or-

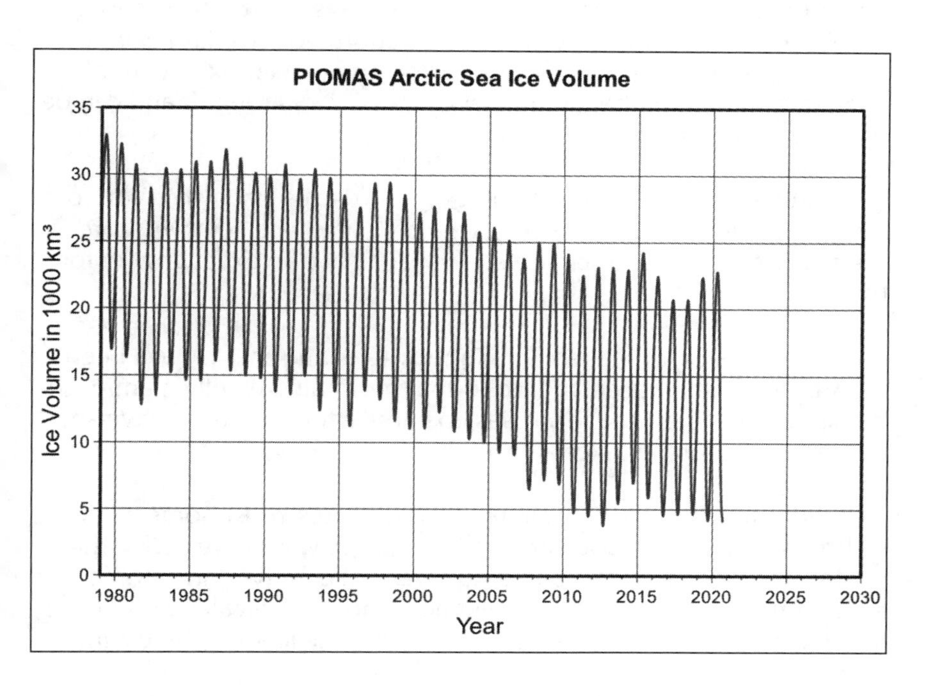

This chart shows the measured decrease in arctic ice over the last 40 years. (PIOMAS is the Pan-Arctic Ice Ocean Modeling and Assimilation System developed by Jinlun Zhang, D. Andrew Rothrock, and Michael Steele at the University of Washington's Applied Physics Laboratory.)

derly crystal structure that's less dense than water because, as water freezes, it loses its density due to the formation of ice crystals and the bonding of hydrogen. While you probably want the ice in your drink to be clear, ice can also be hazy, depending on the impurities within it; it can have varying degrees of blue, or it can even have a greenish tone; and glaciers are generally white but can also be blue or have a light greenish cast.

Scientists recognize nearly twenty different phases of solidifying water; likewise, there are many different kinds of ice, including hail and snow. Indigenous people from Alaska and the northernmost hemispheres do, indeed, have many different names that refer to the various kinds of ice they might encounter in their lives.

When water freezes, it increases in volume, which is why a can of soda will burst if frozen: the ice causes a lack of room within the can, and something's gotta give. When the can thaws, if it's not in pieces, what's inside it returns to its former state and volume. Curiously, if you fill a glass with ice and *water* (or a water-based material, like coffee or soda), the melting ice will not overflow the glass because the water you get from the melting is equal to that which was once ice. Try the same with nearly any other liquid that's denser than water (like milk or corn syrup, for example), and melting ice will likely overflow the glass.

Another oddity: it's a myth that water freezes at 32 degrees F (0 degrees C): sometimes, it requires a lower temp to turn solid; sea-water, for instance, needs lower temps to freeze due to the salt in it. Here's a tip, by the way, next time you're stranded on a dinghy in the Arctic Sea: at that point in the ocean, because of its proximity to glaciers, frozen, then melted, seawater contains very little salt and can be used for drinking.

And for the record, yes, boiling water freezes instantly if you toss a bowlful of it in the air on a very, very cold day. Also, for the record, ice is very slippery, and science still doesn't have an exact answer for that.

While it's a fact that some mighty interesting things have been found, years later, frozen and icebound (mummies, woolly mammoth babies, bodies, plants, and viruses, to name a few), humans have an interesting history with ice.

Prior to the early 1900s in the United States, most homes had iceboxes to preserve food and keep it cold. Ice was harvested in the winter from a local water source, such as a lake or river, and stored in blocks insulated with straw in buildings made specifically for that use. In the warmer months, ice was delivered by "the iceman," who carried

HANDY FACT

From *The Handy Physics Answer Book*, 2nd edition, by Paul W. Zitzewitz, PhD: "The Celsius scale is named after a person whose life work was dedicated to astronomy. Anders Celsius (1701–1744), a Swedish astronomer, spent most of his life studying the heavens. Before developing the Celsius temperature scale in 1742, he published a book in 1733 documenting the details of hundreds of observations he had made of the aurora borealis, or northern lights. Celsius died in 1744 at the age of forty-three."

Anders Celsius

his wares into homes with the help of brute force, a leather shoulder protector, and mean-looking tongs. Blocks of ice were placed in a large, insulated, refrigerator-like device with a tray beneath it, to catch what melted. For geographical places that didn't freeze in the winter, ice was sometimes transported by train, cross-country. Today, you can make your own ice or, like great-grandma, you can buy it: making ice for consumers is a multihundred-million-dollar-a-year business.

BIOLOGY: STICK OUT YOUR TONGUE AND SAY "AHHHH"

Unless you're brushing it, sticking it out at someone, eating ice cream, or, well, reading about it, you probably don't think much of your tongue. But did you know....

IF ASKED, YOUR doctor would say that the human tongue is basically a muscular organ that sits in your mouth between your lower teeth. Contrary to what you've heard since middle school, the tongue doesn't just float in your throat: it's firmly attached in the back by webs of tissue and a bone in the back of your throat called a hyoid bone (yes, the same bone that suggests to medical examiners that you may have been strangled). On the tongue are tissues called mucosa and bumps called papillae, which give the tongue its roughness. As many as 4,000 taste buds coat your papillae.

THE TONGUE IS also home to all kinds of nerve cells that send information to the brain.

ALSO CONTRARY TO every rumor you heard when you were a kid, it is not possible to "swallow your tongue."

ONCE MORE, YOUR childhood buddies were wrong: there are no "zones" of taste on the human tongue. The entire organ is capable of detecting any and all of the five known tastes of sweet, salty, sour, bitter, and umami.

THE AVERAGE HUMAN tongue is nearly four inches long from tip to end, which is nowhere near anything spectacular size-wise. Relative to size, the chameleon and the nectar bat have the longest tongues: twice their body length and 1.5 times their body length, respectively. As for humans, yes, tongues come in longer sizes, but those are outliers.

LIKE FINGERPRINTS, EVERY tongue on Earth has a unique print.

WHEN YOU SEE your doctor and you're asked to stick out your tongue and say, "Ahhhh," the reason is because the tongue is a barometer for many kinds of diseases and ailments, including vi-

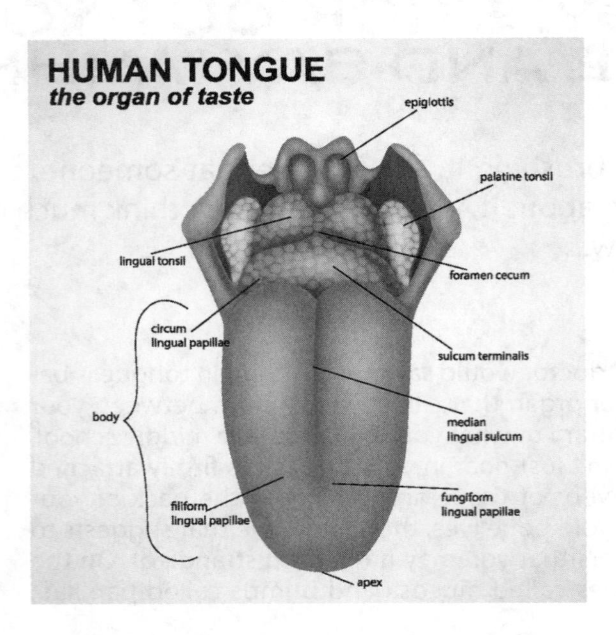

A diagram of the human tongue and its parts.

tamin deficiencies, infections, bacteria overgrowth, and even oral cancer.

BY THE WAY, while sticking your tongue out at someone here in the States is likely seen as some level of insult, many cultures consider that kind of thing to be a sign of approval, respect, or greeting.

YOU CAN PROBABLY already surmise that your mouth is teeming with bacteria, but it gets better: scientists say that more than 700 species live in colonies on your tongue. Fortunately, most of them are good bacteria, so don't worry.

AS FAR AS MUSCLES GO, your tongue is one of the most sensitive on your body. While it's not the most powerful muscle you have, it's one of the more flexible due in part to the fact that the tongue is the only muscle that moves without the support of your skeletal system. Your tongue can twist and curl and move to create speech, mimic sounds, chew, and swallow.

YES, ONLY SOME people can curl or roll their tongue. No, it's really not a matter of genetics or heredity. It's just that some people can't … but research suggests that if you can't, you may be able to learn to do it.

HANDY FACT

The Handy Science Answer Book, 5th edition, says, "The first artificial sweetener discovered was saccharin. It was discovered accidentally by Constantin Fahlbert (1850–1910) and Ira Remsen (1846–1927) at Johns Hopkins University in 1879. According to the story, after a day of working in the laboratory, one or both of the men picked up a roll at dinner. The roll tasted sweet. They surmised correctly that something was on their hands from the coal-tar derivative chemicals they had worked with during the day. They tasted the various chemicals in the laboratory and discovered that the compound, benzoic sulfimide, was a sweet compound."

PHYSICS: ANYBODY GOT THE TIME?

You know what they say about time: it flies when you're having fun. When you're not, or when something's as boring as watching paint dry, then time seems to stand still. Why is that?

To understand, you have to wrap your brain around something that can't be seen, bought, retrieved, or kept. Time, says the dictionary, is a duration of occurrence or a definitive instant when something happens; asking what time it is, therefore needing to know your place in the continuum, is a little of both.

Albert Einstein, in his Theory of Relativity, understood that the speed of light is absolute, can never be varied, and is one of the most irreversible things in the universe. This means that time is a part of the universe, but to accommodate the precision of the speed of light,

One of Albert Einstein's important contributions to physics was pointing out that time was not a constant and that it changed based on the speed at which the observer was traveling.

space and time must flex. It was Einstein's teacher, Hermann Minkowski (1864–1909), who said that proper time was the distance between two events as measured by a clock that passes through both events, depending not only on the motion of the events but also by the motion of the clock itself. Coordinate time is the distance between the two events, measured by a far-off observer. To make it super simple, let's say that the former is the amount of time (using a clock) it takes for a running back to score a touchdown. The latter is how much time you *think* it took.

Though it's a considerable subject to study, today's physicists don't believe that time flows or moves forward, backward, or, really, anywhere. Time just *is*, existing in a four-dimensional space–time continuum that we just need to trust is there, and (to blow your mind a little) the past, present, and future all exist at once within the continuum with no defined direction of travel. As if that isn't enough to absorb, there's this: time can dilate, depending on the movement of the observers of any particular event. If there are two fans watching that running back and one of them is behind a camera on the field, his personal perception of time will be different from that of the couch potato.

For everyday uses, let's just assume that you want to know the time so you don't miss a minute of that game you're excited about. There are websites you can find that will tell you exactly what time it is—but how do they know?

It's important to remember that while humans had sundials and we were aware of the sun rising and setting and the calendar year based on the sun, it was the Babylonians who divided the day into hours, minutes, and seconds, based on a system they borrowed from the Sumerians. The Greeks and Chinese also devised systems of time, too, but since time was overall perceived by an individual's observation, it differed (even if just a little bit) from person to person.

The truth is, in the beginning of clocks, having an *exact* time really didn't matter much. Everyday folks and those who worked the land got up, worked until lunch (if there was one), went back to work until someone fixed the evening meal, then went to bed shortly thereafter. Time was important, but *knowing* the time was mostly just no big deal.

You can't exactly call them precise, but in the fifteenth century, clocks with springs began appearing in Europe, and that had to have been better than merely relying on the skies for the time. More than two hundred years later, British carpenter and clockmaker

411

Tachypsychia: a neurological condition that overly distorts and warps one's perception of time.

Government agencies in the United States and Paris both agree that the measure of a second is determined by how long it takes a cesium atom to vibrate just over nine billion times.

John Harrison (1693–1776) invented the chronometer, which led to safety on the seas by allowing more accuracy when calculating longitude, which gave mariners a set place to calculate their positions when a-sea. Not long after Harrison invented his instrument (and, by the way, collected a pile of prize money from the Board of Longitude), British clockmaker Thomas Mudge (1715–1794) invented the lever escapement, a piece of the clock's guts that keeps internal parts moving. As the early nineteenth century rolled in, Eli Terry (1772–1852) had figured out how to mass-produce timepieces, making them more affordable for nearly anyone. By the turn of the twentieth century, women were wearing watches on their wrists, but men used pocket watches almost exclusively until World War I when soldiers found wristwatches to be much handier in battle.

Ah, but back to your game.

Generally speaking, to determine the *exact* time, the world has agreed on the authority of two different timepieces created since the middle part of the last century: the atomic clock, which uses cesium-133 atoms and has a ratio error of one second lost every 1.4 million years; and the cesium fountain atomic clock, which has a ratio error of one second lost every 20 million years (finessed later to an accuracy of one second lost every 100 million years). You can find the exact time, per an atomic clock, online at www.time.gov.

HANDY FACT

Here's a unique timepiece, courtesy of *The Handy Science Answer Book*, 4th edition: "Carolus Linnaeus (1707–1778), who was responsible for the binomial nomenclature classification system of living organisms, invented a floral clock to tell the time of day. He had observed over a number of years that certain plants consistently opened and closed their flowers at particular times of the day; these times varied from species to species. One could deduce the approximate time of day according to which species had opened or closed its flowers. Linnaeus planted a garden displaying local flowers, arranged in sequence of flowering throughout the day, that would flower even on cloudy or cold days. He called it a 'horologium florae' or 'flower clock.'"

BIOLOGY AND BOTANY: SURE, WHAT COULD IT HURT?

Surely, the following tales can be given a "with best intentions" label. If we could just bring *this* creature to *that* place for *this* reason, it would be great because the creature is so good at what it does best, and that kind of task is needed *there*. And it would all work out because, of course, no harm was ever meant … right?

Of course not, but humans can sometimes make a situation worse by meddling that leads to disastrous endings like these.…

Just before the beginning of the nineteenth century and after Australia's heyday as a penal colony, Europeans migrated to Aus-

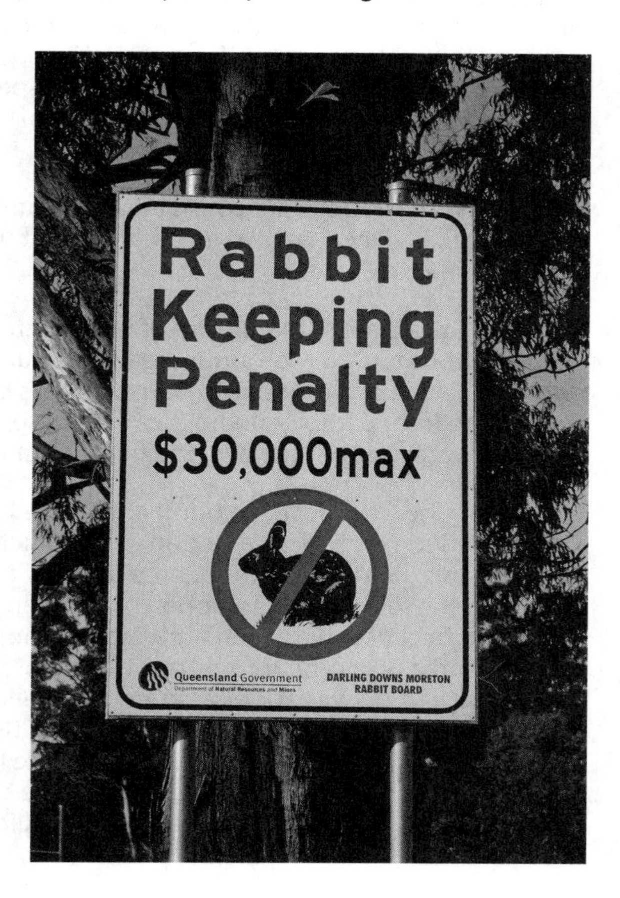

Some states in Australia, including Queensland, have made it illegal to keep domestic rabbits for fear that they might get loose and start another lupine plague.

tralia, drawn to several things: the freedom, the land, and the warm climate were all appealing to someone who had none of the above. It surely helped that nonprisoners were given transportation (usually from Great Britain), as well as tools, seeds, and sometimes even livestock to get them started. Some also came to join family who'd been imprisoned in Australia since the late 1780s.

By most accounts, the Brits loved it in Australia, but there were things about home that they missed. One, Thomas Austin (1815–1871), a man of means, missed the kind of hunting he'd done back home, and so he wrote to his nephew and asked that the lad send a few European hares to Australia so that Uncle might enjoy a bit of sport.

Some say that rabbits had already been introduced to Australia when the penal colonies came to be. So, what harm could come, anyhow, from a few more bunnies?

Austin released his twenty-four rabbits for the hunt, but some of them escaped while he was hunting them. Within a decade, millions of rabbits overran Australia's outback, and while they were doing what rabbits do best, they were also eating every green thing they could find, which threatened crops, destroyed the topsoil, and, ultimately, resulted in decimation or loss of a number of species on the continent, including the greater bilby, two kinds of wombats, and a kind of wallaby.

Today, feral rabbits are considered pests in Australia, and in Queensland, it's illegal to own a bunny as a pet unless you're a bona fide magician.

While there are some species of rat that are native to America, the ubiquitous rats you see around garbage dumps and landfills are common black, brown, and gray house rats. It's believed that the rodents came from Europe hundreds of years ago, although black and brown rats are thought to have originated in China.

Surely, no one *meant* to bring rats to the New World. It likely just happened when ships and transport vehicles arrived at the shores of this continent with rats accidentally aboard. The creatures, always on the lookout for an opportunity, then snuck ashore and multiplied because there aren't many situations that aren't perfect for a rat, and a brand-new country with lots of naive creatures to eat and places to hide was about as good as it gets. It's believed that that first black rat came to America in the 1550s, although it may have happened earlier, perhaps as early as the late 1400s; the brown rat probably arrived at roughly the same time but, because the brown rat is more prolific and adaptable, it quickly overtook its brethren in numbers.

Rats can carry many diseases, including hantavirus, the plague, salmonella, rat bite fever, and tularemia, and wild rats are perfectly happy to bite. Here's why they've survived all these years, despite human efforts to eradicate or at least control them better: rats are agile, fast, smart, can squeeze through the tiniest of spaces, and can move large objects with ease. A pair of rats, if left unchecked, can result in somewhere around 2,000 rats in one year's time, and if you're not shuddering now, you should be (see Science—Animals: Awwww, Rats!).

For more than 2,000 years, the Chinese have known that kudzu had medicinal qualities. It could be used to treat alcoholism, to eliminate a bad headache, to soothe an upset stomach, and to ease vomiting. It helps with inflammation, and it reportedly helps diabetics to regulate glucose levels. What possible harm could come from a plant like that?

Until the World's Fair Centennial Exhibition in Philadelphia in 1876, *Pueraria montana* wasn't known much outside Asia. Its first introduction was intended to appeal to Victorian-era mansion owners for their own gardens, but the ornamental plant didn't take off as expected, and few people asked their gardeners to plant it.

In the late 1920s and early 1930s, when the drought hit the center of the country and soil eroded and blew across the nation, the Soil Conservation Service turned to kudzu. The plants' roots were a natural for holding on to soil and keeping it from eroding, and farmers were highly encouraged to plant it even though horses and cattle weren't fond of eating kudzu. Even so, the plant was embraced in the southern part of the United States by developers and railroad magnates, which tended to convince farmers that kudzu was a mighty nice option (a nice, fat payment per acre planted probably helped). A kind of fervor swept some areas, and there were, at one point, organizations and service clubs set up and running in honor of kudzu.

Post-World War II, though, that all ended. Kudzu was just a plant then, as it may be now—an ornery, pesky, invasive plant that overshadows and kills plants and trees by stealing sunlight from them. It grows all over the southeast quadrant of the United States and in some areas in the northeast; some sources say that millions of acres of the South are covered by kudzu, while other sources say it's not that bad at all.

Part of the appeal is still in its medicinal qualities, and it's edible. Scientists are just now taking another look at kudzu, to see if other uses for it might be possible.

It's also interesting to note that this invasive species is being invaded by an invasive species: a few years ago, the Japanese kudzu bug showed up in Georgia after having hitched a ride on a plane somehow. It—and its progeny—are busy now, chewing up kudzu.

When settlers came to the New World in the 1500s, they had no idea what they were going to see, let alone have for dinner. To stave off starvation, they brought pigs with them to eat along the way and to raise as food once they made it overseas. Due to free-range farming and because pigs can be escape artists, some of the creatures got away from their humans and went feral.

As if that wasn't enough, a little over a century ago, European or Eurasian wild boars were released in the South and south-central parts of the country for sport-hunting purposes. Naturally, some of them got away, too, and when feral pig meets wild boar and both are adaptable, highly prolific, and self-sufficient, you've got a problem that today results in millions of dollars in damage to crops, pastures, and native species and threats to human health and safety.

Feral pigs, by the way, can, at the time of this writing, be found in forty-five of the fifty states, so don't get too comfortable if you haven't seen any up close. The largest population, at more than two million animals, can be found in Texas.

Here's a bit of a surprise: the honeybee as we know it is not native to America. Yes, there existed an ancient honeybee here, but it went extinct millions of years ago; the bee we are familiar with is one of many subspecies, but it's the most common worldwide. That little guy, the Western honeybee, was introduced to this continent sometime in the 1600s by European settlers who brought hives of them to the New World, much to the reported amusement of the indigenous people.

Prior to 1956, there was never much reason to fear the sweet honeybee. She (since most worker bees are female) did what she did best, which was make honey; for the most part, she was docile and easily controlled by beekeepers in the know. In that year, though, Brazilian scientists took bees from Africa, just south of the Sahara Desert, and attempted to cross them with an Italian strain of honeybee to increase honey production. The operation backfired spectacularly: instead, the experiment led to bees that were more aggressive, attitudinal, and less likely to back off from a perceived threat.

411

Just one insect produces food regularly consumed by humans. That's the honeybee.

And in 1957, some of those Africanized bees escaped from the lab....

Twenty-six queens took off, to be exact, and they took a whole lot of worker bees along with them and, as they spread and established their territory, they mixed with other hives both commercial and wild. Still, despite the fact that Africanized bees build their hives and produce offspring quicker than non-Africanized bees, it took years for the Africanized honeybee to completely establish itself in South America.

It wasn't until 1985 that the first few were discovered in California, and in the fall of 1990, the first colony reached Texas. It's found all over the southernmost part of the country now, but scientists believe that if the bee ever acclimates to cold weather, it could reach the northern states easily enough.

As for the bee itself, she offers the typical good news / bad news scenario.

On the "pro" side, Africanized honeybees build hives faster than do other bees, and they produce more little bees, so they need more pollen, and they make more honey if their hive is placed in the right kind of climate. Also good: because they are just bees, the sting of an Africanized bee is no worse than that of any kind of bee.

The bad news is that Africanized bees will swarm and attack with merely the slightest provocation, and they don't give up: some swarms will chase an annoyance—whether it's human, animal, or vehicle—for a quarter mile if they're angry enough. They don't go looking for a fight, but it you accidentally stumble on a hive somewhere, or if the weather is too hot or too cold, or if you smell different to the hive, or if there's a sudden, loud noise that provokes them, or, for

The sting of an Africanized honeybee is not any more potent than a regular bee. The difference—and what can make them lethal—is that they swarm on their victim and keep stinging much longer than the average bee.

heaven's sake, if it's a Wednesday, you won't just endure a single little sting or two. It could be *hundreds* of stings everywhere on your body, stinging that could go on for several minutes, and, well, they don't call them "killer bees" for nothing.

To combat the Africanized honeybee, experts have tried trapping them, forcing a non-Africanized queen into a hive, artificially inseminating an Africanized queen before placing her into a hive, and releasing a large number of non-Africanized drones into a territory. And still, the bees are here....

Sometimes, the extinction of a creature can be traced directly, as if it were a paper map with one road. That's exactly what it was like for the dodo.

For millions of years, *Raphus cucullatus* lived in one place in the world: the island of Mauritius in the Indian Ocean. It had been there so long that it had acclimatized itself to the weather, food, and whatever else the island contained (which wasn't much because of Mauritius's size), and it had lost its ability to fly (it never needed to go anywhere). And the dodo lived there, presumably happily, until very early in the sixteenth century.

That was the year that Mauritius became a handy stopover point for Portuguese and Dutch ships on their way to work the spice trade in the Orient. For the ships' sailors who'd spent weeks with little to eat but what was onboard, a 50-pound bird probably looked mighty good, and the dodo was an easy meal: because it had never encountered humans, it had no fear of them, so sailors were able to basically

A 1626 sketch of two dodos.

walk up to the birds and snatch them. Reports were that the creature didn't taste good, but it probably beat hardtack biscuits any old day. It's been said that thousands of dodos were killed and eaten.

Once people began settling on Mauritius and the island became a penal colony, Dutch settlers brought monkeys, cats, and pigs with them, and rats tagged along on ships. As critters new to Mauritius escaped, repopulated, and became feral, they discovered that dodo eggs were tastier than the dodo (although the latter was easy prey, too).

These new neighbors were the nail in the dodo's coffin. The last confirmed spotting of a live dodo was in the late 1600s; a close genetically related cousin lived on a nearby island and also went extinct soon afterward. Sadly, because *Raphus cucullatus* died off relatively quickly, few records were kept, and there exists no complete dodo skeleton, so scientists really don't know exactly what the dodo looked like.

What we do know, however, is that the dodo's extinction wasn't just a loss for the bird itself. Researchers think the dodo's digestive system was necessary for the survival of a certain kind of tree that only lives on Mauritius.

They're working on this problem. Stay tuned....

HANDY FACT

The Handy Science Answer Book, 4th edition, has this explanation: "An 'endangered' species is one that is in danger of extinction throughout all or a significant portion of its range. A 'threatened' species is one that is likely to become endangered in a foreseeable future due to declining numbers."

SPACE: GARBAGE IN THE GALAXY

On April 22, 1970, Wisconsin senator Gaylord Nelson finally achieved his dream: a day devoted to environmentalism and working to clean up the planet. More than twenty million people participated in that first Earth Day and have celebrated it every year since then—but what about all the garbage in space?

Indeed, the galaxy beyond our immediate atmosphere is slowly becoming as cluttered as Earth. In 2018, there were some 20,000 items of a four-inch-or-larger size catalogued as flotsam and jetsam of space; at the time of this writing, there are undoubtedly more. But just because they've catalogued the larger pieces, don't breathe a sigh of relief yet: the smaller-than-four-inch pieces can do just as much damage in space.

It all started in 1957, when the first satellite, *Sputnik*, was launched from Earth into orbit.

It fell back to Earth and burned on re-entry in very early 1958, but that was only the beginning. In the past sixty-plus years, thousands of satellites have been launched, and some—but nowhere near all—of them have come home again: The Vanguard 1, sent into orbit on March 17, 1958, is still out there.

The vast majority of what scientists say is "junk" or "debris" floats around Earth at low orbit, which is considered to be within 1,250 miles (2,000 kilometers) of the planet's surface. If you were to rocket up to see what's there, you'll find larger parts of man-made craft and satellites that have long been out of commission. You'd spot paint chips, a screwdriver, a spatula, nuts and bolts, a lens cap, astronaut tools, bits

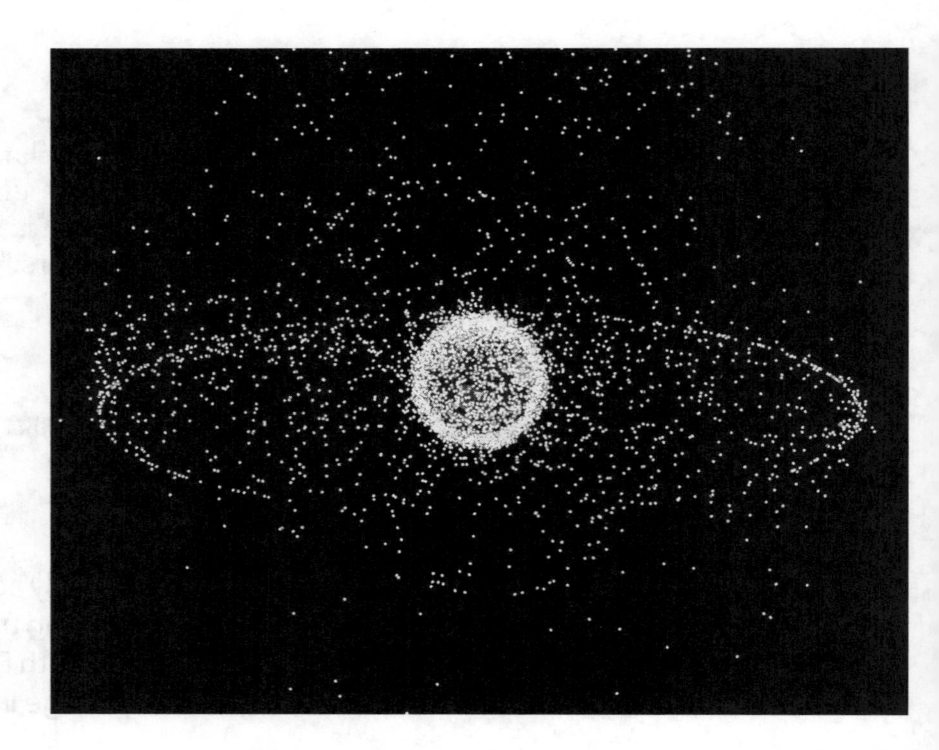

A computer image from NASA indicates the positions of thousands of pieces of space debris currently orbiting Earth.

of rockets and spaceships, and millions of fragments too tiny to capture. You'd also see pieces of nonworking satellites that have crashed or hit working satellites, and then, there are the meteoroids, which also count as space debris. If you took a little side trip to the moon, you'd find some 200 *tons* of trash left over from various manned and unmanned trips there. If you could get to Mars and Venus, you'd find a bit of trash there, too. This is all not just an American mess, mind you; Russia, China, and India must claim their fair share of the blame both on and off the moon. Even amateur satellite makers can get in the game if they've got the money.

So, why worry about all this space trash? It's not exactly like it's in your backyard, right?

True—but it does cause a danger to manned spacecraft and to working satellites. The closer an object is to Earth as it orbits, the faster it goes. Since most space debris is in low orbit (thereby closer to our planet) and so are satellites, it all moves through space at about 17,500 miles an hour, as does the Space Station—indeed, NASA does a lot of repairs on equipment due to damage from microscopic space debris. Now, imagine what a collision would look like between a car-sized bit of space debris and the International Space Station! More relatably, imagine the dent a deck-of-cards-sized flotsam could leave on equipment or a human.

The thing we must remember is that just as we've been told for the past fifty years that we shouldn't litter, we also can't continue launching items into outer space with the idea that there's no harm, no foul. The truth is that the longer unused things are left out there, the more likely it becomes that they'll fragment into *even more* dan-

HANDY FACT

According to *The Handy Science Answer Book*, 4th edition, Landsat maps "are images of Earth taken at an altitude of 567 miles (912 kilometers) by an orbiting Landsat satellite, or ERTS (Earth Resources Technology Satellite).... These scanners can detect differences between soil, rock, water, and vegetation; types of vegetation; states of vegetation (e.g., healthy/unhealthy or underwatered/well-watered); and mineral content."

These maps are useful for geologists, foresters, oil company executives, farmers, and anyone who is employed in the business of land management. You can get your Landsat map from the U.S. Geological Survey.

gerous bits that are capable of inflicting damage. And the more bits there are out there, the less room there is for things we'll rely on for further research and, ultimately, for everyday life.

BIOLOGY: DID YOU KNOW ...?

 ***YOUR BODY PRODUCES*** enough spit to fill a small bathtub once every 40 days.

 ***THERE'S A REASON*** you didn't like brussels sprouts when you were a kid: a child's taste buds outnumber those of an adult.

 ***YOU ONLY USE*** one nostril at a time. (Check it out: try plugging one side and then the other, and you'll see which one you're using at the moment.)

 ***EEEUUUWWW, YOU SHED*** nearly two pounds of dead skin cells each year; in fact, much of the dust in your home consists of dead skin cells.

 ***AROUND THE WORLD,*** our population increases by four babies each second.

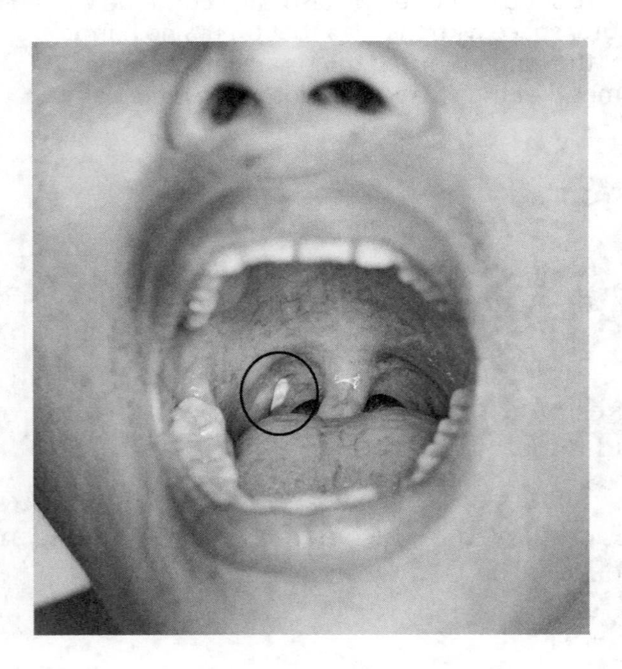

If you still have your tonsils, you might develop tonsil stones (or tonsilloliths). Sometimes, people cough them out easily, but sometimes they need to be nudged out with a cotton swab or rinsed out with gargling.

IN THE TIME between the moment you raised your eyelids this morning and the time you'll lower them tonight, you will have spent a tenth of your wakeful day with your eyes shut. That's because your eyeballs need you to blink, and your brain doesn't even notice. It just picks up where it left off, as if nothing ever happened....

YOU ARE VASTLY outnumbered by bacteria: there's just one of you. There are four quadrillion quadrillion bacteria on Earth.

WITHIN FOUR MINUTES of your death, parts of your body start to decompose. It could take your brain as many as seven minutes to totally cease functioning.

MORE TWINS ARE born today than ever before. From when scientists started keeping records (1915) until 1980, about one out of every 50 live births included twins, a rate of 2 percent. Today, the rate is well over 3 percent, and one out of every 30 live births includes twins. Researchers think the rate is going up because twins are more prevalent with older mothers and more women are waiting to become pregnant.

IF YOUR ANCESTORS came from anywhere outside of Africa, you almost certainly have a small percentage of Neanderthal DNA in your genome.

IF YOU STILL have your tonsils, you might have noticed that there are white bumps on them (or maybe you can even feel them at the back of your mouth). Those are probably tonsil stones, and they're basically just food debris and dead cells that got caught in the folds of your tonsils. Although they can grow to frightening sizes, tonsil stones usually don't cause any problems, but it's a good idea to get rid of them anyhow by swallowing hard, coughing hard, or gargling with warm salt water.

THERE'S MORE TO you than you think there is: your body is made up of between thirty and forty trillion cells.

HANDY FACT

According to *The Handy Science Answer Book*, 5th edition, "the probability that at least two people in a group of 30 share the same birthday is about 70 percent."

TECHNOLOGY: AN ORCHESTRA OF ONE

For about as long as there have been humans, they have made musical instruments out of bone, wood, or reed. The flute, say archaeologists and paleontologists, is the oldest musical instrument known to mankind. The guitar, by the way, is the most popular.

One of the newer instruments is the theremin.

Invented in Russia in the fall of 1920 shortly after the Russian Civil War, the theremin was the brainchild of twenty-three-year-old Lev Sergeyevich Termen (1896–1993), who was tasked by the Soviets to research gases and proximity sensors to detect things that are nearby without any physical contact. Termen was experimenting with sonar when he discovered that moving one's body into an electromagnetic field could alter that field. A little more tweaking, and rather than creating a techy, new instrument for nefarious spy purposes, he discovered that his device made beautiful music.

Lenin was said to have been quite taken by the device, and he sent Termen on an immediate tour around Russia to help convince

Lev Termen (aka Leon Theremin) is shown here playing his invention in 1927.

people to want electricity in their homes. That done, and with audiences so taken with the instrument's music, Lenin sent Termen to Europe to show it off. Audiences there clamored to see the new device and to hear the music that came from it.

Following his success with concerts, Termen (who Americanized his name to Leon Theremin) came to the United States along with his instrument, which he patented here in 1928. Americans loved the instrument as much as the Europeans did, and Theremin was in high demand for concerts. RCA saw the value and beauty of the device's sound and received the production rights to the instrument named after its inventor.

Alas, the Great Depression left concert houses and public venues largely unused, but audiences both here and abroad continued to be interested in the theremin, which RCA called the world's first electronic instrument. That's a pretty big deal, considering that just 10 percent of rural American homes had electricity then. A handful of people even became proficient at playing the instrument.

Proficiency is absolutely necessary for getting music from the theremin.

Basically speaking, the theremin is an electronic instrument with a single oscillator and two antennae on a base that is roughly the size of a desktop computer tower on its side. The base is joined by one antenna on the right and another antenna that loops on the left at the other end of the base (if you're a lefty, it's the other way around). By waving your hands near or away from each of the antennae, you con-

HANDY FACT

Because it takes years to become a whiz at getting beautiful music from a theremin, there haven't been a lot of theremin players to hit the big time. Other than Leon Theremin himself, Clara Rockmore (1911–1998) was probably the first and maybe the most famous of her time; Rockmore was a classical violin player before she mastered the theremin. Lucie Bigelow Rosen (1890–1968) was a trained theremin soloist and friend of Theremin, and Samuel Hoffman (1903–1967) ensured that the theremin was brought to modern ears. Musical engineer Robert Moog (1934–2005) was also a theremin player, as is Led Zeppelin's Jimmy Page (1944–). If you'd like to hear more, look for Carolina Eyck's (1987–) videos online.

trol the pitch and volume. Shake your fingers and get vibrato. Wiggle them and change the song. Move your hand a half inch up or down or sideways, and you get a totally different note. You won't ever touch a theremin to get beautiful sounds from it; it's the only musical instrument you don't touch to play. If you've ever heard "Good Vibrations" by the Beach Boys, you've heard just a fraction of what a theremin can do.

In America, Theremin seemed to be having a ball, but things may not have been as they seemed. Though he loved New York society, was a charming guest at parties, married African American ballerina Lavinia Williams, and found work in the United States (including several high-level projects), something was nipping at Theremin's heels. At any rate, in 1938, he quite abruptly left America to return to the Soviet Union—he left without telling his wife—for reasons that are still debatable. Some said then that he was a spy. Others thought he'd been kidnapped. A large amount of personal debt has been hinted at and may be the likeliest reason.

After World War II, the theremin was still a popular instrument, but that was waning: it was just not very easy to play well, and much simpler instruments were then hitting the market. Still, Hollywood used it for soundtracks now and then, and electronics expert Robert Moog kept the theremin in the public eye to a point, basing his Moog synthesizer on it. Modern rock bands used it on occasion.

In the meantime, Leon Theremin spent time in a Soviet gulag for a few months, then in a prison called a sharashka, a secret place where he was allowed to work as a scientist. At the sharashka, Theremin invented what he called "The Thing," which was a "bug" hidden in a wooden plaque that was given to the American ambassador to Moscow in 1945, 1945 and allowed the Soviets to eavesdrop on conversations held in the ambassador's office for seven years until it was discovered.

HANDY FACT

"Sonar, an acronym for 'Sound Navigation Ranging,' is a method of using sound waves to determine the distance an object is from a transmitter of sound. Sonar is used predominantly as a navigational tool by humans and animals. Dolphins and bats, among other animals, use sonar for navigation, hunting, and communicating. Machines such as depth-finders on boats, distance meters used in real estate and construction, and motion detectors for security devices all employ sonar."

—*The Handy Physics Answer Book,* 2nd edition

In 1967, an American journalist discovered where Theremin had gone when he abruptly left this country some 30 years prior; still, it wasn't until the fall of communism that Russian sanctions eased enough for Theremin to return to the United States in 1991. While here, he played with a few concert orchestras and spent time in New York with his daughter, who accompanied him here.

This time, though, he didn't stay for long. Leon Theremin died in Russia in the fall of 1993 at the age of 97.

ANIMALS: ANIMALS YOU DON'T WANT TO PET

Puppy kisses are the best. And, really, there are not a lot of things on Earth better than the purr of a warm cat on your lap. No doubt, we love our furry babies, but here's a list of animals you probably don't want living in your house and sleeping on your bed.

NEVER MIND THAT Fiona, Cincinnati Zoo baby and star of her own videos, is adorable, you definitely do not want a hippopotamus living in your house. First, there's their size: a hippo can grow to be just over sixteen feet long and more than five feet tall at the

Hippopotamuses might look like they are too fat and cumbersome to be anything but cuddly and sweet, but in reality, there is not a lot of fat on a hippo. They are aggressive beasts that kill more people in Africa than any other wild animal on that continent.

shoulder, and a full-grown male can weigh nearly two tons, which is not going to do your floor any good. Then there's a hippo's mouth, from which emanates noises that can reach 115 decibels, or about as loud as an up-front seat at a rock concert. Inside that mouth, which can open wide enough to swallow a human whole, are tusks and teeth that can grow to twenty inches long. But that's the physical side of a hippo; on the personality side, hippopotami are aggressive, unpredictable, and willing to fight lions to the death. What's worse is that they're perfectly happy to chase and kill a human, in the water or out of it, and they're faster than you. Isn't that all you need to know?

IT SHOULD COME as no surprise that the saltwater crocodile is on this list. Weighing up to 2,600 pounds with a length of up to twenty feet, he'd take up a lot of real estate in your living room. Try ordering him off the sofa, and he'll slither down at up to eighteen miles per hour but he won't be happy: you could be met with the second-hardest bite in the animal kingdom, the same bite-per-square-inch as the great white shark (and crocodiles are not known for their trainability, either). It's not the bite that will kill you, though: the saltwater croc pulls his prey underwater to drown it, then he rolls it in the water to dismember it before he digs in to eat.

THOUGH THERE ARE some that shiver at having a snake in the house at all, there's no denying that snakes can make good pets if you're not an ophidiophobic. Even so, you might want to think twice about having a black mamba around. Highly aggressive, super fast, and prone to attack without provocation, a black mamba generally weighs under three pounds but can reach a maximum length of fourteen feet. If that doesn't put a shiver in your shirt, then consider that the mamba injects up to 100 grams of poison in every bite, which is the very definition of "overkill" since just ten grams will kill you. If you think you can outrun a black mamba, you're probably wrong. You'll be dead before you get far. On the same note, you don't want an inland taipan snake, either, for similar reasons.

JUST THINKING ABOUT having a dragon in the basement would be *so cool,* wouldn't it? Unless, of course, you're thinking about getting a komodo dragon, and then it would be so deadly. It's not the size of a komodo dragon that's so awful: males are larger, as the case is with most creatures, at up to eight-and-a-half feet long and 200 pounds, which really isn't gigantic. They move at about twelve miles per hour, which is slower than the fastest human can run, so no problem there. The reason you don't want a komodo dragon as a pet is that they'll eat *anything*, including humans, dead or alive; their teeth are sharp; and their grip is super tenacious. If those reasons alone don't make you stop a second to reconsider, there's this: scientists used to think it was the bacteria in a komodo dragon's bite that was so lethal; the creatures, after all, dine on carrion and corpses on a regular basis, but no, they've recently discovered that komodo dragons are, indeed, blessed with venom in their saliva. How bad can that get? Well, if you suffer a komodo dragon bite and live through *the bite itself,* the venom will kill you within about an hour.

WHAT POSSIBLY COULD go wrong with a cute, little frog that could just fit in your pocket and that's about the size of a paper clip? Nothing at all ... unless you found him in a tropical forest in South America and he happens to be a certain kind of member of the *Dendrobatidae* family, also called a poison dart frog. You could absolutely be forgiven for picking him up—poison dart frogs sport beautiful, vivid colors on its skin in bright yellows, greens, corals, and black, and they look like jewels. They're tiny and adorable, but you can say "goodbye" pretty much as soon as you pick a wild one up: they won't bite you, but they'll kill you, no problem, right after you come into contact with the slime they release on the skin of their pretty little backs. Just one simple stroke on the topside of one teensy frog is enough to kill ten humans or animals, the latter of which is the reason certain Indigenous people in Colombia hunt with poison dart frog venom on the tips of their projectile weapons. Scientists believe that the toxic matter comes from what the frog eats in the wild; poison dart frogs that are raised in captivity are not poisonous. Now, here's the thing: you don't want a wild poison dart frog in your

Poison dart frogs come in a spectacular array of beautiful colors (reds, blues, greens, yellows, and oranges) set against black spots and stripes. The colors serve as a warning to predators, saying, "I'm poison! If you eat me, you'll regret it!"

house, but you might want him in a local laboratory: researchers are currently looking into the chemistry behind the poison dart frog, and it may lead to lifesaving medicine someday.

EVERYBODY LOVES COWS, right? Yes, but having a cape buffalo begging at the kitchen counter would be no fun at all—as a matter of fact, there'd be no begging: at six feet tall at the shoulder and weighing up to a ton, a cape buffalo would just take whatever he wanted from your countertop. You couldn't tell him no, either: cape buffalo are fearless, with two sharp, curved horns that they're not afraid to use, even against animals that are faster than they are or bigger. Indeed, they may stalk you if they perceive you as prey, and you definitely don't want to get between them and their offspring. Indeed, just give the cape buffalo whatever it wants: the animal is said to kill more hunters than any other land animal on the continent of Africa. On a similar note, don't mess with elephants, either: they remember who crosses them, and they've been known to knock down trainers and step on their heads as revenge for past transgressions.

SO, A CUDDLY puppy isn't for you, and that's okay. Even so, you're not going to want a hyena as a pet instead. Known for a bark that eerily resembles a cackling "laugh," hyenas are slightly smaller than a Great Dane, but they're shaped differently than Fido, with coarse fur, higher shoulders, and longer legs, looking as though a fox mated with a short giraffe. Alas, most pet dogs would be no big deal for a hyena to kill—the hyena's bite is stronger than a dog's bite, and a hyena has patience and tenacity in capturing prey: he's speedy when chasing his dinner and will bite and pick and torture, almost playing with his target as he takes it down. He'll often start eating even before the prey is dead, and if live prey isn't around, he'll happily eat carrion or corpses. What can you expect from a creature that, if it's born first, will kill all its siblings as soon as they emerge from the womb?

A "PRETTY BIRD" in a little cage likely doesn't trigger your sense of danger, but a cassowary really should. These are not cute songbirds, to say the least. As the second-largest bird in the world (only the ostrich is larger), the cassowary stands up to six-and-a-half feet tall, weighs up to 150 pounds, and is as bad as it looks: with a head topped by a keratin-rich casque that looks like a fierce helmet, the bird is covered with feathers that are not for flight but rather for protection from the elements of its rain forest habitat and for protection against predators. If you're determined to have one as a pet, and if its scowl isn't enough of a deterrent to approachment, know that cassowaries are elusive and athletic: they can run more than thirty miles an hour, they can jump as high as

Large birds native to New Guinea, cassowaries are as large as an adult human and dangerous! They can gut a man with their powerful legs and claws. Don't mess with them!

a talented basketball player can, and they're known to be excellent swimmers. Still intrigued? Then this is important, too: the cassowary has three toes on both of its feet, with the middle toe sporting a daggerlike claw that can measure around four inches in length. Because of the size of the claw and the size of the bird, that dagger can slash and kill in a heartbeat, and cassowaries have been known to disembowel an enemy in a flash, guilt free. Is it any wonder why the bird is considered the most dangerous in the world?

IF YOU SQUINT, you might be able to convince yourself that a honey badger is somewhat like a cuddly, little kitty. Never mind the honey badger's superstrong jaw, its sharp teeth, the incredibly long claws on its feet (perfect for digging!), or the loose skin that helps protect the honey badger from danger and is nearly impenetrable by teeth, fangs, or claws. Never mind its infamous bad temper, its willingness to take what it wants at all costs, and the inclination to attack anything at any time, no matter its size. Never mind that it's one of the wiliest animals at the zoo. So, yes, if you can ignore all these very unappealing things about it, the honey badger might be kind of like a big kitty, at just under a foot in length and weighing up to thirty-five pounds, with impossibly cute, little ears on its pointy skull and a plumey tail that beneath it sports the same kind of stinky scent gland its cousin the skunk has. The diet, however, is what may give you pause about the honey badger as a pet: about 25 percent of the honey badger's diet consists of snakes, and if they're poisonous, that's fine because the creature's skin protects it from most bites. They'll also eat birds, small mammals, some berries and plants, amphibians, fish, insects, tortoises (they bite through the shell), and carrion, and they've been known to raid human cemeteries. Time, perhaps, to stick with domesticated animals, maybe?

Okay, but you're not into sharing your sofa with anything, and your pets are a little different. Here are the things you don't want in your saltwater aquarium:

OH, MY, BUT wouldn't the blue-ringed octopus look beautiful in your fish tank? At a little larger than a baseball but lighter—about five inches in diameter and just an ounce in weight—the creature looks like any other octopus until it gets frightened. At that point, it lives up to its name by morphing into an octopus with tiny, blue rings all over its body, which serve as a warning to any creature thinking of eating it. Heed that warning: the blue-ringed octopus is considered to be one of the most dangerous, deadly creatures in the ocean because it will bite, and each chomp contains enough neurotoxin to kill more than twenty-five adult humans in a matter of minutes. Scarier still: because of their size, these guys are easy to step on in the wild, their bite is generally painless, and there is no antidote.

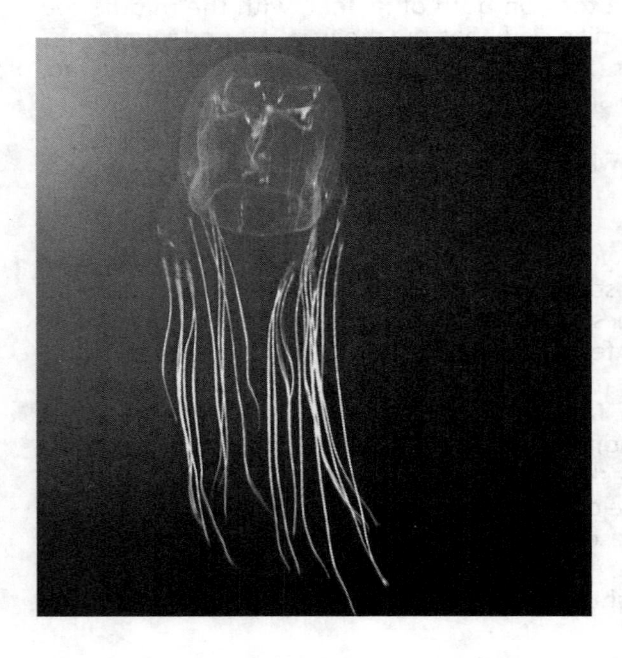

*There are more than 50 species of box jellyfish living in the South Pacific around Australia and the Philippines. Not all of them are poisonous, but those that are, such as the wasp jelly (*Chironex fleckeri*), will cause agonizing pain and even death.*

HERE'S ANOTHER BEAUTY for your home aquarium: the box jellyfish. You'll want to make sure you've got a lot of room for it, though: with a body that's right around five inches in diameter, the creatures are ethereal, nearly transparent, and lovely to watch as they move around at speeds of up to four knots (a little over four miles per hour), with tentacles of up to ten feet in length in tow. They're even more impressive in the oceans surrounding Australia and the Philippines, but don't go trying to catch one. While a handful of people have endured the excruciating pain that follows a brush with a box jellyfish, which can linger for weeks and weeks, death comes within minutes to most people unfortunate enough to be entangled in those deadly tentacles because the venom is just that horrible. And, by the way, just getting stung by a regular jellyfish is terribly unpleasant, and no, urine is not going to help.

TOSS A STONEFISH into your aquarium, and you might think that you didn't: the creature is a pro at camouflage and looks like an ordinary rock on a good day (hence the name stonefish). It'd be a good-sized rock, though: normally found off the northernmost parts of Australia, the creature weighs up to five pounds and can survive out of the water for a little while, which means you can actually invite him to watch TV with you if you're careful. But don't stroke him on the back, and don't even think of stepping on a stonefish: that's where the creature keeps his sharp dorsal spines, each of which sports a venom sac packed with nastiness. One small poke causes excruciating pain that lasts for days, even weeks, and can result in paralysis, loss of limbs, and even death. There is an antidote for stonefish venom, but don't take chances.

IF YOU HAPPEN to be looking around for something pretty to add to your aquarium, be careful what you reach for. The cone snail's shell is gorgeous, with its bright colors and patterns, but the second you touch that shell, you'll be sorry. The cone snail doesn't look too fondly on anybody messing around, and it has a long proboscus with a sharp barb on the end that it's more than happy to use on your hand. It'll happen so quickly that you might not even realize you were stung; not even diving gloves are a good

HANDY FACT

On dangerous creatures, *The Handy Science Answer Book*, 4th edition, offers this: "In a study of 570 shark attacks, it was found that the most shark attacks occur near shore. These data are not surprising since most people who enter the water stay close to the shore."

defense against a cone snail's sting and the fact that analgesic properties in the venom help to maintain your innocence ... at least for a little while. You definitely don't want to wait around: the analgesic lasts for an indeterminate minutes or days, and the pain is horrific. If you're lucky, it'll be accompanied by redness and swelling; if not, you could die because the venom of some cone snails is so deadly that one drop can kill ten adults.

So, maybe it's time to stick with the fish in the pet store, then, eh?

BOTANY: WHAT A FUN GUY!

So, here's something you don't want *on* your skin, but putting it in your mouth is a good thing. You eradicate it from your front yard, but you welcome it in the woods. It's eliminated from the bathroom but also put there for a reason. It's fungus, and without it, we humans literally could not live.

Up until about 1.5 billion years ago, fungus was just another part of life on Earth, closer to animals than plants, but nothing uncommon in the scheme of things on the planet. Then, it diverged to become its own creature, which is exactly what fungus is: it's a living thing, a eukaryote with encapsulated DNA in its cells, just like you, so don't mistake fungus for a plant. Plants need chlorophyll to survive through photosynthesis, but (again, like you) fungus needs to seek out food from things that are living or have lived and died. It's just been in the past few decades that scientists have realized this; before the late 1960s, fungi were classified as plants, but now, they have their very own classification. Overall, though, we humans have generally known about fungus for at least 30,000 years, but possibly longer.

Yep, fungus has been around awhile: science tells us that the first fungus probably appeared more than two billion years ago. From there, it reproduced much as it does now, asexually or sexually (the latter through a complicated process called meiosis), which ultimately results in spores that are spread far and wide, each to become its own colony of fungi. Early fungus then specialized into basically what fungus does now, which is to help decompose dead things.

That isn't as gruesome as it sounds. Actually, the opposite sounds worse: imagine a world in which nothing—including dead human bodies—breaks down into the soil.

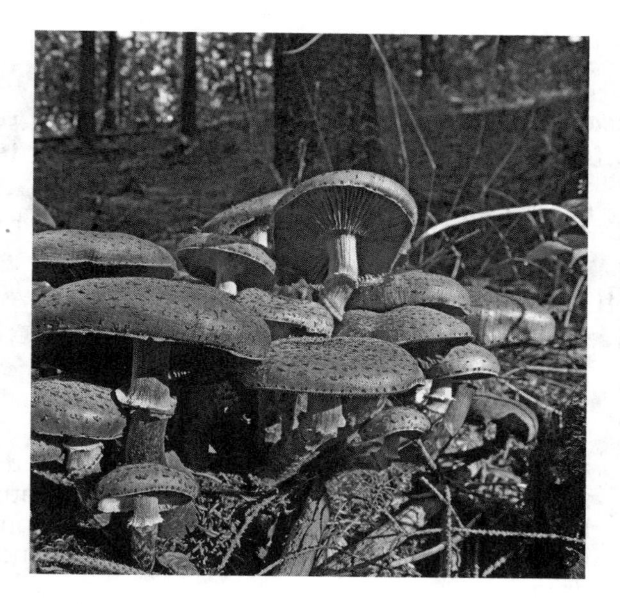

*The largest single living organism found in the world is not a mammal but, rather, a humble fungus. A single honey fungus in the Malheur National Forest in Oregon covers 2,200 acres (8.9 square kilometers) and is estimated to weigh over 600 tons. A runner-up to the fungus: Pando, a colony of quaking aspen (*Populus tremuloides*), which is not a single organism but multiple trees connected by a root system that lies beneath about 106 acres on the western side of the Colorado Plateau in southcentral Utah.*

Fungi can come in many colors and sizes from the microscopic to unimaginable sizes. It can be single-celled, or it can contain a multitude of cells in one organism. It can live on land or on water. One version of honey fungus (*Armillaria ostoyae*), an organism that could be nearly 9,000 years old, has basically taken over a four-mile-square patch in Oregon. Conversely, fungi can be so tiny that a pinch of garden soil can hold thousands of fungus spores and pieces of fungus, each able to start its very own colony. Overall, there are at least 144,000 different species of fungi, although scientists have hinted that there may be many more we either haven't yet found or that we just haven't yet noticed.

On one side of the fungal coin, having fungus around can be very bad.

Pathogenic fungi are just what they sound like: They're they're parasites that can cause disease or even death of the host they're feeding from. This can be a plant, like rye grass (ergot) and chestnut trees (chestnut blight fungus); or it can be your toes when you get athlete's foot; or your head if you get dandruff; or your arm if you get ringworm; or your mouth if you get thrush. Fungus destroys crops that have already been harvested through mold and mildew, and it can poison a human who ingests damaged crops.

Without fungi, however, we'd be in a world of hurt.

Fungus is what allows us to have bread and wine since both use yeast, and yeast is a fungus. Mold is a fungus, and penicillin is made of

HANDY FACT

The Handy Science Answer Book, 4th edition, says: "Fungi grow best in dark, moist habitats, but they can be found wherever organic material is available. Moisture is necessary for their growth, and they can obtain water from the atmosphere as well as from the medium upon which they live. When the environment becomes very dry, fungi survive by going into a resting stage or by producing spores that are resistant to drying. Fungi also thrive over a wide temperature range. Even refrigerated food may be susceptible to fungal invasion."

a certain kind of mold. One fungus, ergot, is bad for grasses but good for pregnant women since it's a part of the treatment to prevent early labor and to control hemorrhage. Fungus is found in statins to prevent heart attacks and in some cancer treatments and immunosuppressant drugs.

Fungus can be used in insect control or to benefit insects that have been infected with viruses; it's sneaky, and it kills certain species of amoebas and roundworms by stalking them and killing them; some fungi will even trap small animals like nematodes. The presence of fungal spores can help scientists solve crimes. Mycorrhizal fungi work in tandem with plant roots through a symbiotic relationship, giving the fungi food through photosynthesis and furnishing the plant nourishment through fungal nutrition. Lichens (another kind of fungus) can help researchers determine the health of glaciers. And fungus can be yummy: mushrooms, of which about 200 species are edible, are fungus; and some fungi are necessary for certain kinds of cheeses. Indeed, it's a better world when there's fungus amongus.

TECHNOLOGY: DID YOU KNOW?

IF YOU COMBINED all the computers that NASA had in 1969 (the year they put the first men on the moon), your smartphone still has more computing power.

A "JIFFY" IS a real measurement of time in astrophysics and quantum physics. First described by Gilbert Newton Lewis (1875–1946), it's defined at 33.3564 picoseconds, or the time it takes light to travel a centimeter in a vacuum. Since Lewis defined it, it

Early versions of the VCR were bulky, expensive, and could only record TV shows for later viewing because Blockbuster Video did not exist yet.

has been altered slightly for different areas of study, but suffice it to say that you can't move that fast.

HIGH-TECH MEETS low-tech: Google's headquarters in California is lawnscaped by goats. Every so often, a goat herder brings a couple hundred goats and a border collie to the grounds, allowing the animals to do double duty: they munch the grass down to acceptable levels while they also fertilize the soil.

THE COST OF the first VCR (videocassette recorder) was well over $1,000 when it sold in the late 1970s. Back then, you could only record TV programs; prerecorded movies and such were not yet available.

TECHNOPHOBIA IS THE fear of technology. Cyberphobia is the fear of computers. Nomophobia is the fear of not being near your cell phone.

AN "ANDROID" IS a figure representing a man. A "gynoid" is a figure representing a woman. And now you know.

SOME 15 BILLION spam emails are sent to someone in the world every day. Don't ever "unsubscribe" to them; doing so only indicates to spammers that yours is an active inbox. Instead, sequester them into your "Spam" file and then delete. For what it's worth, those nasty spammers statistically have less than a one in twelve million success rate.

HANDY FACT

According to *The Handy Science Answer Book*, 5th edition, "Coal, oil, and natural gas are all considered fossil fuels since they were formed from the buried remains of plants and animals that lived millions of years ago. All fossil fuels are nonrenewable sources of energy since their supply will diminish eventually to the point of being too expensive or too environmentally damaging to retrieve for use. Uranium, essential for the production of nuclear energy, is also considered a nonrenewable source of energy since a limited supply of uranium exists. However, it is not a fossil fuel."

COMPUTER SECURITY DAY is celebrated every year on November 30.

SCIENTISTS ARE STUDYING algorithms to predict outbreaks of violence and hate group activity.

THE ENVIRONMENT: THE SUMMER THAT NEVER HAPPENED

If you live in the Northern Hemisphere—or if you're even just slightly familiar with it—you know that it can get mighty cold way up north in the winter, but there are some glorious summers that follow. Even the more temperate areas have thermometer swings, but peek back in time and see that it almost wasn't so.

Every year, we expect that the snow will melt, the soil will thaw, and spring will arrive. That was the expectation for most of the country in 1816, and there were a few warm days that made folks want to air out their homes and maybe start planting crops. On April 24 in Salem, Massachusetts, it was 74 degrees. Thirty hours later, it was far below freezing, and that's never good, but it was spring. Though similar things happened across the United States and in Europe and Asia as well, folks noted the oddity but weren't surprised; that kind of spring happens up north.

About three weeks later, things got worse, though: on May 12, a clipper from Canada brought frigid temperatures, which killed the buds on fruit trees in New England and New York, and *that* was something unusual. It was so cold for so long that ice formed on rivers and ponds in the middle of May (normal low temperatures average around 50 degrees), and corn wasn't growing anymore. The ground refroze in the beginning of June in New York. As far south as Virginia, residents left diary entries and notes on the unusual temps.

And then it got *really* cold....

On June 6, a half foot of snow fell on New England; the following day, it snowed in Boston, and on June 8, parts of Vermont were pelted with eighteen inches of the white stuff. Day after day after day, the temperature struggled to get above the freezing point in New England, and most days, it didn't. Birds froze on fence lines; sheep, which had been shorn that spring, froze to death from exposure. And yet, the temperature reached over 100 degrees in Salem on June 22.

By the beginning of July, though, it was back to the deep freeze; even Norfolk, Virginia, had frost that month, and by then, it was basically all over for food crops. It didn't help that there was a drought that year, either. August rushed in with warm, sunny days, and farmers had hope that there might be a small chance for a window of crop growth, but the bad news wasn't over: cold weather was back by the beginning of the second week, and it came with more snow. Widespread killing frosts put a period at the end of this particular sentence, but The Year Without a Summer wasn't done yet.

Despite the snow and because of the drought, wildfires were a danger to New Englanders and New Yorkers. With nothing to eat, livestock died off and famine hit the upper part of the country; the same thing happened in Europe: food riots broke out, and many died for lack of food. When ice broke through dams, others drowned.

Some parts of Europe didn't rebound from this summer until two years after the calendar said it was over. Scientists estimate that the world's climate dropped by three degrees that summer, which doesn't seem like a lot ... *but it is.*

In the meantime, the disaster began to change the country in all corners: New England farmers, hopeful that the cold hadn't reached other parts, migrated to the Midwest. Politically, many members of Congress voted themselves a pay raise and then promptly lost their jobs at the following election due to outrage from voters. Modern, regular observations and record keeping for climate and temperature began in earnest, and the freeze contributed to a boost in Arctic ex-

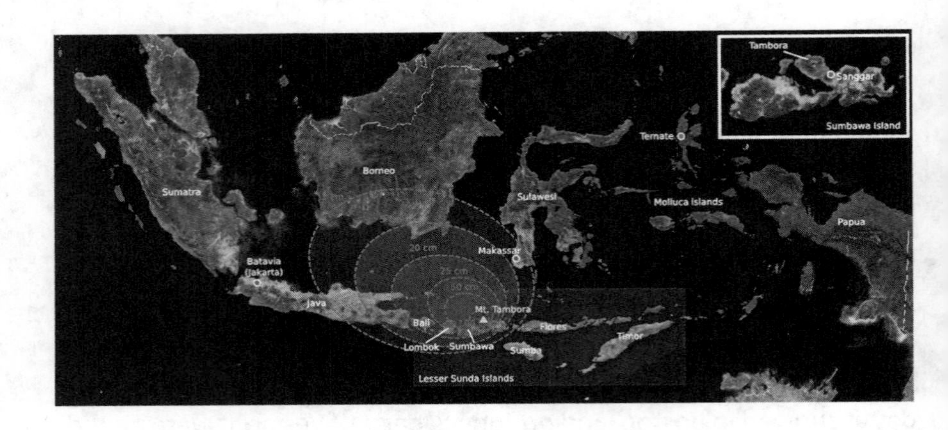

A NASA photo of Indonesia and surrounding islands shows the location of Mt. Tambora. The circles around the volcano indicate ash depths that resulted from the 1815 eruption.

ploration. And, escaping the gloom and rain in Europe, Percy Shelley, Mary Wollstonecraft, Lord Byron, and others sequestered themselves in a room and told scary stories to one another—one of which became the masterpiece *Frankenstein.*

So, what happened to the world that summer of 1816?

Scientists are sure now that Earth itself was at fault: on April 10, 1815—a whole year before the Nightmare of 1816 began—Mt. Tambora, a volcano on an island in what is now Indonesia, exploded with ferocious strength that rivaled Mt. St. Helens and even Krakatoa. It erupted for nearly two continuous weeks, spewing gases, ash, and debris high into the atmosphere and for a hundred miles around, which killed thousands instantly. The volcano led to tsunamis, ash piled to a reported depth up to a dozen feet and, ultimately, a volcanic winter due to sulfur dioxide gases in the atmosphere that retarded sunlight's ability to reach the planet and the blocking of the sun by all the volcanic debris.

Okay, wow, you're thinking. But can this happen again?

Possibly, but not likely. Our modern system of monitoring volcanic activity is so much better than what was available in 1816, which was basically nothing; this, by the way, is all relatively new because the link between volcanic eruption and temperature change hasn't been known but for the last half century or so.

Bear in mind, however, that even a small volcanic eruption can wreak havoc on Earth's climate. Should another Big One happen, however (it's unlikely that it would happen quietly and suddenly without notice), the general agreement is that the resulting explosion, debris, and its aftermath would cause worldwide destruction and complete, total chaos for months and months.

HANDY FACT

According to *The Handy Science Answer Book*, 5th edition, Earth has "four kinds of volcanoes. *Cinder cones* are built of lava fragments. They have slopes of 30 degrees to 40 degrees and seldom exceed 1,640 feet (500 meters) in height.... *Composite cones* are made of alternating layers of lava and ash. They are characterized by slopes of up to 30 degrees at the summit, tapering off to 5 degrees at the base.... *Shield volcanoes* are built primarily of lava flows. Their slopes are seldom more than 10 degrees at the summit and 2 degrees at the base. The Hawaiian Islands are clusters of shield volcanoes.... *Lava domes* are made of viscous, pasty lava squeezed like toothpaste from a tube."

PHYSICS: THE COLOR BLACK

Trick title. According to physics, black is not a color because of its lack of specific wavelengths. Black, says physics, is an absence of light, so despite your tendency to say it's a color and despite all the websites that say it's so, it's not (see Science—Physics: Color My World). Green is a color. Red is a color. Blue is a color. Black is ... black.

And there you are. But you didn't think that was the final word on black, now, did you?

To understand black, it's perhaps easiest to start with what's *not* black: the human eye is so finely attuned to the presence of light that it is, indeed, possible for you to spot a single lit candle more than a mile away on a clear night. Although your pupils were made to expand or contract to allow in the right amount of light to see, your sight is pitiful compared to your cat, which has twenty-five rods per cone in its eyes. You, by comparison, have a measly four rods per cone; even so, when the light is low, your eyes will likely adjust well enough that you can see to navigate around your living room.

But there's black ... and there's black.

To explain that, you need to understand black in all its facets.

Scientists at the National Institute of Standards and Technology (NIST) are constantly hard at work to find the most ultra-black of blacks by manufacturing carbon nanotubes that can capture an astonishing amount of light—the best ones to catch not only the light you see but also all the light across the electromagnetic spectrum. At the time of this writing, scientists at the NIST have manufactured a black that catches 99.99 percent of light, but researchers at MIT have done them one better with a black that catches 99.995 percent of light beams. Technology like this is not, by any means, completed, and its uses will be found in solar products, satellites, and telescopes, to name a few items that are fast becoming consumer common.

One of those technology-created blacks, Vantablack, is the blackest black known so far. Vantablack is exclusively licensed to artist Anish Kapoor (1954–), so you can forget about using it for your painting class.

And while we're thinking about physics, science doesn't seem to have hit upon a definitive answer to the question of what would happen to you if you fell into a black hole. The best advice is not to go looking for a black hole (see Science—Space: Falling Forever).

Psychologists know that while black is a perception taken by the eye, it's also imbued with a kind of personality: black signifies power and fury, elegance and wealth, humility and commonness, magic, anger, leadership, and force. Furthermore, all colors have been shown to evoke emotion; for instance, black can be scary, or it can remind one of decadence, diamonds, and mink. Even the youngest of children have been shown to perceive black as having these attributes.

In history, black was the very first tone used by our prehistoric ancestors, who scooped it up from fire and mixed it with minerals to make art. Ancient Greeks painted black art on other-colored backgrounds, or vice versa, and medieval artists used black to paint many of their demons on canvas. Later, Johannes Gutenberg used black ink to print his Bible. By the 1300s, black began to point to a certain social status (or lack thereof): Italian bankers first embraced black for their clothing, and kings followed suit; conversely, in certain social strata, black indicated poverty because the individual wearing it couldn't afford clothing with more expensive dyes. That still holds to a point: laborers often wore black in times past, and many servants, waiters, and housekeepers still do.

The term "Black Friday" may have several meanings. It's commonly believed to signify the day when retailers' financial bottom line goes "black," or on the side of making a profit. It could also be so-named because it

411

More than a dozen bands, groups, and musicians have released an album called *The Black Album*.

was a day when early workers called in sick in order to have a four-day weekend, leaving the lights off at the workplace; or it's said to describe the traffic and long lines that shoppers endured on that day.

During the Middle Ages, clerics wore black robes as a sign of humility and penance; indeed, the Bible says that there was dark before there was light, and biblically, black is used to indicate sorrow, mourning, sin, and death. Though it still signals humility, black is a symbol for death in Catholicism and is the liturgical color for Good Friday. For Buddhists, it's the color of primordial darkness and hate and can transform hate into love. In Judaism and in some Muslim cultures, it's the color of modesty.

Fashion knows that black is basic and classic, but it's also the color of mourning in the West. Wearing all black can seem menacing, or it can seem elegant (we might thank Coco Chanel [1883–1971] and her "little black dress" for that), but it really depends on the wearer and what's worn. Black is also known to shave pounds off its wearer, which could explain why black is the most popular shade of clothing. For the person who's dressing their home, painting your child's room black can incite calm and creativity, and black accents are good feng shui.

In nature, scientists have discovered that some birds and insects sport black markings that are able to capture up to 95 percent of the light in a nifty physics trick, although most animals that are black are not hued quite so deeply. Black, to varying degrees of shade, can also be found in minerals, soil, rocks, plants, clouds, and the sky.

And then there are black things you should avoid. Superstition, for instance, says that you should never let a black cat cross your path. Don't mess with someone who has a black belt in judo. And never, ever ask to hold a black mamba snake.

HANDY FACT

The Handy Physics Answer Book, 2nd edition, says, "It would seem that a color inkjet printer mixing the three primary pigments of yellow, magenta, and cyan should be able to produce all the other colors, including black. When all three primary colors are combined, however, the mix looks more like a muddy brown color than black. Although these are the primary colors from which other shades can be created, they do not represent all the colors of the spectrum needed to form black. That is, there are gaps in the wavelengths that those pigments absorb. Therefore, most color inkjet printers have a cartridge with yellow, cyan, and magenta ink and another separate ink cartridge of just black."

CHEMISTRY: GIMME A LITTLE SHAKE

Who can have French fries without salt? Or potato chips? Or any kind of snacky crunch? Not you, but did you know that the shaker on your table holds grains of a very fascinating thing?

At its very least, salt is a mineral. That's all it is, just sodium chloride that leads to one of our most basic taste senses. At some point, your doctor may have told you to watch your intake of it, but you can't eliminate salt, nor would you want to. Without salt, there would be no you. Historically speaking, you could also say that without salt, there would be no civilization.

Ancient peoples knew that salt made food taste better, that it worked well to preserve food, and they knew exactly where it came from because ancient civilizations dug their own salt mines thousands of years ago. Those who couldn't get salt from the ground traded it with merchants in caravans. Those with no access to caravans got salt in war, where they also used it to punish an enemy for future growing seasons by salting the earth.

The Bible mentions salt, of course, but the mineral was known far and wide. Egyptians used it in funeral practices. Abyssinians used it as currency. Europeans who lived near the ocean captured their salt from

The salinity concentration of the Dead Sea is 34%, making the waters there far more dense than that of the human body. The result is that people float very easily on the Dead Sea!

the sea and traded it since ocean salt was said to be far superior than anything that came from mines. As trade routes flourished, salt became a major commodity for food and food preservation, and governments occasionally used it for income-generating and citizen-control purposes by instituting salt taxes.

Government interference with the citizen's salt generally backfired, however. American colonists imported salt from Europe and the Caribbean, which was fine for a while until, in the spring of 1776, they established a salt works in North Carolina for use during the Revolutionary War. Mahatma Gandhi led a revolt over a British salt tax.

The Dead Sea, in the Jordan Rift Valley in the Middle East, has a salinity of over 33 percent, which is ten times the salinity of seawater. The salinity of Great Salt Lake varies, but it's up to 27 percent salt in some places.

We don't like our salt messed with.

Generally speaking, the stuff in your saltshaker has been mined from the ground. It's purified by dissolution, evaporation, and recapture and often iodized—which is to say that potassium iodide, which has been an ingredient of table salt since 1924, is added to help prevent goiter and other afflictions caused by a dietary lack of the mineral that our bodies cannot synthesize. It's also possible that stabilizers are added to the table salt you shook on your dinner tonight.

Another kind of salt you may consume is sea salt, which contains small amounts of calcium and magnesium as well as possible sediment, giving sea salt a bit of a different taste. Generally speaking, though, sea salt and table salt taste the same when used in an average meal. You may occasionally find Himalayan salt on the menu (it's pink, so don't be alarmed) and kosher salt, with a larger grain. Black salt may have traces of carbon and lava in it! Rock salt is large-grained and is used in the making of ice cream.

At any rate, be sure you don't go wild with your saltshaker, nor do you want to completely eliminate the salt you put in your mouth: consuming too much of it (hypernatremia) can lead to kidney disease and cardiovascular problems; too little of it (hyponatremia) can lead to death, which is why we're urged to hydrate when the weather gets hot. Literally, an imbalance of salt can be a killer; in fact, in ancient China, the nobility used to commit suicide by ingesting too much salt water.

But if you've had to put down the saltshaker lately, don't toss out all the salt in your house. Because a mere 6 percent of all the salt processed in the world ends up on someone's table, that leaves 94 per-

HANDY FACT

How salty is seawater? According to *The Handy Science Answer Book*, 4th edition, "seawater is, on average, 3.3 to 3.7 percent salt. The amount of salt varies from place to place. In areas where large quantities of freshwater are supplied by melting ice, rivers, or rainfall, such as the Arctic or Antarctic, the level of salinity is lower. Areas such as the Persian Gulf and the Red Sea have salt contents of over 4.2 percent. If all the salt in the ocean were dried, it would form a mass of solid salt the size of Africa."

cent for things like water conditioning, de-icing in colder climates, and in agriculture (cows and horses love their salt licks!). Religions use it for some rituals. Salt is also used in making chlorine for your pool. It's used in the manufacture of paper and plastic products. It's an ingredient in the making of aluminum. You'll find it in soaps and medicines, synthetic rubber, pottery, and textile dyes.

Keep your eyes open and check those ingredients. You never know where you might find salt!

BIOLOGY: YO MAMA!

From the lowliest insect to the highest being in the world (that's you!), everybody's got one. So, here are some facts about our mommies.

THE YOUNGEST HUMAN mother ever recorded was a five-year, seven-month-old Peruvian girl named Lina Medina, who gave birth in 1939. Not much is known about the circumstances except that Lina's mother thought that her daughter was near death since Lina's belly kept growing and growing. The family finally took the girl to a doctor, where it was determined that she had gone through puberty at age three (something that's so rare that just one child in 10,000 experiences it), and she was most definitely about to become a mother. Years later, Lina went on to marry and have another child, but little else is known, and few know who fathered Lina's six-pound son, Gerardo. Since it's uncertain whether or not Lina's still alive, the mystery may never be solved.

IT IS POSSIBLE for a woman to become pregnant with two babies who are not twins at the same time. Uterus didelphys, or having a double uterus, is a congenital defect that begins in the womb when the two tubes (Müllerian ducts) that usually join to create one uterus in normal females don't join completely and, instead, end up becoming two separate uteri. This defect, which occurs in about one out of every 3,000 women, results in two uteruses, two cervixes, and, sometimes, two vaginas. Technically speaking, therefore, and though it's extremely rare and often results in premature labor or miscarriage, a woman can be pregnant with two fetuses conceived at two different times and even from two different fathers.

SO, HERE'S WHAT you did to your mom when she was carrying you: you gave her heartburn, stretch marks, maybe varicose veins, and a sore back. You put her through hours of pain, and you stole her heart. But you also left behind small amounts of your fetal cells, a process known as microchimerism, which happened through the placenta. If you looked at Mom now, you might not see evidence of microchimerism—it doesn't last forever—but scientists discovered in 2012 that fetal cells may have traveled as far north as Mom's brain.

PRIOR TO ABOUT the early 1800s, motherhood was perceived more as a biological function than something to revere. You can point to the beginning of Queen Victoria's reign as somewhat of a catalyst of change. After she took the throne, ideals of romance, motherhood, and domesticity began to be what the then-modern

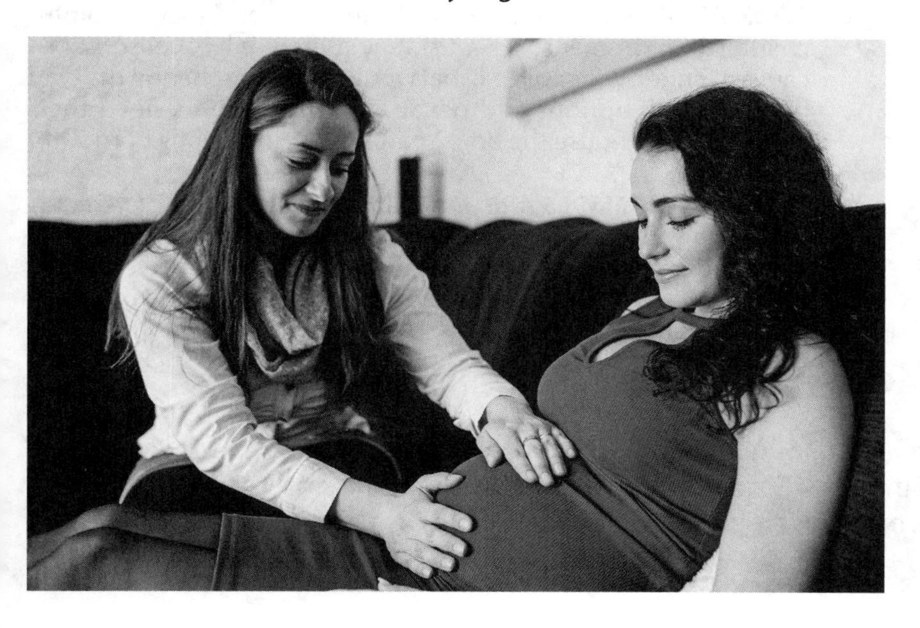

You can still choose a midwife to help with your pregnancy. Studies show that midwife prenatal care can have many benefits, including reduced need for caesareans, higher rates of breastfeeding, fewer perineal tears during birthing, lower rates of induced labor, and less need for anesthesia.

woman aspired to have; women were seen as unfulfilled and unwomanly if they eschewed home, babies, and family. Although it's famously been said that Queen Victoria hated pregnancy and wasn't a fan of small children, as compared to other women of the eighteenth century, women of her era were more likely to enjoy their children, to interact with them, to want them educated, and to have closer long-term relationships with their offspring.

UNTIL THE LATTER part of the 1700s, most American babies were delivered by midwives; by the middle of the eighteenth century, women whose families could afford it began asking for doctors to help deliver their babies, mostly in the comfort of their own homes; for middle-class women, slaves, and poorer women, midwives continued to help bring children into the world. Again, it was Queen Victoria who made it okay to give birth with anesthesia; in her younger days, ether was very uncommon, but just a few decades into the twentieth century, the rate of doctor-assisted childbirth was 85 percent and anesthesia was commonly used. At the same time, new mothers had begun embracing the idea of having their baby at a hospital, in part because doctors insisted it was safer and because methods to quash pain were better available at medical facilities. The only problem? Sometimes, women died of the rudimentary anesthesia, and sometimes, with the mother asleep, doctors used unnecessarily harsh and hasty methods to get babies out. It wasn't until the 1950s that the pendulum swung the other way and new mothers were allowed to think of childbirth as a natural thing that didn't necessarily require intervention.

SO WHY HASN'T evolution fixed the issue of difficult, painful childbirth for humans? Scientists say there are two reasons: because a longer, wider birth canal would seriously change the way humans walk upright and because human infants need their larger heads to house the brain matter we're inherently born with.

PLATYPUS MOMMIES ARE unique mammals in that the females will lay eggs up to three weeks after fertilization, storing the eggs in an underground burrow that can reach more than twenty-five feet down and which is kept very humid using mud and plant de-

HANDY FACT

According to *The Handy Science Answer Book*, 4th edition, you had footprints and fingerprints (in the twenty-fourth week of your mother's pregnancy) before you had hands (which developed in the twenty-fifth week of your gestation).

bris. After a period of less than two weeks, platypus pups hatch but are nowhere near ready for independence yet; instead, for four to five months, the pups enjoy Mom's constant care, under which they receive milk from secretions on her abdomen.

THE MAMMAL WITH the longest pregnancy: the elephant, with a gestation that can last for 95 weeks. That's nearly two years, y'all. Newborn elephants routinely weigh more than 200 lbs. An elephant is no match for a frilled shark, though: poor Mama shark is pregnant for about three and a half years.

THE CREATURE WITH the shortest pregnancy: the North American opossum, which is pregnant for as little as 12 days. After her gestation, she'll give birth to up to 20 babies, which she'll carry in her pouch until they're able to be independent.

RABBITS DESERVE EVERY bit of the reputation they have when it comes to reproduction: a rabbit kit reaches sexual maturity as early as its third month of life. Pregnancy lasts a mere month, and the doe can get pregnant again in as little as a few hours. If she bears the maximum twelve baby rabbits once a month, and if all those babies live to have 12 litters a year, and *their* offspring are so lucky, and so on until they die (which can be at ten to twelve years of age), you'll have to add on to your warren because you'd have literally billions and billions of bunnies.

ANIMALS: DID YOU KNOW (PART I)?

REPORTEDLY, GERALD FORD taught his beloved golden retriever, Liberty, to purposefully distract White House visitors so that Ford might escape small talk at events.

DOGS CAN LEARN to count and have a rudimentary idea of "less" and "more."

OF THE 10,000 different species of birds we know of on Earth, only about 1,800 migrate in the spring and fall.

MICE CAN'T VOMIT. Contrary to popular belief, however, horses *can* vomit, but if your horse does, call the vet immediately: it could be caused by a life-threatening issue.

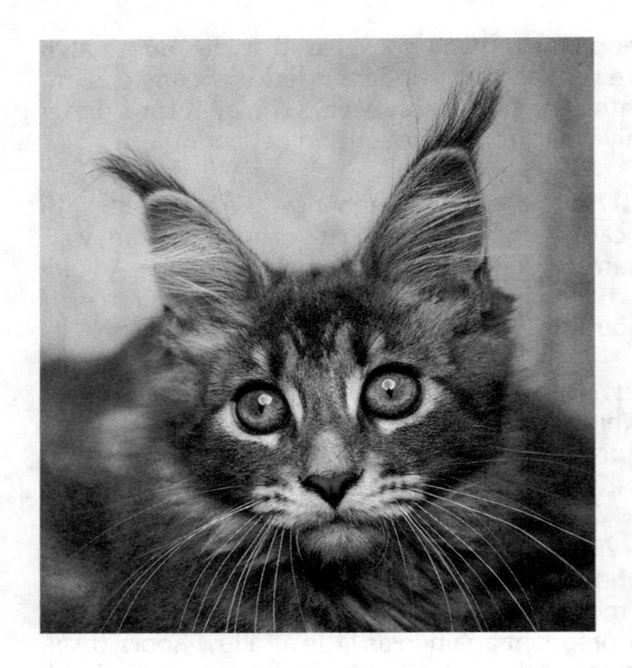

Next time you mention the fur tufts at the end of a cat's ears, you know what to call them: "ear furnishings."

WHILE MANY ANIMALS give birth lying down, a giraffe does not. This means that the giraffe calf will fall from great heights at birth—sometimes as high as five feet and sometimes on its head. Not to worry. Baby giraffes are resilient and can stand within 30 minutes of being born.

THE TUFTS OF hair that stick out of a cat's ears are called "ear furnishings."

THE OLDEST HORSE on record was "Old Billy," who was verified to have been more than sixty years old when he died in 1822 in England. The average horse lives up to thirty years.

AT THE TIME of this writing, the American Kennel Club (AKC) recognizes fewer than two hundred different breeds of dogs. The Kennel Club in the United Kingdom recognizes 218 different dog breeds.

OFFICIAL RECORDS VARY, but it's relatively common to read about captured alligators that get uncomfortably close to 17 to 19 feet in length. To put this into a bit of perspective, think of the size of your bedroom (the average is 11 by 12 feet). A 'gator wouldn't fit stretched out.

FEMALE KANGAROOS ARE pregnant for as little as three weeks and up to almost five weeks. At birth, a joey can be smaller than

HANDY FACT

The Handy Science Answer Book, 5th edition, says, "The Clydesdales were among a group of European horses referred to as the Great Horses, which were specifically bred to carry the massively armored knights of the Middle Ages. These animals had to be strong enough to carry a man wearing as much as 100 pounds (45 kilograms) of armor as well as up to 80 pounds (36 kilograms) of armor on their own bodies. However, the invention of the musket quickly ended the use of Clydesdales and other Great Horses on the battlefield as speed and maneuverability became more important than strength."

a kidney bean; his mother licks a path from her birth canal to her pouch and the joey must pull itself up and into the pouch, where he attaches to the mother's nipple. There, he'll stay for as long as fifteen months. A mother kangaroo can care for two joeys in her pouch in different stages of maturity; while she's nursing them each with milk that has different nutritional content, she may be pregnant with a third embryo.

ACCORDING TO A 2003 study, penguins have impressive bathroom habits: in order not to dirty their beds or nests, when an Adélie penguin needs to eliminate, it backs its bum to the edge of the nest, lifts its tail, and shoots feces as far away as possible, a distance of up to four feet.

BOTANY: EAT YOUR VEGGIES AND FRUITS

Mama always said it best: eat your vegetables and grow up strong. But *why* would you want to nibble on a part of a plant? Well, first of all, they really are delicious.

CARROTS

IF THE AVERAGE newly harvested carrot is eight inches long, it would take nearly 8,000 carrots to make a mile of the veggie—and that's if you can stop yourself from eating even one or two. If you can't help yourself, that's okay: the calories in three carrots will be burned off from walking that mile.

CARROTS ARE EASY to grow, particularly if you have sandy soil; in fact, the best, straightest carrots are from gardens in which the soil is not too heavy or rocky.

HISTORICALLY, CARROTS WERE grown first as a medicine and then as a vegetable for every day. The reason is clear: one regular-sized carrot contains a little over twenty calories, plus about five grams of carbohydrates and a little bit of fiber, but it's absolutely packed with beta-carotene, which our bodies convert to vitamin A. Just one carrot gives you 200 percent of your daily requirement of vitamin A.

THERE IS NO such thing as a "baby carrot" except in marketing. What you know as "baby carrots" are actually the broken bits of regular-sized carrots, tumbled and smoothed for snackers who prefer them.

EAT YOUR CARROTS—or wear them. Scientists and entrepreneurs have discovered that carrot fibers are stronger than carbon fibers by a factor of two. Curran is the brand name of the result of this experiment, and it's being used in helmets and other items.

411

People often get fruits and veggies confused. Most people think that if something is sweet, it's a fruit, and if it is not sweet, then it's a vegetable. This is not always the case. A good example is the tomato, which is a fruit and not a vegetable. Fruits are edibles that grow from the flower of a plant, while vegetables come from other parts of the plant, such as leaves, stems, and roots. An easy way to tell the difference is to look inside. If there are seeds inside the edible, then it is a fruit; if not, it's a vegetable. Some other comestibles that are mistaken for veggies include squash (zucchini, acorn squash, pumpkins), avocados, peas, green beans, bell peppers, eggplant, okra, and even olives.

CORN

ASIDE FROM HAVING it by the spoonful or on the cob, there are many hidden ways that you consume this tasty vegetable: it's a great food for large animals and is often a major ingredient in pet food. Cornmeal and corn flour come from corn. Corn starch works as a great alternative to talc. Corn can be used in gasoline; it can become an ingredient in sweet food and drink, soaps, cosmetics, adhesives, candles, and any product that requires starch; and it's used in the plastic industry. All in all, more than 3,500 products are made wholly or partially with corn.

IF YOU TRAVEL abroad, you might not find *corn* anywhere. That's because it's often known as maize, and it's grown on every continent except Antarctica.

BEFORE YOU BITE into that next cob, consider this: the average corn cob has roughly 800 kernels lined up in sixteen rows of goodness. You will always find an even number of rows on a cob of corn. As for the stalk itself, the average height at maturity is eight to ten feet, but the record for a single stalk of corn is thirty-three feet. In the United States alone, some eighty to ninety million acres of land is dedicated to growing corn.

BIOLOGICALLY SPEAKING, CORN is both cereal and vegetable. A single cup of pure corn is under 200 calories and for that, you get over forty grams of carbohydrates and nearly five grams of fiber, plus vitamin C, folate, and thiamine. Bonus: corn is gluten-free.

BEANS

IF YOU LOVE beans, you're in luck: there are somewhere around 40,000 different kinds of beans for you to try and countless dishes to try them in: beans can be fried, boiled, or baked, and green beans can be eaten raw from the garden. We know that because we've known about the goodness of beans for nearly 10,000 years; in fact, beans were one of humanity's first cultivated crops.

NUTRITIONALLY SPEAKING, YOU almost can't beat a bean. Beans are full of Vitamin B6, iron, magnesium, potassium, thiamine, riboflavin, and other vitamins and minerals, plus antioxidants. Because they're high in protein, beans are good substitutions for meat. They're also high in fiber, which, in addition to the complex

Beans are increasingly being used as healthy alternatives to meat, such as with this black bean burger.

sugars that beans contain, can cause flatulence in some humans. Despite what Granny says, though, there's really no foolproof way to eliminate flatulence from eating beans; the good news is, not all beans will do it to you.

YES, BEANS ARE legumes. So are peas, lentils, peanuts, and, for that matter, clover.

YES, AS MENTIONED above, you can eat *green beans* raw, but as for other kinds of beans, well, just don't. Because of a kind of protein called lectin, consuming most beans raw will give you major gastrointestinal problems, and they can even kill you. Nope, beans contain much more nutrition after they're cooked, so don't be tempted to do otherwise.

PEAS

WHEN YOU READ this, you may think that peas are a little boring: there are only three basic types that you can grow in your garden. It's easy to get confused about the differences between English peas, snow peas, and sugar snap peas because they all look basically the same. For most people, the only important thing is that snow peas are not good for shelling but *are good* for cooking, pod and all. If you look, you'll find other varieties of peas to grow, but the basic three are what you'll find in the average grocery store.

LIKE BEANS, PEAS have been cultivated by humans for about 10,000 years, but peas are thought to have originated in Middle Asia. It's known for sure that Christopher Columbus brought peas to the New World, and the love for them spread across the continent.

ACCORDING TO BRITISH experts, the proper way to eat peas is to mash them with your fork and then scoop them up. The record for eating peas was set by Janet Harris of Sussex, England, in 1984, when she ate nearly 7,200 peas, one by one, in sixty minutes, using chopsticks.

SHOULD YOU SAY, "Yes, peas"? Of course you should: peas are rich in vitamins A, C, and B1; they're high in fiber and carbs and low in fat. As for protein, about five teaspoons of peas have more protein than one teaspoon of peanut butter.

SQUASH

AS A PLANT, it's believed that squash was domesticated up to 9,000 years ago, probably somewhere in North America. There

Did you know that squashes are actually fruits and not veggies? That's because the squash comes from the flower of a plant. As a fruit, then, zucchinis are great for the garden. They are easy to grow and bear a lot of fruit!

are two basic types of squash: summer squash, with thin skins that can be eaten whole (think: zucchini); and winter squash, which have tough shells when mature (like butternut or acorn squash) and seeds that can be dried and eaten later. Pumpkins, by the way, are considered to be a kind of winter squash; melons are also related to garden squash.

WINTER SQUASH AREN'T named because they're grown in the winter but because they store easily over the coldest months of the year.

ON A LIST of vegetables, squash probably isn't on a lot of Top Three lists, and that's too bad. Aside from the fiber that a cup of orange (winter) squash gives you, it's absolutely packed with vitamin A—so much, in fact, that it contains four times more vitamin A than most people require per day! The seeds of those orange squash are also packed with vitamin E. Summer squash (not orange) offers vitamin C, niacin, and protein, among other good-for-you benefits.

THERE'S A REASON for the old jokes about forcing zucchini on your friends and neighbors: one zucchini plant will yield edibles for many months. You may get upward of ten pounds of zucchini *per plant*. That's a lot of bread, cakes, and stir-fry.

BECAUSE THEY CAN, some gardeners compete to grow very large pumpkins. As of this writing, the largest pumpkin ever grown came from Belgium when Mathias Willemijns brought forth to the world his 2,600-plus-pound behemoth in 2016. That's just slightly smaller than a compact sedan.

POTATOES

AS FAR AS vegetables go, potatoes are relatively new to the table: the Inca people began cultivating them a mere 2,200 years ago. When Spanish conquistadors came to South America, they snagged a few taters to take back with them to the Old World—although few would actually eat them. Potatoes are related to the nightshade, which is deadly, and potatoes got a bad rep by extension. Once the tubers were introduced to Ireland, however, they became popular in Europe, and they came to North America in the 1500s.

NO DOUBT YOU already know a few things that come from potatoes. Legend has it that Thomas Jefferson introduced French fries to America during his tenure in the White House. Potato chips were invented in 1853 because Commodore Cornelius Vanderbilt insulted Chef George Crum in Saratoga Springs, New York, when the commodore complained that his fried potatoes were too thick; in response, Crum shaved potatoes, fried them, and served them, and the rest is history. By the way, if history could design a better name than "Crum" for the guy who invented potato chips, I don't know what it could possibly be.…

IF YOU'RE AVERAGE, you'll eat more than 100 pounds of potatoes this year—and that's a good thing. Taters (without embellishment) are low in calories, and they're naturally gluten-free, fat-free, sodium-free, and contain no cholesterol. One potato contains about half of your RDA of vitamin C, plus it's packed with potassium, vitamin B6, and antioxidants.

HANDY FACT

This may confuse any arguments you might have: according to *The Handy Answer Book for Kids (and Parents)*, "a fruit is the part of a plant that has developed from a flower and contains seeds. Fruits are generally fleshy, sweet, juicy, and colorful.… Tomatoes, peppers, pumpkins, and even nuts are all technically fruits.

"Vegetables are simply other parts of plants that are eaten, like roots, stems, and leaves. Carrots and sweet potatoes are roots. Asparagus is a stem. Cabbage and spinach are leaves.…"

IN THE UNITED STATES, Idaho grows the most potatoes for the consumer market, with Oregon, Washington, and Wisconsin following. The largest potato ever grown came from Great Britain in 2010. It topped the scales at more than eight pounds of white tastiness.

TECHNOLOGY: DON'T PUT THAT IN YOUR MOUTH!

Actually, your dentist wants you to put these things in your mouth. And so you do, possibly every single day, at least once, better twice. But beware, and maybe *don't* open wide....

First of all, people everywhere should herald the invention of the toothbrush.

Ancient Egyptians and Babylonians who wanted to preserve their pearly whites chewed on a stick, the end of which was frayed in order to cover more surface. In the very late fifteenth century, the Chinese began attaching coarse boar's hairs to a stick or bone or piece of ivory to clean their chompers; a boar, for you nonfarmer types, is a big pig, so let that sink in. Nylon was invented in 1927, and ten years later, it was finally used to make toothbrushes. The minor shock about this is that the concept of oral hygiene was still somewhat new at this time in American history; in 1900, less than 10 percent of the residents in American households brushed their teeth. It took the concerted efforts of several dentists in several areas of the United States to change that, and a campaign to promote oral hygiene during World War I helped. By the time toothbrushes were available for the masses, brushing one's teeth was a widely accepted idea.

But here's where the danger lies: toothbrushes wear out and need to be replaced about every six months or so. Worn-out toothbrushes don't clean your teeth very well, and they could harm the enamel on your teeth, *and* they could lead to gum disease. Other cautions: you could swallow a toothbrush bristle (generally harmless, although that's been reported to have caused

Nearly half of the toothbrushes sold in the world sport red handles.

appendicitis), and a surprising number of people have swallowed entire toothbrushes (which requires a doctor, stat).

As for toothpaste, it's been around for longer than toothbrushes have. The Egyptians used it some 7,000 years ago, and the technology spread to China and India about 2,500 years ago. It was not the minty-fresh stuff we know, though: some of the more common ingredients included the powdered hooves of oxen, ashes, pumice (yikes!), eggshells, and bones. The Chinese added salt and mint. Later, ground nuts made their way into toothpaste, as did charcoal, and the whole concoction was left as a powder. All a user had to do was add water. It wasn't until after World War I that toothpaste as we know it outsold tooth powder (and yes, you can still buy tooth powder).

The stuff you brush your teeth with today consists of a majority of safe abrasives, to get the stains off your pearly whites. There's likely fluoride in your toothpaste as well as a sweetener and some type of detergent.

As for the safety of your toothpaste, it's like anything else: too much of a good thing can be a bad thing. Because of its active ingredients—which are generally safe in small amounts—swallowing large amounts of toothpaste can be poisonous, so you want to keep it away from kids without supervision, and never give it to pets. Your dentist can offer you more information on the dangers of toothpaste and fluoride overuse.

Some people enjoy making their own tooth powder to clean their teeth. Some of the ingredients commonly used are baking soda, calcium powder, bentonite clay (a binding agent), sea salt (soothes gums), sage (for whitening), ground cloves or cinnamon (natural antibacterials), and peppermint.

Finally, no discussion of tooth-hygiene dangers is complete without the mention of the plain old, common toothpick.

It's a safe bet that people for millennia have been using whatever tool was at hand to get pesky annoyances out from between their teeth; bones, ivory, porcupine quills, and sticks are but a few of the things that history tells us were used as toothpicks. Toothpicks are even mentioned in the Old Testament.

Like everything that's handy, though, toothpicks were made better when they became mass-produced: entrepreneur Charles Forster of Maine invented a way to manufacture toothpicks, first by hand and then by machine, although both were laborious processes. By 1860, his factory was large enough to keep up with growing demand for this modern-not-modern product, which was made (some twenty million toothpicks *per day*) by slicing birchwood into strips, then into thousands and thousands of wooden picks, which were then smoothed and polished for consumer use. Because Forster did this in Strong, Maine, the town was known as the Toothpick Capital of the World until early in this century, when the factory closed. Alas, the use of dental floss has overtaken the once ubiquitous wooden pick.

Here's why you need to be ultracareful with a toothpick: they're very easy to swallow. And because your body can't absorb or dissolve a wooden toothpick, the pick remains largely intact, which means it can poke a hole in the intestines, stomach wall, bowels, or an artery. Don't mess around: if you've swallowed a toothpick, head for the ER.

HANDY FACT

According to *The Handy Science Answer Book*, 5th edition, "Most mammals have two sets of teeth during their lives. They are born without teeth and remain toothless while they are nourished by their mother's milk. They develop deciduous teeth around the time they are weaned. The deciduous teeth are sometimes called baby teeth or milk teeth. As they mature, the deciduous set of teeth are lost and replaced by a permanent set of teeth."

ANIMALS: AWW, RATS!

Here's a solid fact for you: when it comes to rats, people either love them or hate them. There is no middle ground. Several different kinds of animals fall under the genus *Rattus*, including the long-haired rat, the bush rat, and the extinct bulldog rat. For us, the black rat (*Rattus rattus*) and the brown rat (*Rattus norvegicus*) are the most familiar outside the laboratory; the latter is most often the kind you'll find in a pet store. Both types of rats originated in Asia and came to America aboard ships in the seventeenth and eighteenth centuries.

It might sound like the stuff of nightmares, but all it would have taken was one single male–female pair of rats to colonize the country.

Let's assume that the mother rat (the doe) was pregnant when she skittered ashore with her mate (the buck). Rat gestation is less than a month, so she might have immediately sought a place to give birth to up to a dozen baby rats. Six weeks later, the babies would reach sexual maturity; meanwhile, the doe would have given birth again already, and she could be pregnant for a third time. Do the math—if you dare.

Physiologically speaking, although the two main kinds of rats are related, there are main differences: brown rats (which can be brown, black, or gray) are considerably larger than their black rat brethren and can grow to be twenty or more inches tail tip to nose. The brown rat has smaller ears than does the black rat, which is generally only eight inches in size from tip to tip. Another marked difference is that the brown rat is omnivorous in a major way and will eat nearly anything that remotely resembles food (and a number of things that don't), while a black rat (which is black or dark brown) is happier with

HANDY FACT

Science learns from the darndest things: from ancient, fossilized pack rat urine, a substance that crystallized into a cementlike bond with surrounding material as it dried, researchers have discovered DNA from things that the animals brought into their nests. This tells scientists what the rats ate and collected; it also offers a picture of the surrounding landscape, climate, and other animals and fauna that existed in the area millennia ago.

organic fodder, small birds, and pet food that you might leave out for Fido.

The CDC says that a rat can enter a home through a hole slightly larger than a quarter.

As for where you'll find these critters, the answer is "almost anywhere," although black rats prefer warmer climates. Brown rats can be found in large urban areas (think New York City's infamous rats) and aren't bothered by cold weather. In either case, you don't want to go making friends with a wild rat. You don't even want to try to get along with him. The reason why is twofold: disease and destruction.

Rats are fastidious about their fur and will groom themselves and others both as a way of bonding and of ridding themselves of dirt—but that doesn't necessarily make them clean: they are major carriers of leptospirosis, toxoplasmosis, hantavirus, salmonella, and a host of other organisms that can make you sick. That's not to even mention the fleas that live on the rat, one type of which caused Europe's bubonic plague, a disease that occasionally crops up here and there in the United States and still kills hundreds of people each year in this country alone. Yes, rats will bite, and they will eat living human flesh, if the victim isn't swift or mobile; thankfully, it's rare for rats to carry rabies.

As for the destruction wrought by rats, it's considerable: rats burrow, which could harm crops from beneath and could undermine buildings. They'll chew through cement (partly to keep their constantly growing incisors trimmed), sheetrock, drywall, wood, and hard-packed dirt. They'll chew through roofs, HVAC systems, and electrical wire. A rat will happily live in your house, leaving feces and urine anywhere it happens to be, which can cause millions of dollars in needed repairs and extermination fees around the country. As for hoping they'll go away, if there's plenty to eat, a rat will stay within the same basic area for life because ... why leave?

And yet, aficionados don't *want* their rats to go.

Pet rats can be highly affectionate and will groom a person as if they're part of the rat's own family. They're intelligent and may be happy to learn good behaviors; they're calm, clean, quiet, and can be the perfect pet for apartment dwellers or people with time constraints. It should be mentioned that lab rats aren't meant to be pets; they are specially bred, often with genetic alterations, and are meant for entirely different purposes.

The worst part about a pet rat is the best part about a wild rat: the average life span is usually from one to three years.

HANDY FACT

According to *The Handy History Answer Book*, 3rd edition, "The most egregious outbreak of the bubonic plague was the version called the Black Death, which occurred in the middle of the fourteenth century. Between 1348 and 1350, it is estimated that the bacterial disease killed more than one hundred million people. Scientists believe that a bacterial strain developed from rats and fleas that traveled from China to Europe."

BIOLOGY: HURTS SO GOOD?

So, let's say you just sliced your finger making dinner. Not much, a little blood, but the reaction was instantaneous: you yanked your hand away from the knife. Even before your reaction, though, something else happened: in a fraction of a millisecond, sensory receptors in your skin sent messages through nerve fibers (A-delta and C fibers), which hit your spinal cord and brain stem, which sent them on to the brain, which registered the information, processed it, recognized the pain, and caused you to say things you wouldn't say to your mama.

Not a lot of people like pain, and you have to believe that it's handy to know how big a world of hurt you'll be in if you handle a certain kind of insect. For that, Justin O. Schmidt (1947–), an entomolo-

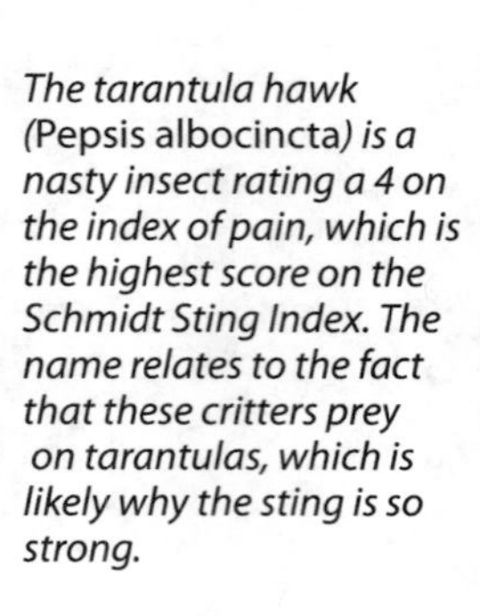

*The tarantula hawk (*Pepsis albocincta*) is a nasty insect rating a 4 on the index of pain, which is the highest score on the Schmidt Sting Index. The name relates to the fact that these critters prey on tarantulas, which is likely why the sting is so strong.*

gist at the Carl Hayden Bee Research Center in Arizona and author of *The Sting of the Wild,* has taken one (or, as it turns out, many more than one) for the team by allowing all kinds of stinging insects to have a go at his hands, fingers, and arms. Of course, because one must keep track of such information, he created the Schmidt Sting Index.

Starting with the mildest, perhaps almost annoying, stings you can endure (a sweat bee, which rates a 0), Schmidt moves up to fire ant (a mere 1.2, although it should be noted that you never get stung by *just one* fire ant), to velvet ants that'll make you scream, and all the way up to the top of the scale. Tarantula hawk wasps, warrior wasps, and bullet ants reside there, and that's information you don't want to test.

If *Hymenoptera* aren't your thing, Coyote Peterson (1981–) has gone up against all kinds of animals for his YouTube channel, *Brave Wilderness*. Like Schmidt, his interest in wildlife began in childhood; Peterson launched his first show with plans for an encounter with a bear, but he ended up with a North American porcupine that he let brush his hand, thus embedding a number of quills in his skin.

On Peterson's pain scale, porcupine quills and scorpion stings hit Level 2 out of four. His personal worst encounter, however, is with the executioner wasp—also *Hymenoptera*—although the bullet ant is right up there on the Peterson scale, too.

Don't try this at home....

HANDY FACT

The Handy Science Answer Book, 5th edition, says that "the average North American porcupine has about thirty thousand quills or specialized hairs, comparable in hardness and flexibility to slivers of celluloid and so sharply pointed that they can penetrate any hide. The quills that do the most damage are the short ones that stud the porcupine's muscular tail. With a few lashes, the porcupine can send a rain of quills that have tiny, scalelike barbs into the skin of its adversary. The quills work their way inward because of their barbs and the involuntary muscular action of the victim. Sometimes, the quills can work themselves out, but other times, the quills pierce vital organs, and the victim dies."

MATH: HOW LONG HAS THIS BEEN GOING ON?

Imagine this: you've been tracking a woolly mammoth for days, and you finally killed your prey. Dinner's on you, but when you get back to your pals to tell them you need help bringing home the woolly bacon, they want to know how far they'll have to walk.

It's not hard to see why we needed measurements of distance, weight, length, height, and other ways of seeing our world. We needed those numbers—and we all needed to agree on them. But that took awhile.

In the beginning, regions, if not individual cities and villages, had their own ways of measuring things, and because of that, it's often difficult to determine the final answer on many of the origins of the measurements we are familiar with and use most often. We do know that basic measurements in ancient times included the length or width of a man's foot; the width or length of his finger, arm, or hand; the size of common crops; or the distance someone could walk in a predetermined time, all of which would have varied widely. No doubt, this mishmash of measurement was likely the reason for many ancient arguments, and it didn't do much for trade, which highlighted the need for an agreed-upon way of determining such things.

For early humans, length and distance were determined by whatever the Celts, Greeks, Romans, and Egyptians had nearby. The Egyptian cubit, the unit of measurement that Noah supposedly used to build his ark, was the distance from a man's elbow to his fingertips. In Europe, inches were determined by lining up barleycorn or poppy seeds, a method the Anglo-Saxons devised before they began calling this new measurement by a modified Latin word for part of a Roman foot. Originally, as legend has it, the Scots said an inch was the width of their king's thumb but that obviously was prone to change, so they decided to measure the thumb, of three men of three different sizes and then they averaged them to determine the measurement. And then there were Swedish inches. And French versions and Chinese adaptations, and nobody said it was easy. You see where this is going, right?

In 1959, an international agreement was *finally* reached to set—once and for all—a foot at 0.3048 of a meter and an inch at 25.4 millimeters.

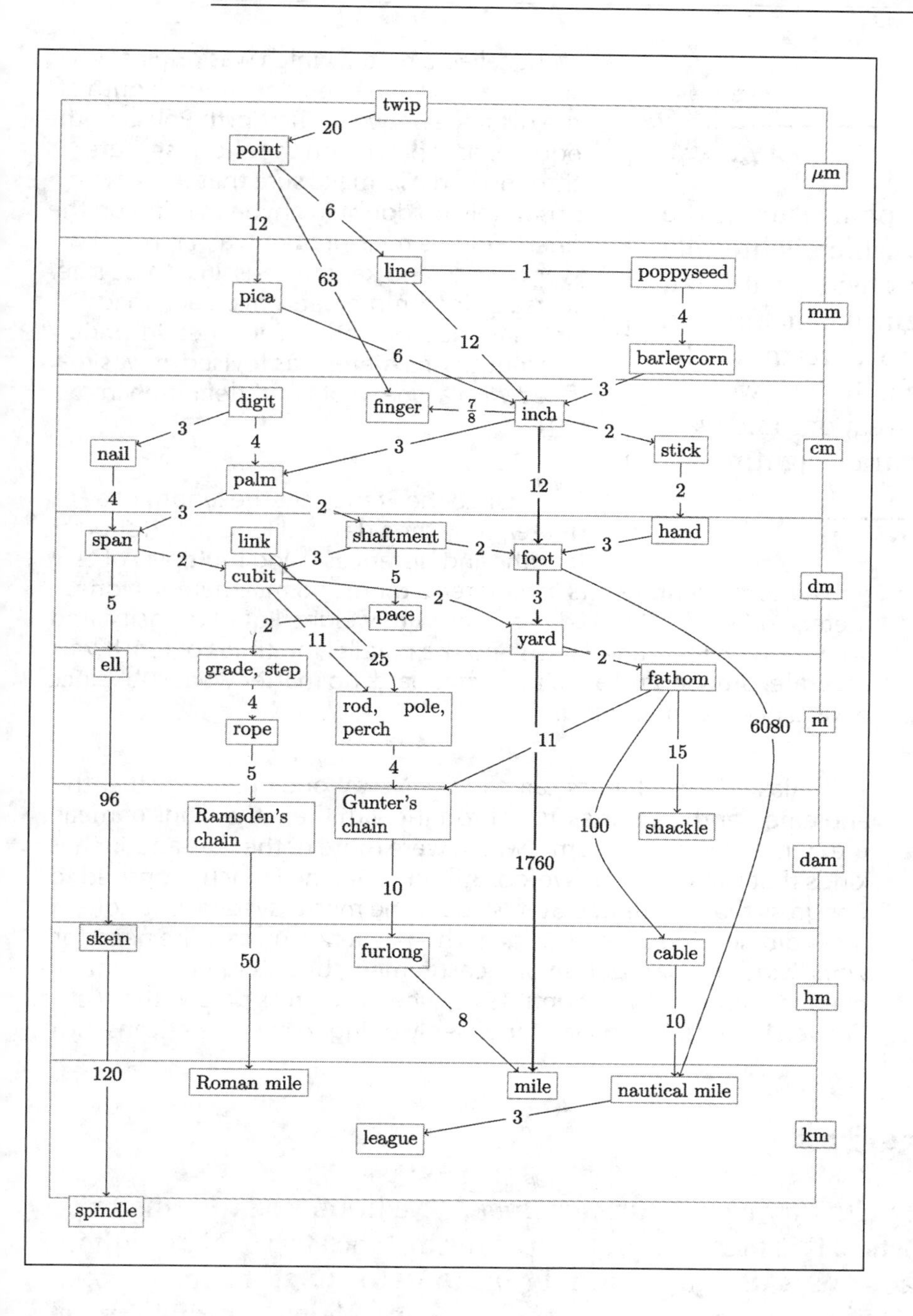

This handy-dandy chart shows the relationships of different units of length in the English Imperial system to one another (metric system is at right). For example, you can follow along to see that there are 20 twips in a point, 6 points in a line, 12 lines in an inch, 3 inches in a palm, 3 palms in a span, 2 spans in a cubit, 2 cubits in a yard, 2 yards in a fathom, 11 fathoms in a Gunter's chain, 10 Gunter's chains in a furlong, and 8 furlongs in a mile (whew!)

Which begs the question: *So, whatever happened to the metric system in the United States?*

In the midst of all the perplexing lack of standards, French tradesmen quietly made their own standards based on a measure-

The smallest unit of measure is the "hair's breadth," which is literally measured in how wide a human hair is. The origin is ancient and once was used to determine the width of an inch; today, we use hair's breadth in weapons and warfare for more accurate sighting.

ment called a meter, which was originally supposed to have been one ten-millionth of the distance between the North Pole and the equator as it passed on the shortest route through Paris. To make sure that there was absolutely no doubt in anyone's mind on the finality of this measurement, a platinum bar was created and kept in Paris, just to be safe. In years since, more bars have been made and are housed in other countries. In 1983, the length of a meter was revised; it was refined with a "speed of light" definition in 2019.

Once the French set the length of a meter, they then devised longer and shorter lengths and distances using multiples of ten: one hundred centimeters in a meter, a thousand millimeters in one meter, and so on. They did the same with weight, liquid capacity, and volume, with liters and grams and the parallel multiples of ten. It was accurate, precise, and absolutely standard, no matter what substance or distance was measured.

Alas, when colonists came to the New World in the eighteenth and nineteenth centuries, they brought with them the kinds of measurements they had at home, which were more of the feet-and-inches kinds that Great Britain favored. Still, in 1795, the French proposed to Congress that the United States adapt the metric system, but Congress did nothing about the idea. Thomas Jefferson saw the need for worldwide standardization of measurement, but his interest went unheeded—and so it went until 1975, when Congress passed the Metric Conversion Act, which was supposedly going to spur Americans into

HANDY FACT

According to *The Handy Science Answer Book*, 5th edition, today's "meter is equal to 39.37 inches. It is presently defined as the distance traveled by light in a vacuum in 1/299,792,458 of a second. From 1960 to 1983, the length of a meter had been defined as 1,650,763.73 times the wavelength of the orange light emitted when a gas consisting of the pure krypton isotope of mass number 86 is excited in an electrical discharge."

embracing metric measurements in union with much of the rest of the world.

No timetable was ever set for this, however, and the idea seems to have quietly gone away—at least in the minds of most Americans—but not in their language. Today, you can purchase a two-liter bottle of soda, go on a 5K run, or use a metric wrench, which shows that a metric thread is indeed sneaking into our culture.

TECHNOLOGY: FIRST COMPUTERS AND OTHER TECH FORGETTABLES

As you read this book, you may occasionally find something interesting, even irresistible, and you know you've got to investigate further. Without thinking much about it, you might grab your smartphone or your laptop and do a quick search. It's easy to do, but in the not-too-distant past, you wouldn't have had that speedy option.

While computers have been in existence for many decades, they didn't appear in homes on a wide basis until the mid-1970s. Even then, only hobbyists had computers since they were generally build-it-yourself electronics that came in kit form; some kits required that you add your own keyboard, power supply, and monitor, which was often just an extra TV you had around. In 1977, two years after the introduction of the MITS Altair 8800, the first kit sold, Apple introduced the Apple II for home use. Radio Shack also weighed in with its TRS-80, Commodore had an available device, and the first computer store opened for business.

It wasn't until a year later that computers became different things to different people when programs could turn your computer into a word processor, a calculator, or a gaming device. By 1980, it's estimated that about a million desktop computers were in use in American homes.

Back in 1976, the 5.25-inch floppy disk was introduced, and it revolutionized the home-computer industry then. In case you never used one, the name of the product fit its description exactly: the

The first commercially successful personal computer was the Altair 8800 from MITS. Released in 1974, it was sold mostly through electronic hobbyist magazines as a kit you had to assemble. The cost was about $2,000 in today's dollars. It featured a 2 MHz CPU and a 4K memory board.

square disk was made of thin plastic, and it flopped in your hand as you grasped it to insert into the slot on your computer. The earliest "floppies" held about 80 kilobytes of info (for single density; double-density capacity was roughly 115k), or a tiny, tiny fraction of today's common 8GB thumb drives, and they sold for around $60 for ten disks; later floppy disks—the ones that were smaller, hard-bodied, and not so floppy—were 3.5 inches in size and held about 1.44 megabytes of information (see: Science—Technology: It's No Flash in the Pan!).

To actually use one of those newfangled computers to its fullest capacity, though, you'd have to find access to a dial-up connection to jump on the World Wide Web (from which we get the "www" part of website addresses). Perhaps the best part of dial-up connections was that all you needed was an available phone line: you'd plug one end of a cord into your computer, the other end into your phone jack, and with a few clicks, you were in business. What happened was that your computer waited for a dial tone; once it got the tone, it then dialed a phone number and waited for it to connect, much like you do when you make a call. The bad news was that that could take anywhere from a few seconds to a few minutes or longer if you had a bad phone line; in the meantime, users had to listen to a series of tones and screeches to get online, and if you ever heard that music, you'll never forget it.

Two major downsides to dial-up connections: they were terribly, annoyingly easy for the connection to be lost, and while you were on the computer, regular calls couldn't come in or go out.

The first fully portable computer was released in 1981. The Osborne 1 was larger than a briefcase, weighing in at more than 23 pounds, without a battery. As one of the first all-in-one computers, the Osborne 1 included a tiny monitor, a keyboard that flipped down from the top, and two slots for your floppy disks. You could buy an Osborne 1 for around $1,800. The Epson HX–20 was also released in 1981; considered to be the first laptop, it was about the size of a piece of notebook paper and roughly 2 inches thick. Data for the HX–20 was stored on a small cassette.

Hard to believe, but the computer mouse predates the home computer kit!

Rudimentary computer mouses were invented in the 1940s, but they were largely kept secret; they had trackballs or wheels to move the cursor but were otherwise quite different from the mouse on your desk now. The first computer-controlling mouse was publicly demonstrated in 1968 in Germany; those early "mechanical" mouses were attached to the computer by a wire and had the rolling ball inside the underside of them; users might remember that they had to occasionally take the mouse apart and clean it with a puff or air or breath, to rid the device of lint and debris. Though they were available a few decades ago, the use of "optical" mouses became more widespread at around the turn of this century.

The word "mouse," as used to refer to the device we use to move our cursors, by the way, first appeared in the mid-1960s. It likely sprung from the very appearance of a computer mouse, its size, and its long "tail" that attached it to the computer.

Another rather amazing thing: gaming systems predate computer kits, too. In 1972, Magnavox released the Magnavox Odyssey, a machine for consumers that could be hooked to any TV, which, at that point in tech history, was generally a large boxy thing with an enormous tube that surely made hooking and unhooking a hassle, at best. Even before Magnavox launched its product, though, electronic gam-

The first computer mouse (shown here) was invented by Douglas Engelbart (1925–2013) in 1964.

ing was available if you knew where to look for it; Magnavox snagged the first official rights to electronic gaming and sold around 100,000 units before Atari's Pong moved into the top slot. If you don't remember Pong, talk to someone who does. Oh, we have stories....

The story of the telephone is one that's been told many times; suffice it to say that we've been enjoying the benefits of phones for nearly 150 years. We've been enjoying the benefits of an answering machine for just about that long.

At the Exposition Universelle in Paris in early 1900, Danish inventor Valdemar Poulsen displayed what he dubbed a "telegraphone," which recorded sounds with a magnet. He astounded exposition-goers with it, and the *London Daily News* gushed that it would be a great device for businessmen. But here in the States, AT&T, which had a monopoly here, immediately blocked the device for several decades—except for its own uses. If you'd been around in 1934, you could have left your messages for AT&T employees by calling their offices. It wasn't until 1951 that the technology was available to consumers through a little box that hooked up to a home's phone system. Today, you can get your voicemail without a box, probably digitally.

And when you're watching your flat-screen TV (one that's vastly slimmer and lighter than your grandma's old television), keep your eyes open for 1990s-era programs. Feel free to laugh at the cell phones we used then. Early cell phones were introduced for consumers in early 1992, but they were about the size of bricks, and about all they could do was take and make calls and retrieve voicemail. There's no way they'd fit in anybody's pocket.

How far we've come....

HANDY FACT

"The first weather satellite, the Television and Infrared Observation Satellite (TIROS I), was launched by NASA on April 1, 1960. Although the images were not of the same resolution as we have now, they were able to reveal the organization and structure of clouds and storms. One of its accomplishments was to see a previously undetected tropical storm near Australia. The information was conveyed to the people so they could prepare for the approaching storm. It operated for seventy-seven days until mid-June 1960, when an electrical fire caused it to cease operating."

—*The Handy Science Answer Book*, 5th edition

PHYSICS: TURN UP THE HEAT

Until Columbus visited America and returned to Europe with all kinds of nifty new things, the only spicy spice that Europeans knew was plain old pepper. Pepper was an important spice for European palates, to be sure, and South Asia enjoyed a brisk trade because of it, but plain old black pepper was nothing like the chilies that Columbus took from the South and Central Americans that he met here.

The people of Columbus's "New World" had known of chilies for a long time by then. The chili was perhaps the first plant domesticated in Central America, having been cultivated since about 7,500 B.C.E. Once Columbus opened the door for chilies in Asia, Spanish and Portuguese explorers brought the plants from Mexico and north Central America to South Asia for cultivation there. Today, you'll find just as many spicy, chili-based dishes in Asia as you will in South America.

As members of the Capsicum family, chilies can range from hardly spicy to burn-the-roof-down hot, depending on their capsaicinoids, which are a class of compounds found in the chilies. The most common one is capsaicin, and it, like all the other capsaicinoids, is not water soluble; if you consume a superhot chili, drink milk to rid your mouth of the burn. That helpful hint leads to what is perhaps the most curious thing about capsaicin: aside from the fact that it does some strange things to human cells similar to burns from a fire, if you take capsaicin on a regular basis, you can build up a tolerance to its heat, which lowers overall sensitivity to pain. This is why some spicy chilies suddenly don't taste quite as spicy. It's also why capsaicin is sometimes used in medicine.

HANDY FACT

You may be asking yourself now: will eating superspicy chilies kill me? The answer is: probably not. The first bite of a Carolina Reaper will hurt, and that's no lie. You'll turn red, you'll sweat, and you'll think you're being incinerated from the inside out. But in order to die from consuming more, you'd have to eat *a whole lot more,* and it's likely that your body would save you from your own idiocy. Most experts say that while death is theoretically possible, you'd likely vomit and/or pass out long before eating too many hot chilies would kill you. But seriously—don't try to find out, eh?

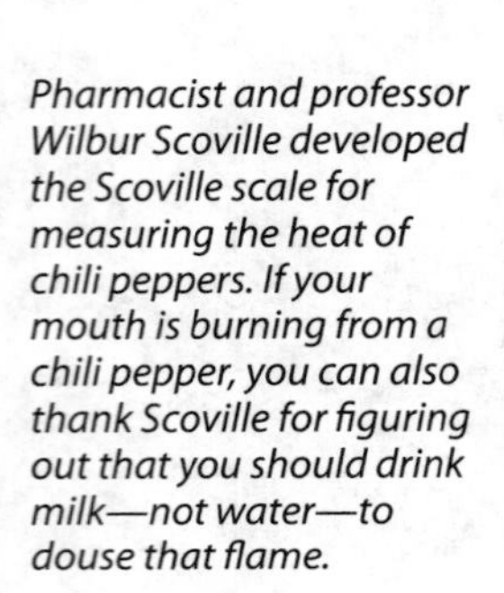

Pharmacist and professor Wilbur Scoville developed the Scoville scale for measuring the heat of chili peppers. If your mouth is burning from a chili pepper, you can also thank Scoville for figuring out that you should drink milk—not water—to douse that flame.

All this history begs one big question, though: how spicy is spicy?

Enter the Scoville scale.

Wilbur Scoville (1865–1942) was an accomplished author of a handbook on pharmacology and had worked as a professor at the Massachusetts College of Pharmacy and Health Sciences when he started working at Parke-Davis pharmaceutical company. He had already known that milk was a good remedy for chili burn, so he began working on a heat ranking of popular chilies of his day. He originally called it the "Scoville Organoleptic Test," which measured spiciness based on the testing of humans and the chilies they consumed in a reverse-dilution method. A number was then given to the chili based on how much dilution was needed before it didn't burn anymore. Scoville went on to win several awards for his work as well as an honorary doctorate of science from Columbia University.

His results are now known simply as the "Scoville scale," which ranks the heat of chilies and peppers in Scoville Heat Units (SHU), measured in 100 units per number given on rankings. So, for instance, a sweet bell pepper gets a SHU ranking of zero. Cayenne pepper gets up to 50,000 Scovilles. Ghost peppers get slightly over a million SHU. Carolina Reapers (the hottest pepper known at the time of this writ-

HANDY FACT

On Christopher Columbus, *The Handy American History Answer Book* has this to say: "History wrongly billed Columbus as 'the discoverer' of the New World. The native peoples living in the Americas before the arrival of Christopher Columbus truly discovered these lands. It is more accurate to say that Columbus was the first European to discover the New World, and there, he encountered its native peoples."

ing) have up to 2.2 million SHU, but there are many contenders (like the Dragon's Breath) in the wings.

Even so, that's nuthin'. Police-grade pepper spray contains up to 5.3 million SHU.

Pure capsaicin contains up to 16 million SHU.

BIOLOGY: WHAT MAKES YOU TICK?

Chances are, what's going on inside your skin isn't high on your priority list of things to think about all day. That doesn't mean you shouldn't know these things about your innards....

YOUR HEART

CONTRARY TO POPULAR belief, your heart does not sit entirely beneath your left breast. It's actually closer to your breastbone, just left of center.

THE HEART OF an infant weighs about twenty grams, which is less than the weight of one light bulb. Once you become fully grown, your heart is roughly the size of two small hands clasped or a large fist, roughly eight to twelve ounces, depending on your gender. The creature with the smallest heart is a kind of fruit fly that measures fractions of a millimeter in length; it should be no surprise that whales—the largest creatures on Earth—have the largest hearts.

YOUR HEART WILL beat a little more than 100,000 times today, and it'll pump around 2,000 gallons of blood through your veins. If you live an average lifetime, that ticker will beat some two-and-a-half billion times.

DANIEL HALE WILLIAMS (1856–1931), a Black cardiology surgeon, performed the first open-heart surgery in 1893. The youngest person to survive heart surgery wasn't even born yet; the technology that enables doctors to do cardiac surgery on babies in utero arrived just this century.

WHILE A HEART attack can happen at any time, statistically speaking, Mondays are when most heart attacks happen. Be doubly aware if that Monday happens to be Christmas Day; that's the calendar day when the most heart attacks are reported.

SMILE! RESEARCHERS SAY that happy people have 22 percent fewer heart attacks than people who report less satisfaction with life.

YOUR BLOOD

IF, SOMEHOW, YOU could magically spread your blood vessels out in a single, continual line, they would take you around the world at least four times.

THE CORNEAS IN your eyes don't have a blood supply and are the only organ in your body that doesn't. Corneas get oxygen directly from the air that hits them.

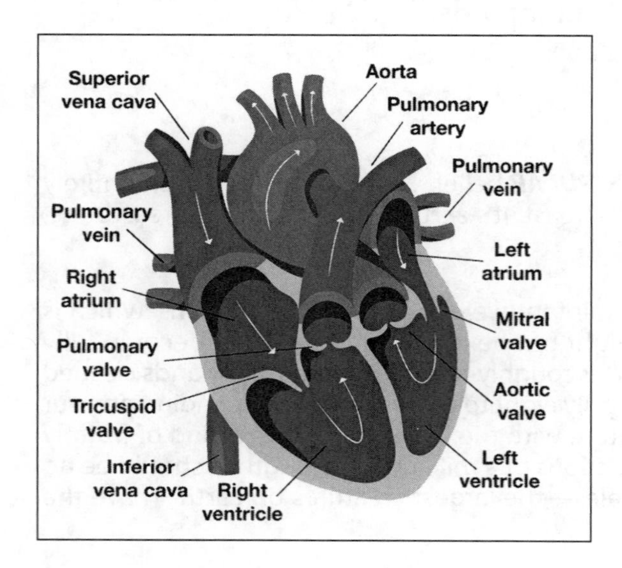

The human heart is a remarkable, four-chambered muscle machine that pushes 2,000 gallons of blood through your body every day!

ONE OUNCE OF blood contains 150 billion red blood cells. If you donate blood, you can do so without fear: your body manufactures millions of new blood cells per second. Those red blood cells come from bone marrow and could live up to four months; white blood cells are also made in the bone marrow but may only last a few hours to a few days.

EVERY TWO OR three seconds, someone in the United States uses donated blood.

YOU'RE GOLDEN. No, seriously: tiny trace amounts of gold are found in human blood.

IN THE UNITED States, O-positive blood is the most common blood type. In Japan, A-positive blood is most common. The least common blood type in America is AB negative.

WILLIAM HARVEY (1578–1657), the man who discovered how blood circulates in the human body, was heavily involved in the investigation of witches in the early 1600s.

YOUR BRAIN

THAT THING IN your noggin weighs roughly three pounds. It consists of well more than half fat, making it one of the fattiest organs in the human body. (That changes the insult power of the word "fathead," doesn't it?)

IN CONSUMPTION OF oxygen, your brain comes in second, at not quite 20 percent of the body's usage. The liver wins, with slightly more than 20 percent.

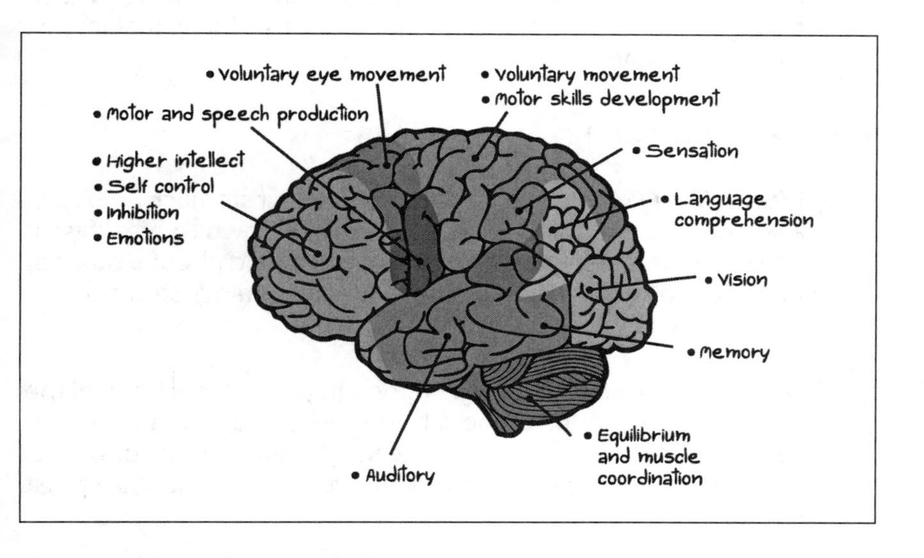

This somewhat simplified diagram indicates which regions of the brain are responsible for various life functions.

EVEN BEFORE YOUR mother knew she was pregnant with you, your brain grew at about 250,000 new brain cells per minute. In the 12 months after you were born, your brain doubled in size.

YOUR BRAIN FEELS the pain sent to it by receptors in the body, but the brain itself is incapable of feeling pain.

BY THE TIME you are in your late twenties, your capacity for memory has started to fade. Memory is affected by alcohol use, traumatic events, and dehydration. Even so, scientists suggest that the things you remember aren't directly the things you remember, but that you're remembering the last time you recalled that memory—meaning that, over time, what you think you remember could be skewed by your own brain. It should also be noted that if you're not familiar with a setting or a person, your brain may not hold details well, which is why witnesses' memories can be fallible.

SOMEWHERE IN MIDDLE age, the size of your brain will start to shrink.

YOUR BRAIN STEM (at the base of your skull) controls breathing, heart rate, swallowing, consciousness, and alertness. You can't, therefore, live without it, but science has noted several instances of people who have lived normal lives with very little brain matter in their heads. CT scans of those individuals show normal or slightly enlarged craniums filled with fluid and not much else.

BECAUSE WE ARE social creatures, humans are wired to see faces where they do not exist. That's called pareidolia, and it's not just faces we may see. Pareidolia also includes seeing objects in clouds, hearing hidden lyrics in music, or hearing voices in everyday sounds like furnace noises or table fan hums. Pareidolia used to be labeled as a psychosis, but now scientists know that it's just us being human.

YOUR STOMACH

WHEN YOU FIRST get up in the morning, your stomach holds just a few ounces of fluid and bile. Once you sit down to breakfast, if you're an average person, it can expand to hold about a quart of food. For comparison's sake, a one-day-old infant's stomach can hold about a tablespoon of food.

ONCE YOU'VE EATEN enough, nerves in your stomach signal the brain that you're done, and a hormone called ghrelin simultaneously decreases. In between meals, ghrelin does the opposite, stimulating your appetite by telling your brain that it's almost chow time.

HANDY FACT

From *The Handy Physics Answer Book,* 2nd edition, by Paul W. Zitzewitz, there's this about your ears: "The ear allows humans to hear frequency ranges between 20 hertz and 20,000 hertz, but it is mostly sensitive to frequencies between 200 hertz and 2,000 hertz.

"The lower and upper fringes of this bandwidth can be difficult to hear, but many people—especially younger people—hear these frequencies quite well. As people age, their sensitivity to high frequencies diminishes. Damage to the hair cells caused by exposure to loud sounds also reduces the ear's sensitivity to high frequencies."

YOUR STOMACH IS lined with epithial cells, which produce mucus to keep a barrier between the walls of the stomach itself and the acid and food mixture that will ultimately nourish you. Because the enzymes and acids that allow digestion are caustic, the lining of the stomach is replaced by the body on a regular basis. And because you're wondering now, no, stomach acid wouldn't tear a hole in your hand if you held some of it, but you would feel a little bit of a burn.

EMETOPHOBIA IS DESCRIBED as an extreme fear relating to vomit, whether seeing it, doing it, watching someone else do it, or reading about it. (Sorry.)

CHEMISTRY: I CAN MAKE IT MYSELF!

Remember the first time you realized that mixing a little of *this* and a little of *that* made something entirely different? In the 1950s and 1960s, a lot of kids with that kind of experimentation in mind rejoiced to see a chemistry set beneath the Christmas tree.

They were in good company: it's possible that their great-great-grandparents might have had that same yen to experiment and the fun chance to do it.

The first chemistry sets were manufactured in Europe in 1791 and were meant for highly educated but curious adults to purchase for themselves. The kits came with blowpipes, glassware, and chemicals needed to study the budding science of chemistry. In 1797, James Woodhouse, a professor at what would eventually become the University of Pennsylvania, authored a booklet on conducting experiments with chemistry kits, but by then, the kits were meant not just for the educated professional but also for adults in well-to-do households. Other scientists weighed in with similar books, and by the early 1800s, everyday people who were interested in the new science of chemistry in both the United States and Europe could learn more from public demonstrations and lectures, which were held in larger cities.

By 1860, the chemistry set fad exploded (no pun intended). Consumers had many options for the purchase of chemistry cabinets, including small sets for those who had just a slight interest in the science to chests that were made to answer requirements for advanced education. Adults reveled in conducting experiments and showing their friends what were literal parlor tricks.

As for child as chemist, kids hadn't quite yet entered the picture here. That happened near the beginning of the twentieth century when toys that happened to use chemistry to do "magic" tricks of color change and light fire-play were marketed. This new perception of chemistry sets as toys surely changed a lot of marketing plans as

A circa 1942 A. C. Gilbert Company chemistry set. Alfred Carlton Gilbert was a medical student at Yale and an amateur magician in the early 1900s, when he came up with the idea of selling such sets to children.

more companies with different kinds of chemistry sets entered the picture.

Because Germany was the largest manufacturer of chemistry kit components, and because of restrictions on materials and that country's need to divert resources to the war effort, sales of kits in Europe waned during World War I. Here in the United States, however, two enterprising brothers, John J. Porter and his brother Harold Mitchell Porter, started a chemical company in 1914 and began selling toy chemistry kits, specifically marketed for younger boys. As of 1920, the Porters had serious competition from Alfred Carlton Gilbert (1884–1961), who also invented the Erector Set and who expanded his array of hands-on toys to include chemistry kits.

In the aftermath of World War II, after the technological and chemistry advances of companies like DuPont, Goodyear, and others who worked in the war effort, interest in chemistry rose and expanded to include sets for girls.

To be sure, most of what a kid would find inside a chemistry set was harmless, but very early sets contained radium, and even post–World War II kits might include radioactive materials or poisons that did cool things. In 1960—aw, gee whiz—the government put a stop to that kind of "fun" by passing the Federal Hazardous Substances Labeling Act of 1960, which mandated labels on kits with dangerous chemicals inside as well as the removal of certain equipment and acids. The Toy Safety Act of 1969 removed lead, and two more acts in the 1970s made chemistry kits even more tame. Even the media began to warn parents of the dangers of letting kids play with chemicals. New and growing awareness of environmentalism pretty much sealed the end of the chemistry set as it was known for nearly two centuries.

HANDY FACT

According to *The Handy Science Answer Book*, 5th edition, "The chief odorous components of the spray [of a skunk] have been identified as crotyl mercaptan, isopentyl mercaptan, and methyl crotyl disulfide in the ratio of 4:4:3. The liquid is an oily, pale-yellow, foul-smelling spray that can cause severe eye inflammation. This defensive weapon is discharged from two tiny nipples located just inside the skunk's anus—either as a fine spray or a short stream of rain-sized drops. Although the liquid's range is 6.5–10 feet (2–3 meters), its smell can be detected 1.5 miles (2.5 kilometers) downwind."

Once upon a time, a chemistry set could be had for as little as a dollar, but today they're considerably pricier and a whole lot more docile than they used to be. Still, if you've got a science-minded kid handy, you can find a decent enough set.…

ANIMALS: DID YOU KNOW ...? PART II

ELEPHANTS HAVE FOUR knees. Because of that, they are physically unable to jump.

BOTH KOI AND goldfish are carp, but they are two distinct kinds of carp. Koi can grow up to three feet in length, and their bodies are more cylindrical than their bowl brethren. Koi also come in a larger variety of colors. Koi can live for up to 60 years if properly cared for; goldfish can live for less than half of that.

A GOAT'S PUPILS are rectangular and horizontal, a trait they share with sheep, octopi, and toads. It's thought that the pupil allows for a wider range of sight, so they know if a predator is sneaking up on them.

TIGERS DON'T JUST have striped fur. Their skin is striped, too. Another tiger fact: the stripes on each tiger are unique to that cat, acting like a sort of "fingerprint" in the tiger world.

BEWARE: A GRIZZLY bear's bite is so strong that it can bust a bowling ball.

YOU MAY KNOW dogs to be good scenting animals, but studies show that African elephants have twice as many olfactory recep-

HANDY FACT

The Handy Science Answer Book, 4th edition, says: "The average orb-weaver spider takes 30 to 60 minutes to completely spin its web. These species of spiders (order Araneae) use silk to capture their food in a variety of ways, ranging from the simple trip wires used by large bird-eating spiders to the complicated and beautiful webs spun by orb spiders. Some species produce funnel-shaped webs, and other communities of spiders build communal webs."

A bear's jaw muscles are strong enough to literally crush a bowling ball!

tor genes as do dogs. Bears may also be better sniffers than dogs. Whales, conversely, completely lack olfactory organs.

CLOWNFISH ARE ALL born male. A number of them will "become" female in order to mate and propagate the species.

RING-TAILED LEMURS use their unique (and, it's said, rather pungent) scent not just to mark territory but also to signal a readiness for mating, to avoid inbreeding, and to fight with one another.

SCIENCE IS STILL not entirely sure whether or not bats pass gas.

BIOLOGY: THE MONSTER WITHIN

If you've read this far, you know that the human body is a wonderfully made, beautiful thing. But sometimes, it can be a wonderfully weird, very unusual thing, too. Your body can accomplish amazing feats—but it can also veer into downright odd territory.

Teratomas are usually benign (meaning not cancerous) tumors, most commonly found in the coccyx (tailbone), ovaries, and testes but that can develop anywhere in your body. If that doesn't sound bad enough, there's this: a teratoma often contains fully developed tissue, organs, teeth, bones, hair, even entire little limbs.

Yes, there's a reason the Greeks gave them their name: "teratoma" derives from words meaning "tumor" and "monster."

If you're diagnosed with a teratoma, pay close attention: it will be described as "immature" or "mature"; the former can develop into cancer, while the latter includes dermoid cysts and can be removed, although mature teratomas have a higher chance of reoccurrence. Also, if you're diagnosed with one, scientists think that it's probable that you've had it since before you were born. While teratomas are most common in female patients, men can have them, too. They should not be confused with....

A parasitic twin, which occurs when the eggs for identical twins split improperly in their mother's womb, leaving behind a generally healthy human and a partially developed "other" body that's physically attached in some way. Parasitic twins differ from conjoined twins in that at some point during gestation, one twin's development stops, while the other continues as normal.

There are several variants of parasitic twins, including three variations indicating the head and one that indicates a lack of heart or head. The latter variant is most common and can, because of the way the circulatory system works in the twins' bodies, mean life-or-death problems in the larger, developed twin. Fortunately, parasitic twins occur in fewer than one in a million live births, and the situation is often able to be surgically dealt with; unfortunately, doctors don't know much about the phenomenon.

Parasitic twins also differ from vanishing twins, in which two heartbeats are indicated on a gestational ultrasound test, but at some early point in the pregnancy, one of the fetuses is absorbed either by

HANDY FACT

The Handy Science Answer Book, 5th edition, says, "The twin birth rate in the United States in 2015 (data released in 2017) was 33.5 twins per 1,000 live births, or 3.35 percent of live births. The birth rate for triplets or higher-order births was 103.6 per 100,000 live births."

the mother, by the placenta, or by the fetus's more dominant sibling. Partial absorption changes a vanishing twin into a parasitic twin.

A lithopedion, or a "stone baby," was first recorded back in the tenth century when one unfortunate mother's pregnancy went wrong: the baby died in utero, and rather than absorbing it (as would happen in the early weeks of pregnancy) or expelling it (as would happen if the miscarriage was slightly later in the pregnancy) or inducing labor (as might have occurred in modern times), her body allowed the fetus to calcify into a solid, stonelike mass.

Although extremely rare—they have been recorded a mere 300 times in the last four centuries of medical record keeping—lithopedion formations generally happen in pregnancies that develop outside the mother's uterus and inside her body cavity. Despite this waterfall of medical happenings, records indicate that many women have gone on to have children while they carry a lithopedion formation inside their bodies; many, in fact, are unaware that they even carry a stone baby. On the flip side, lithopedion formations are actually rather beneficial things: by calcifying the fetus, the body protects the mother from the dead tissue inside her. Even so, the truth is that stone babies are exceedingly rarely seen today in places where modern medicine is practiced because, should miscarriage occur, measures today would be taken to avoid calcification long before tissue would ever reach that point.

GENERAL SCIENCE: THIS AND THAT AND RANDOM FACTS

SODIUM IS A highly explosive chemical when it comes in contact with water. Chlorine is what sanitizes a swimming pool. Put them together, and you've got plain old run-of-the-mill table salt.

THERE'S A NAME for the person who trains circus fleas. That would be "pulicologist." You're welcome.

ACCORDING TO DOCTORS, the number-one complaint from patients who walk into the emergency room is abdominal pain.

SCIENTISTS SAY THAT babies see the color red before they're able to discern the other colors well. It takes five months before they are able to see the full spectrum of colors.

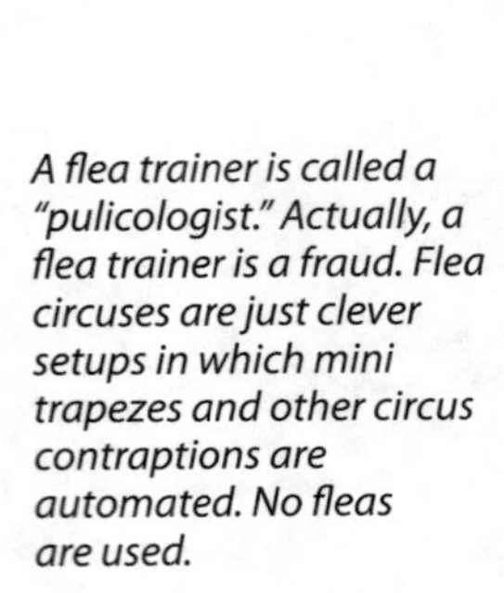

A flea trainer is called a "pulicologist." Actually, a flea trainer is a fraud. Flea circuses are just clever setups in which mini trapezes and other circus contraptions are automated. No fleas are used.

THERE ARE MORE bacteria in your gut than there are other cells in your entire body.

THE AVERAGE PERSON passes gas around twenty times each day. Many times, you wouldn't even know that you've done so.

IN PLACES ALONG the Gowanus Canal in New York at the turn of the last century, the bacteria was tested at more than 600,000 bacteria per cubic centimeter. Today, it's not recommended to swim in water that contains more than twenty-four bacteria per cubic centimeter.

THE TIMING AND cadence of conversation between two humans is basically the same across cultures. Each "turn" lasts an *average* of mere seconds, and there's about a 200-millisecond gap between turns, which is how long the brain needs to determine that the speaker is done. Indeed, the brain is busy: it's actively listening, while it comes up with possible answers to keep the conversation going.

A GOOGOL IS the number 1 followed by 100 zeroes. Google does not list a googol of websites yet; in fact, it's mathematically doubtful that there will be a googol of websites available for perusal when your grandchildren have grandchildren.

IN THE ANIMAL world, there are few species that are matriarchal (which is to say, they are run by females). Those include lions, honeybees, elephants, and spotted hyenas.

HANDY FACT

The Handy Science Answer Book, 5th edition, says, "Light travels at 186,282 miles (299,792 kilometers) per second or 12 million miles (19.3 million kilometers) per minute." You can't even blink that fast.

"FOUR" IS THE only number that, spelled out, has the same number of letters as the number it represents.

IT IS ENTIRELY possible that human victims of the guillotine were alive for several seconds after being separated from their heads. As for animals, Mike the Headless Chicken lived for 18 months after losing his head (but not his brain stem) to a farmer who intended to have Mike for dinner. Mike choked to death in an accident. It should also be noted that the heads of decapitated rattlesnakes can still bite many minutes after the head is severed, although that is a matter of dying nerve reflexes and not an indication of life.

TECHNOLOGY: PUT A LITTLE SONG IN YOUR HAND

Back in the early 1960s, a transistor radio was a near-mandatory accessory for teens and many adults. With a transistor, you could take your tunes (via radio station) with you wherever you went. But what if reception was bad or you just plain didn't like the local stations?

In 1963, Philips introduced a new technology called compact cassettes, which used twenty-year-old technology originally devised for reel-to-reel recording machines; like original reel-to-reel machines, cassettes weren't meant for home use. Originally intended for dictation machines in business settings, the sound quality wasn't the best, but it worked for the office. Once the technology improved, ever-evolving computer manufacturers chose the cassette for data storage, and cassette players became more popular with music lovers, who could either buy prerecorded music, or they could record their own music on blank tapes.

About a year after cassettes were introduced, a group consisting of the founder of the Learjet, William Lear (1902–1978), and several other large tech firms and automobile manufacturers got together to create what Lear called a Super 8 cartridge. It was an adaptation of a development made by someone else; in fact, inventors had already come up with a four-track player, and others had been tinkering with endless-loop recording methods and cartridges. They'd tinker with it again.

The innovation of the Super 8 cartridge (later, more commonly called an eight-track tape), as indicated by its name, lay in the number of tracks: an eight-track cartridge offered four pairs of stereo tracks presented as "programs" (rather than sides of a record), which were meant to give listeners a better experience. Inside the plastic case, the tape itself was spliced with thin strips of metal that kept the cartridge's contents playing continuously unless users either skipped a song or popped the cartridge out of the machine. There was no need to rewind, therefore, less stress on the tape; the tape was also supposedly less likely to tangle.

While eight-track tapes may seem, in our collective minds, to be completely tied to muscle cars and lava lamps, the truth is that Ford introduced the eight-track players in 1965 for its 1966 model Mustangs, Lincolns, and Thunderbirds as an option. RCA released some of its back catalog so that consumers could use their newfangled eight-track players. The following year, eight-track players were standard on all Ford models, and home players were next, introduced so listeners could take their tunes from car to couch.

But by 1975, the eight-track tape was relegated to the back room for myriad reasons: the tape was awfully unreliable, and people hated

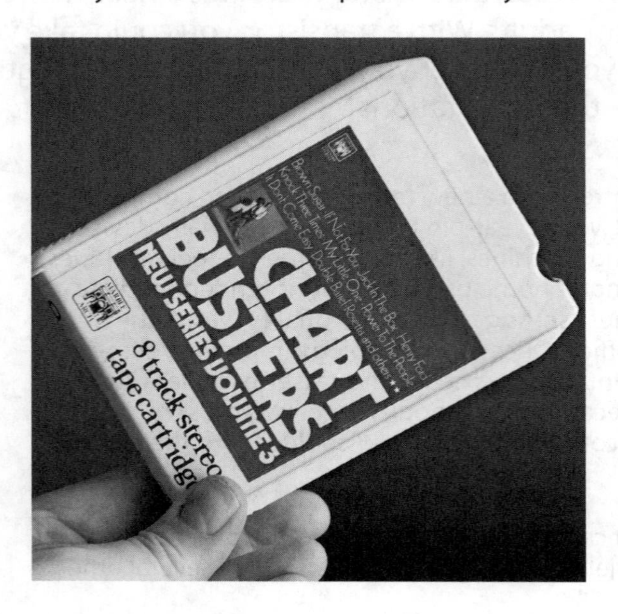

The eight-track tape came out in the mid-1960s specifically for use in Ford automobiles, but it would become a popular way to play music in the 1970s before being supplanted, along with cassette tapes, by compact discs in the 1980s.

when a track would switch because any song that was playing would fade in and out (which really messed with a good sing-along!). That no-rewind feature got to be more of a hassle than a boon, and even though the tape was meant to be protected, it *did* tangle. As for the technology, it also had its problems, including the constant need to clean the player heads and the degradation of the tape inside the plastic case.

By the mid-1980s, eight-track tapes were all but obsolete. Yes, you can still find them, but they're rare, having been taken over first by its close cousin, the cassette, and then by CDs and digital audio players.

While there is some room for argument, many collectors agree that Fleetwood Mac's *Greatest Hits*, released in late 1988, was the last "mainstream" album released on eight-track.

THE ENVIRONMENT: ABRAHAM LINCOLN, ENVIRONMENTALIST

On April 22, 1970, Americans celebrated their first Earth Day, meant to remind us that a clean Earth is essential to the survival of … well, almost everything. But what many Americans don't realize is that environmentalism was a 110-year-old notion.

So much has been written about our sixteenth president, Abraham Lincoln (1809–1865), that you may feel as though you know all there is to know about him. Lincoln's thoughts and actions during the Civil War have been well documented, and his assassination is a whole historical event unto its own. But there was another aspect of Lincoln's

HANDY FACT

From *The Handy Science Answer Book*, 5th edition: "The frequency of a sound is the pitch. Frequency is expressed in hertz (Hz). Sounds are classified as infrasounds (below the human range of hearing), sonic range (within the range of human hearing), and ultrasound (above the range of human hearing)."

interest that isn't widely known, and that's that he may have been the nation's first environmentally minded leader and one who was certainly interested in science.

Lincoln was born in Kentucky and spent much of his childhood and youth on a farm surrounded by woods in southern Indiana. After the death of young Abe's mother, Nancy, Abe's father was the sole food provider; it's likely that Abe, too, eventually had to hunt for dinner, and though it's been said that Abraham was a bookish boy, it's a surety that he knew a lot about wildlife and the trees and plants that made up his world. He even once noted knowing that there were bears living in the woods near his home.

In his young manhood, Lincoln joined the Whig Party and worked hard in the Illinois state legislature while in his twenties. The Whig Party was trying to flip America from a nation of farmers to a nation of industrialists, and Lincoln was adamant that it was the right path for this country, even though industrial progress wasn't always kind to the natural resources that it tamed. This harm (at best) and destruction (at worst) continued through the time when Lincoln became the country's first Republican president and the beginning of his time in office.

After the Civil War started, however, he seemed to have second thoughts. It's believed that the war and the ruin it left behind—the impact battles had on the environment, the destruction of forests, and pollution, not to mention potential problems with war dead on the soil itself—suddenly became clear to Lincoln, and with the urging of

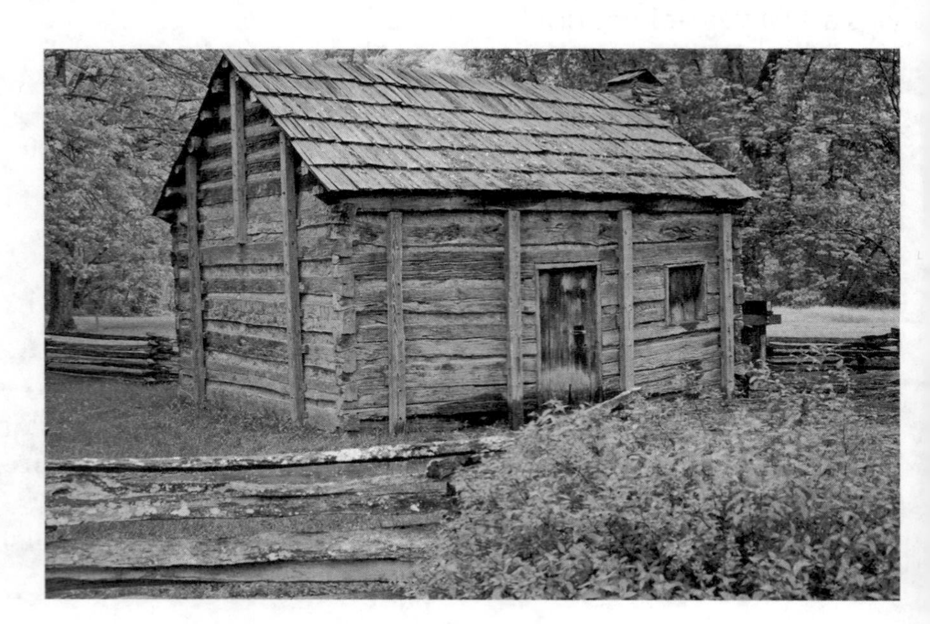

The childhood home of Abraham Lincoln in Knob Creek, Kentucky, was humble, to say the least.

HANDY FACT

From *The Handy History Answer Book*, 3rd edition, by David L. Hudson Jr.: "The Battle of Stones River (in Murfreesboro, Tennessee) is sometimes called the bloodiest battle of the Civil War because of the high percentage of losses on each side. Union commander William Rosecrans (1819–1898) and Confederate leader Braxton Bragg (1817–1876) engaged in a series of conflicts that left nearly 13,000 Union casualties and nearly 12,000 Confederate casualties. The conflict left no clear winner, though historians have said that the Union gained morale by driving the Confederates from Murfreesboro into a retreat southward."

nineteenth-century scientists, he began advocating for the environment. Nature—one of the things Lincoln appreciated in his boyhood—returned to his consciousness.

In 1862, Lincoln established the U.S. Department of Agriculture, which helped, among other things, to ensure the health of American soil to maximize growing capacity. In 1863, he signed a bill that was introduced by Massachusetts senator (and, later, Ulysses S. Grant's vice president) Henry Wilson (1812–1875), to incorporate the National Academy of Sciences. Shortly after that, Lincoln signed the Morrill Land-Grant Act, which ensured that land donated to colleges and universities was available for study, so long as those schools offered classes in agriculture and engineering. On June 30, 1864, Lincoln signed the Yosemite Grant Act, which set aside California's Yosemite Valley and the Mariposa Grove of giant sequoias for use as public land for the enjoyment of future generations, giving them to the state; this action set the stage for the later creation of the national parks system.

Less than one year later, Abraham Lincoln was shot by an assassin's bullet.

Even today, scientists wonder what other benefits Lincoln's environmentalism would have bestowed on this country.

BIOLOGY: I'M SO BLUE

By now, there should be no doubt in your mind that people are different, no matter where you go. Some of us are a little more unique than others.

A little over two hundred years ago, when this nation was still largely unsettled, Martin Fugate, an orphan from France, moved to a small patch of land near Troublesome Creek in a little town called Hazard in the Kentucky Appalachians. There, he settled down, met and married his wife, Elizabeth Smith, and they set out to start a family. That part of Kentucky is a beautiful, peaceful area, but it was remote in the nineteenth century. Roads were iffy at best and nonexistent at worst. Not even the railroad went to that part of the country until well into the twentieth century.

Records disagree on what Martin looked like, but family legend says his skin was unique; Elizabeth was a mountain woman, likely sturdy and used to hard work, but she was also said to have been fair-skinned and lovely. What neither Martin nor Elizabeth knew then when they were preparing for a family, but what we know now, is that they *both* carried a recessive gene that made it possible for statistically half of their children to have methemoglobinemia.

Indeed, of their seven children, four of them had blue skin.

The original gene for methemoglobinemia seems to have come from French Hugenots, long before Martin was born. Here's how it happened: in red blood cells, hemoglobin, which carries oxygen to all parts of the body, is bound to iron ions that are bound to oxygen ions, a process called oxidation, which gives blood its red color. If hemoglobin is damaged, iron ions are unable to bind to the oxygen, but an enzyme called diaphorase (or methemoglobin reductase) steps in and saves the day—otherwise, there would be a decrease in the amount of oxygen available throughout the body due to a buildup of met-

HANDY FACT

"Blood entering the right side of the heart (right auricle or atrium) contains carbon dioxide, a waste product of the body. The blood travels to the right ventricle, which pushes it through the pulmonary artery to the lungs. In the lungs, the carbon dioxide is removed, and oxygen is added to the blood. The blood then travels through the pulmonary vein, carrying the fresh oxygen to the left side of the heart, first to the left auricle, where it goes through a one-way valve into the left ventricle, which must push the oxygenated blood to all portions of the body (except the lungs) through a network of arteries and capillaries. The left ventricle must contract with six times the force of the right ventricle, so its muscle wall is twice as thick as the right."

—*The Handy Science Answer Book*, 5th edition

hemoglobin; in fact, the blood is so oxygen starved that it can appear purple. An overabundance of methemoglobin in the body is (ta-da!) methemoglobinemia, which can lead to life-threatening issues involving cardiac problems or, at the very least, sufferers' skin could turn blue. While methemoglobinemia can be acquired, in the case of Martin and Elizabeth Fugate, they both carried genetic material that suppressed the diaphorase in their bodies. As for their descendants, they may genetically inherit methemoglobinemia when two individuals who carry the met-H gene marry, giving their children a greater chance of having the affliction.

Through the years, because of the isolation of the family and the community, the Fugates and the Smiths interacted and intermarried again and again. Some reports say they were embarrassed by their blue skin and willingly isolated. Others say they were just sticking to themselves because that was the way things were. At any rate, as others moved in and out of Hazard, Kentucky, through the years (especially after the railroad arrived in 1910) and rural areas became much less isolated in general, Fugate genes slowly moved out of the area, there were fewer chances for intermarriage, and fewer and fewer "Blue Fugates of Kentucky" were born.

And yet, the mystery was still there, waiting to be solved.

In the early 1960s, hematologist Dr. Madison Cawein (?–1985) went with his trusted staff to that area of the Appalachians to try to find Fugate descendants to study. Though the family was reticent to put themselves into the hands of someone they didn't know, Rachel and Patrick Ritchie—both blue-skinned—and a small handful of other descendants ultimately agreed to allow Cawein to study and perform tests on them. Because of his past work in medicine, it took surprisingly little time for Cawein to determine that methemoglobinemia was the cause of the Fugate descendants' blue skin, and it was he who discovered a "cure" for the malady: oddly enough, injecting methylene blue restored the iron levels in the blood of methemoglobinemia sufferers, but the effect didn't last long. Later, Cawein learned that regularly ingested pills of methylene blue did the trick and ended the Fugate family's blue symptoms.

Cawein, in case you're wondering, went on to do research on treatment for Parkinson's disease.

Today, doctors and researchers say that methemoglobinemia cases are rare, and it's known that larger pockets of Fugate family descendants live in Virginia and Arkansas. Even so, anecdotal evidence indicates that the gene is still out there: many people have tales to tell of relatives with blue skin or hands. At any rate, you wouldn't necessarily know you have methemoglobinemia unless you're tested or you have a child who's blue.

CHEMISTRY: GO BOOM!

Few things signal "summer" (or, for that matter, New Year's Eve) more than fireworks. Whether it's done ironically or authentically, we look forward to every "ooooh" and "aaaaahhhhh" over them. The history of and chemistry behind those boomers, well, it sparkles.

Most historians believe that fireworks were invented by the Chinese in about the second century B.C.E., but evidence exists that fireworks may have come from India or the Middle East. Either way, before the year 1, the Chinese were using roasted bamboo (which explodes) to BANG away evil spirits.

At some point after the seventh century, an early and enterprising Chinese chemist learned that combining potassium nitrate, sulfur, and charcoal made a lot of noise if the mixture was dumped in a hollow, bamboo tube and lit. That mixture, by the way, is what we'd call "gunpowder." By roughly the end of the tenth century, that mixture was poured into paper tubes and sold by vendors to everyday people. Also, by this time, the powder was made into a sort of weapon attached to an arrow that would explode in midair and literally rain fire down upon enemies.

By the thirteenth century, fireworks had spread to Europe, not as a weapon but as a thing to make noise and enhance a celebration, both religious and secular; rulers who had them particularly liked how their opulent castles looked when framed by spectacular shows of fireworks. The Italians learned to make crackerjack firecrackers by the Renaissance, and schools were set up to teach the craft; in later years, that was a good thing because the Italians learned, in the early nineteenth century, that other substances, including metals, made fireworks splash the skies with color.

411

China is still the world's number-one manufacturer of fireworks.

Folklore has it that Captain John Smith brought fireworks to the New World in the early 1600s, and it didn't take long for America to embrace them with fervor: it's said that in Rhode Island in the 1700s, too many pranksters with fireworks annoyed the wrong neighbors, and use of the boomers was banned there.

While sparklers may have been your introduction to fireworks way back when, you arguably can't get much tamer, fireworks-wise, than the weirdly fascinating "black snake." Even so, always have an adult around when those snake tablets are lit.

By the time America was one year old, fireworks were firmly meant for celebrations, and they naturally became tied to Independence Day, but as much as they were beloved, an equal number of people hated them over the years. In 1906, the Society for the Suppression of Unnecessary Noise formed, initially because of river traffic and its inherent sirens and whooshes; when the organization's founder, Julia Rice (1860-1929), realized that she was gaining traction on her idea, she focused on all loud, sudden noises, including fireworks.

It was after Rice's work that fireworks—which are undeniably dangerous if used improperly—became regulated, and they largely remain as such today in most states, especially with the larger pyrotechnics, which often require a license to detonate. Even so, and despite the warnings and legalities, thousands of Americans are injured each year, some severely, by the mishandling of fireworks.

You're wondering how fireworks work. Here's the skinny on the big ones.

Fireworks consist of an aerial shell, gunpowder, and lots of "pods," which hold fuel, a binder, oxidizing materials, and metals or salts that add color, plus a fuse to light the whole thing off. Fuse lights fuel, fuel lights gunpowder, ka-boom. As the metals or salts are heated, their atoms absorb energy that creates light as the energy dissipates; various colors are made by different chemicals when they absorb and diffuse energy at different rates.

And then ... ka-BOOM!

HANDY FACT

"Magnesium [in fireworks] burns with a brilliant, white light and is widely used in making flares and fireworks. Various other colors can be produced by adding certain substances to the flame. Strontium compounds color the flame scarlet, barium compounds produce yellowish green, copper produces blue-green, lithium creates purple, and sodium results in yellow. Iron and aluminum granules give gold and white sparks, respectively."

—*The Handy Science Answer Book*, 5th edition

PHYSICS: THE SCIENCE OF MAKING A SNOWMAN

It's too irresistible: it snows, you must go outside and play in it—especially if you're under age twelve or live with someone who is. But who was the first person who thought, wow, that would make a great temporary statue?

In his book *The History of the Snowman*, Bob Eckstein tried to determine the answer to that question but couldn't quite do it. An exact determination just can't be made because such things weren't specifically recorded for future generations, but it is certain that fourteenth-century snow lovers (at least) were building snowmen and that making snowmen came from overseas along with America's earliest European settlers—that is, if those cold white guys weren't here already.

At any rate, snowmen are nearly as ubiquitous as Santa during the Yuletide season, and they are seen even in the most tropical climates. Hundreds of books and movies have been made about them. Lots of folks collect them. There are literally tens of thousands of videos about them available for purchase or online.

So, how do you build the best one?

Kids know that the best snow for a snowman is not too dry but has just the right amount of wetness to it for good packing.

The first ingredient is the medium: you have to have perfect snow to make a perfect snowman. Scientists have determined five simple ways to categorize snow: dry, with 0 percent water; moist, with less than 3 percent water; wet, with between 3 and 8 percent water; very wet, with 9 to 15 percent water; and slush, which is more than 15 percent water. This is all determined by the temperature outside: if it's too high, the snow will be too wet and melty for crystals to adhere together, and it's going to make a larger snowball awfully heavy to lift when you're adding height. On the other hand, if it's too cold, the snow will have less moisture, it'll be drier, and it won't stick together well enough to build much of anything, much less a snowman's base.

As it turns out, then, the best temperature for building the perfect snowman is just below freezing, at about 30° or 31° F (–1° C) and the best snow is moist to wet on the scientific scale. Those temperatures allow for dendrite snow crystals (like the ones you think of when you think of snowflakes), and the temps allow for better stickiness, if you will. To further perfect your snow, you'll want to be sure it's free of twigs, leaves, dirt, and other debris.

Once you've determined that the temperature is just right, you need to determine where your snowman will stay. Choosing concrete or asphalt almost guarantees that he'll melt quickly since surfaces like that tend to collect heat faster than bare ground (which would likely be frozen). Also, don't choose a hill; while it might look cool to have a fierce snowman guarding your home, hillside snowmen are not only harder to build (you have to push or carry that snow uphill!), they're also prone to being tipsy. Nope, choose a flat patch with at least a few inches of snow base, preferably closer to your home, so mischief-makers won't be tempted to destroy him. Make sure it's not too windy there or too sunny.

Next, make sure you have enough snow. A six-foot-tall snowman is going to take close to twenty square feet of compacted snow for the best building. If you want a bigger snowman, you'll have to have a bigger yard (or know someone who'll truck the white stuff in for you). The best proportion for a six-footer is in a three-foot-diameter snowball on the bottom, two-foot-diameter in the middle, and a one-foot-diameter for the top, but you may want to play with those measurements. You'll need that little extra snow for your creation's feet and to smooth out the area between the three "body" parts.

There are reasons, by the way, that snowmen are made of basically round shapes: because spheres are easy to shape by mittened hands and because a sphere is stronger than would be a column. Stacked spheres also expose less surface area to solar warming agents (also known as "the sun").

Start with a snowball of average size (generally, whatever size you can fit in your hand), and pack it tight. Place it in the snow at your feet, and start rolling; pack as needed, but eventually, the weight of the snow will pack itself. Be sure to roll all ways; if you only roll your snowball one way, it'll look a little like a cinnamon roll rather than the base of a snowman. Roll the bottom snowball until you achieve the magical three feet, and place it where you want your snowman to stay. Make another snowball for the middle section, and roll it in the same manner until you've achieved the right proportion (hint: roll it close to the base, so you can gauge properly), then lift it onto the base. If it seems wobbly, pack snow into the crevasses between the segments. Make another snowball for the top section, lift high, and pack the crevasses as needed. Finish with twigs for arms, rocks for eyes, mouth, and buttons, an old hat for jauntiness, and a carrot nose. Embellish as you wish, to make your snowman personalized.

Bonus: scientists say that because the snow in a snowman melts slower than the snow that remains on the ground where it fell, building a snowman may slow the rates of flooding.

HANDY FACT

To put the old "Are any two snowflakes identical?" question to rest once and for all, *The Handy Science Answer Book*, 5th edition, says this: "Some snowflakes may have strikingly similar shapes, but these twins are probably not molecularly identical. In 1986, cloud physicist Nancy Knight believed she found a uniquely cloned pair of crystals on an oil-coated slide that had been hanging from an airplane. This pair may have been the result of breaking off from a star crystal or were attached side by side, thereby experiencing the same weather conditions simultaneously. Unfortunately, the smaller aspects of each of the snow crystals could not be studied because the photograph was unable to capture possible molecular differences. So, even if the human eye may see twin flakes, on a minuscule level, these flakes are different."

While two snowflakes can be very similar, there is really no chance they will be absolutely identical down to the molecular level.

THE ENVIRONMENT: HOW'S THE WEATHER DOWN UNDER?

Imagine temperatures of well over 100 degrees Fahrenheit all day, every day during the summer; temperatures below freezing all night, every night; very little rainfall; and not a lot of greenery because, well, it's just too hot for things to grow. But never mind the effect of that kind of environment on plant life. Seventy- or eighty-degree daily swings like that are rough on the human body, and without a good indoor system, it could be downright brutal.

The superhot area of South Australia was initially explored by John McDouall Stuart (1815–1866), who in 1858 became the first European to see land previously used only by Aboriginal people for hunting, gathering, and celebrations. What really put the area on the map, however, was the discovery of opals in 1915 by a 14-year-old boy named Will Hutchinson, who was accompanying his father, Jim, on a gold-prospecting mission.

Like every other organic gemstone, the creation of an opal didn't happen yesterday. More than a hundred million years ago, after ocean

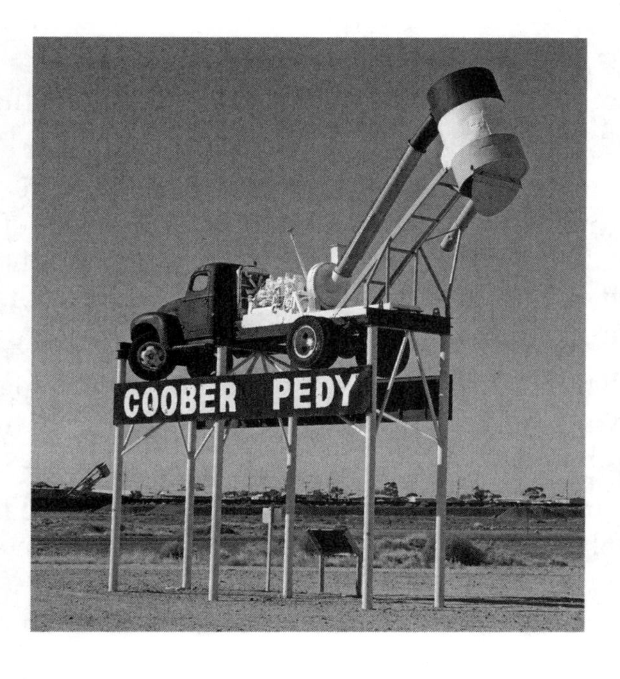

The sign at the edge of Coober Pedy, Australia, sports opal mining equipment. The city was at the epicenter of opal mining in the country. It was here that the largest opal ever mined was found in 1956: the "Olympic Australis," a 17,000-carat gemstone weighing 7.5 pounds (3.4 kilograms).

water receded from the continent of what is now Australia, the lessened water altered the water tables and pulled silica down deep into the ground. That's where silica settled in earthen cracks and where silicon dioxide and water ultimately became opals.

Although it didn't take long for an opal claim to be filed or a town to spring up, the actual mining of opals on a larger scale didn't happen until after World War I, and at about that time, the town that had been created by mining had to be named in order to get a new post office. The original name of the area gave a nod to the first white explorer; it was called the Stuart Range Opal Field, but that was too close to a similarly named mountain range nearby. Instead, the name Coober Pedy was chosen for the mining town, based on the Aboriginal words "kupa-piti," which loosely translates, according to the Coober Pedy website, to "white man in a hole."

Because of its remoteness, it was nearly mandatory that miners live, rather than commute, to Coober Pedy as mining increased following World War II, when Europeans fled their war-torn homelands to move to Australia. This was when the miners did what humans have done for tens of thousands of years: they accommodated by moving into chambers under the inhospitable ground.

Today, mining is still the number-one industry in Coober Pedy; nearly seven out of every ten of the world's opals come from there, in

HANDY FACT

As for sand, *The Handy Science Answer Book*, 5th edition, has this to say: "Soil is the weathered-out layer of Earth's crust. It is a mixture of tiny rock fragments and organic matter. Soils are categorized broadly into three types: clay, sandy, and loam.

"Clay soils are heavy, with the particles sticking close together. Most plants have a hard time absorbing the nutrients in clay soil, and the soil tends to become waterlogged. Clay soils can be good for a few deep-rooted plants, such as mint, peas, and broad beans.

"Sandy soils are light and have particles that do not stick together. Sandy soil is good for many alpine and arid plants, some herbs such as tarragon and thyme, and vegetables such as onions, carrots, and tomatoes.

"Loam soils are a well-balanced mix of smaller and larger particles. They provide nutrients to plant roots easily, they drain well, and loam also retains water very well."

fact, and many of the town's population is employed by one mine or another. As for the city itself, Coober Pedy is home to around 2,500 folks, mostly descendants of Europeans but also representing multiple nationalities. The underground part of the town sports a beautiful, underground church, a museum, bars, restaurants, a casino, a hotel that's both above and below ground, shops and assorted service businesses, and an underground water tank that serves the population's water requirements. Modern methods of building have allowed more aboveground housing today, and about half the town's residents live on top of the soil, while the rest live beneath it in dwellings that run from simple, underground places to veritable mansions that sprawl like labyrinths, all of which are, by virtue of being underground, kept by Earth at a comfy seventy-five degrees. The vast majority of Coober Pedy's subterranean homes are carved out of the soft sandstone in nearby hills; just a few are located in old mine shafts.

And if this sounds like a great place to live, check it out. It will literally take you hours to get from your home in America to a hotel in Coober Pedy.

Or check out the movies *Mad Max Beyond Thunderdome* and *Pitch Black* and watch the background. Yep, they were filmed in and near Coober Pedy, Australia.

BOTANY: MOW, MOW, MOW YOUR LAWN

For most people, it's desired. In some communities, it goes beyond that to "requirement," demanding that you spend a good chunk of your free time on cultivating a gorgeous, green expanse of lawn that won't feed you. It makes no sense. So, why do we devote so much land to the growth of grass?

To understand, let's take things back. Way back.

In the earliest days of humanity, grass was a very good thing.

The grass then wasn't like the verdant lawns we know and love. Grasses then were wild and unruly, most likely longer bladed, and often dry. But those grasses served two purposes: they fed the creatures that early humans preyed upon, and they worked well for cover when on said hunt. To extend upon that, grasses also likely made fine bedding once humans got that far in the evolutionary process.

Toward the middle of the twelfth century, Japanese gardeners wrote *Sakuteiki*, a guidebook on gardening that included instructions on growing and using Zoysia grass, particularly as sod. It's possible that around this time, carefully tended, grassy areas were used for very early sports, games, and competitions. Such expanses likely doubled as pasture for livestock then since the only lawn mowers that existed were the kind that mooed or baaaaa-ed.

It was during the Renaissance that the general use of grass became something altogether different: as somewhat of a status symbol, rulers and the wealthy began to use green, lackey-tended expanses of grass surrounding their estates and castles to prove that they had the money to waste on a crop that wouldn't feed them. Beautiful gardens might have joined the verdant lawns on the grounds, all for the sole purpose of looking nice and all tended by an appointed gardener or by servants with scythes. Still, this indulgence by the rich and powerful heralds the birth of the modern lawn.

As immigrants began moving from the Old World to North America, they found plenty of native grasses here, but they also brought grass seeds with them both on purpose and accidentally in their clothing, shoes, and suitcases and in livestock food, bedding, and manure. Having an expanse of grass for no purpose other than looks was

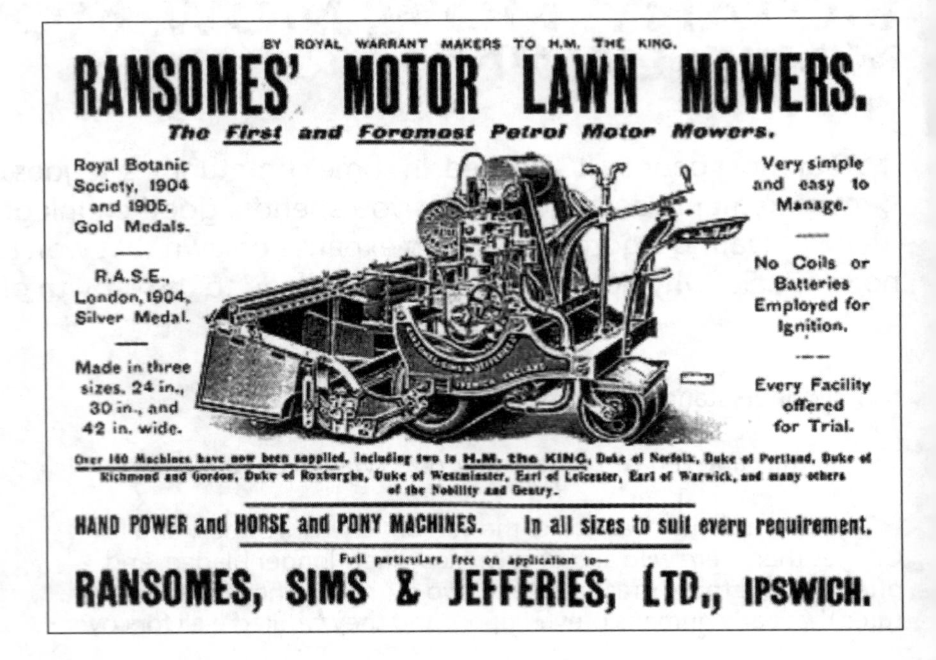

The first gasoline-powered lawn mower was available for sale in 1902. As you can see by this ad, it wasn't exactly the most elegant machine!

still mostly something the wealthy aspired to have (Thomas Jefferson was said to have promoted the cultivation of a lawn), but that started to change when citizens realized that outdoor games were just as fun here as they were back home.

411

Lawn grass is the number-one irrigated crop in America.

Surely, such leisure pursuits were enhanced when Edwin Beard Budding (1796–1846) invented the first mechanical lawn mower in Great Britain in 1830. His invention was somewhat like those sold in some hardware and specialty stores today: it was a reel-type machine with blades in a cylinder mounted on wheels, run entirely on manpower. Twelve years later, Scotsman Alexander Shanks (1801–1845) invented a lawn mower that was horse-drawn. Within a few years, landscape architect Frederick Law Olmsted (1822–1903) promoted the use of green space in both parks and homes, while the patent for the lawn sprinkler was registered in 1871. Other patents for lawn mowers followed in the United States, including a steam-powered mower at the end of the nineteenth century and a lawn mower run by a gasoline engine shortly after the twentieth century began.

And there we were: having a green lawn surrounding your home was no longer something that only the rich and powerful could have. Lawns didn't require livestock or servants anymore, and they could be maintained quite easily by someone with push power. The first riding lawn mower entered the plush picture in the mid-1950s, and by then, it

HANDY FACT

"When Europeans first visited the American grasslands, they saw a territory that had essentially no trees but instead a sea of grasses that could grow up to ten feet tall in the wetter tallgrass prairie. Prairie soil was some of the most fertile soil on Earth. Almost all of the tallgrass and mixed-grass prairie in the United States were eventually plowed under for cultivation, while the shortgrass prairie is now used for crops and cattle grazing. Prior to the coming of Europeans, twenty-two million acres of Illinois were prairie. Today, that figure stands at approximately two thousand acres. From the former prairie, however, the United States feeds itself and some other areas of the world. Illinois, Kansas, and Nebraska are still grasslands, but the grasses that grow there now are not the native big bluestem or needlegrass but corn and wheat."

—*The Handy Science Answer Book,* 5th edition

was almost mandatory for Eisenhower-era homeowners to have magnificently flawless, green lawns surrounding their new suburban homes.

Considering what early lawns consisted of, today's lawns are masterful things. Lawn-care scientists know what kind of grass will grow best in a certain kind of soil; both can be engineered for maximum color, sturdiness, growth in various conditions such as shade, and drought resistance.

And if you really want a low-maintenance lawn, you can buy a robotic mower (invented in 1995) and never lift a finger on your field of grass. Or buy some goats: fertilizer and mowing at the same time.

PHYSICS: HOW TO SURVIVE AN ATOMIC BOMB BLAST

Let's hope you never need to know the following information. The simple fact is, however, that if there's ever a nuclear blast heading our way, there are no right answers on avoiding problems or staying alive. Each situation will be different in the event that nuclear war is imminent.

But okay, let's say that the sirens have gone off. You've been glued to the news, prepping yourself and your family, and you've anticipated this. Maybe you've even got a bug-out bag ready to grab. If you don't, according to experts, there are several basic things you'll need for a fighting chance at life, ranging from a source for clean air to a water purifier and food that will withstand the radiation to come. You should absolutely know that carrying a bug-out bag won't be an easy thing—some necessary items could be rather heavy—but you can surely spread the necessities out between the others in your party.

While you're preparing, know the locations of nearby and long-term shelters so that IF leaving is the best choice, you'll know exactly where you're going and how to get there in an expedient manner.

There are three basic zones of destruction when a nuclear bomb is dropped.

The Severe Damage Zone is the closest to the center of the blast. If you are there when the hypothetical explosion happens, you will see a flash for a fraction of a second, then you'll feel the shock wave before the heat of the blast (millions of degrees of heat) probably incinerates you. Having said this, though, there were people who lived through the first-zone damage in Nagasaki and Hiroshima by taking shelter in thick, concrete shelters, but with today's nuclear know-how, the truth is that you likely won't live long enough to care about the other two zones.

The Moderate Damage Zone is survivable but only if you have the right kind of shelter. Think: concrete buildings. Belowground shelter. Vaults.

The Light Damage Zone is absolutely survivable. But you need to be smart.

The first thing to do is: nothing. If your shelter is intact, don't go anywhere. Don't set out on foot. Your car is not a decent shelter. Stay where you are; if your shelter kept you alive through the blast, it'll keep you safe for at least a little while longer. You may have temporary blindness from the blast, and contamination of your surroundings may not be immediately apparent, which makes it an even better idea to stay put inside.

If you were outdoors when the blast hit, shed your clothing as soon as you get inside, and try to find some way to take a shower or wash with soap and water. Radioactive debris is dangerous, and the quicker you rid yourself of anything such as dirt, dust, debris, or particles that may have fallen on you—by bagging it and placing it as far as possible away from you—the better your chances of avoiding radiation poisoning.

HANDY FACT

The Handy Physics Answer Book, 3rd edition, regarding the first "atomic device," states: "'The Gadget' was a test version of the plutonium bomb. It was installed on the top of a 30-meter (100-foot) tower in the New Mexico desert at a location 35 miles southeast of Socorro, New Mexico, on the White Sands Proving Ground near Alamogordo. The explosion, called 'Trinity,' occurred on July 16, 1945. The energy yield was about 20,000 tons of TNT, more than twice what had been expected. The implosion-type bomb was much safer and more effective than the cannon-style 'Little Boy' design. It became the standard method used for all other nuclear bombs."

Try to wait at least three days before moving locations. Gamma rays decay relatively quickly, and scientists say that the outdoors should be safe enough to leave after about three days. Remember, though, the longer you wait, the safer it is to go.

Lastly, try to find a radio signal or some sort of communication. Try to make your way to a community shelter or place where you can get some news.

The bad news: a nuclear blast is serious stuff.

The good news: experts say that it's mostly survivable.

BIOLOGY: HUMANS ARE SO WEIRD!

You are unique. You're a product of generations of genetic material, but still, you're the only *you* that's ever been.

Even so, you might not be as unusual as you think you are. Here are some common odd-human things and a few reasons why.

Scientists think whether you're right-handed or left-handed is a language thing. Language enters the left brain, which controls the right side of the body. The theory is that the extra brainpower required to speak gives the left side more wiring, so to speak, and makes the right side dominant, thus we are mostly a species of righties. But brains have been discovered that disprove this idea, so the bottom line is, we don't really know why some people are lefties and some are right-hand dominant.

We are not unique in this because other species do it, too, but we lie. Researchers say that we do it for myriad reasons: to make ourselves feel better, to elevate our importance, to avoid hurting others, or to avoid punishment of some sort.

We shave. It's a cultural thing that no other animal does.

Chances are, you feel a little bit *frantic* when you're at a store, you're handed your change, and you're trying to get out of the way of the person in line behind you while you also try to put that change away. Or when you get a free sample of a new product in the store, and even if you hate it, you act like you might actually purchase it sometime. We fake laugh when we have absolutely no clue what someone said or meant. If a stranger sits down near us, we might wait a few minutes to leave, so we don't hurt anyone's feelings. Reason for all this: It has to do with us being social creatures and wanting to avoid embarrassment—ours or someone else's.

Humans are the only creatures with chins that jut out beyond our upper teeth, and while there are theories, scientists don't know exactly why this is.

Admit it: You have to turn the radio down when you get lost driving or are looking for a specific address. You do this because you need the extra concentration and fewer distractions to think specifically.

Another odd thing we do is act as if everyday objects are out to get us; for instance, the computer is acting wonky because it hates you and, *of course*, that means the microwave would choose this day to go on the fritz, too. Likely, this is just an example of you venting, but in its most extreme form, this altogether human action may indicate a rare version of synaesthesia called personification, which causes sufferers to firmly believe that inanimate objects have personalities or genders.

We talk to ourselves. The reason is simple: researchers say we do so to help us organize our thoughts, feelings, and plans; to give ourselves a pep talk; or to focus better when we've lost our way. Talking to ourselves is basically a way for us to control ourselves.

It's likely that you have a "place" at the table where you always sit, and everybody in your household knows it. You probably also take the same seat in the living room every time you relax, and if you cosleep, you have "your side" of the bed. Psychologists say that's a nonaggressive territorial issue.

Even radio and TV personalities go through this sometimes: we hate to hear the sound of our own voice. That's because you're used to hearing yourself through your own head, which is subtly different from what others hear, and that may mess with the self-image you carry or the perception you hoped to convey. Eventually, if you're exposed to the sound of yourself enough, you may get used to it, but this distortion can be problematic with people who have gender or body dysmorphia.

We use paper tubes as swords or light sabers. We give our pets nicknames. We dance when nobody's watching. We play air guitar or

HANDY FACT

"If they could be laid end to end, the blood vessels in the human body, including the arteries, arterioles, capillaries, venules, and veins, would span about 60,000 miles (96,000 kilometers). This would be enough to encircle Earth more than two times."

—*The Handy Science Answer Book*, 5th edition

air drums. We eavesdrop on strangers' conversations. We yell at the TV or at our favorite team in a totally crowded stadium, as if the person we're yelling at will hear us. We hit "save" multiple times. We backspace to fix a word three words back.

And why? Because, well, we're weird.

THE ENVIRONMENT: "FALL" AND OTHER SEASONAL THINGS

Ask anybody about their favorite season, and you'll probably get some firm opinions. We love fall's colors, winter chills, summer sunshine, and spring's sense of renewals. But at the bottom of it all, what's so great about the seasons, anyhow?

Before we get to the fun stuff, we need to know why we even have seasons in the first place.

The easy-answer reason for the seasons is because Earth tilts—23.5 degrees, to be exact. No one knows for sure why this is so; it's believed to be because something crashed into Earth eons ago, back when it was first born, and set the whole planet off-kilter.

Of course, our planet spins on its axis, which gives us night and day, but the spin of Earth doesn't change its angle. The angle of the planet as it orbits the sun is always the same, so parts of it are more exposed to the sun at different times of the year, resulting in what we know as seasons. When Earth's axis (the North Pole) points away from the sun, the Northern Hemisphere has shorter days and winter weather because the tilt away from the sun allows for less sunlight;

when the axis points toward the sun, we get summer in the Northern Hemisphere and longer days. As a transition between this 365-ish-day orbital cycle, we have autumn and spring, with daylight that is roughly equal between dark and light.

So, why do we have a preference? The reasons are too numerous to list, but for many people, having four distinct seasons isn't for them; the change in both warmth and wardrobe is more than they can handle. To be sure, each season has its benefits and its detractors. Each can be counted on to be different from the others....

WINTER

WINTER, OF COURSE, often means snow and cold weather (see Science—Physics: Ice, Ice, Baby; Science—The Environment: The Summer That Never Happened; or Science—Physics: The Science of Making a Snowman). Psychologists say that winter also heralds a difference in personality: people born in the winter months tend to be more optimistic and patient with others.

THERE IS SUCH a thing as thundersnow, but it's relatively rare. It's caused when the air closer to the ground is warmer than the air above it but not warm enough to stave off a snowstorm.

SNOW ISN'T ALWAYS white. It can be pink, purple, or brown, depending on what substances it might absorb as the snowflakes

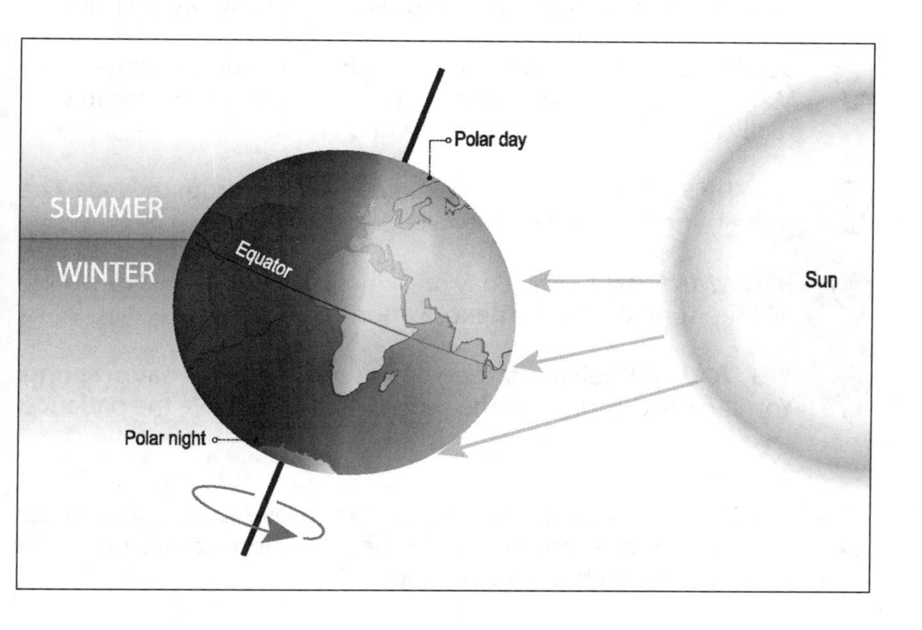

Because Earth is tilted on its axis, we get seasons as our planet orbits the sun.

are forming—which could be dust, algae, or any other unsavory substance. The bottom line: don't eat any snow that's not pure white.

THE WIND CAN make its own snowballs if conditions are just right.

BELIEVE IT OR not, someone has determined that the number of snowflakes that fall on Earth each year is a "1" followed by twenty-four zeroes.

ROUGHLY TEN TO thirteen inches of snow equals one inch of rain. That's because snowflakes have a lot of air between them.

BEWARE: THE DEEP-COLD temperatures of winter kill more people than do the hot temps of summertime. On average, it takes thirty minutes or less to get frostbite.

SPRING

THE SPRING EQUINOX can happen on either March 19, 20, or 21. Spring itself (meaning the seasonal spring) can happen in the Northern Hemisphere at any time between February and May. Meteorological spring begins on March 1.

THAT EQUINOX IS officially called the "vernal equinox." "Vernal" and "equinox" are both Latin words—the first meaning "spring" and the second meaning "equal night." Indeed, there are twelve hours each of daylight and darkness on the first day of spring.

BLAME BEN FRANKLIN for the daylight saving time change every March. He suggested it back in the 1700s, though it wasn't actually put into effect until after World War II.

ON THE SPRING equinox, the sun rises exactly from the east. The same thing happens on the fall equinox.

BEGINNING ON THE first day of spring, residents of the North Pole will see sunlight for the next six months.

YES, YOU CAN balance an egg on its end on the first day of spring. You can do it any time of the year, in fact, if you have the patience.

SUMMER

IT'S TRUE THAT the Eiffel Tower in Paris grows up to a few inches higher each summer due to the heat causing expansion of the iron from which the tower is made.

THE SUMMER SOLSTICE is the day when the Northern Hemisphere gets the most sunlight; therefore, we say that the "day is longer." (It's really not, but with more daylight, it seems like it is.)

SUMMER IS A time when more thunderstorms tend to happen. Pinpoint any single time or day on Earth, though, and there are more than 2,000 thunderstorms in progress.

THE DOG DAYS of summer have nothing to do with your pooch. The days between July 3 and August 11 refer instead to the time when the sun is in the same region as the Dog Star, Sirius, which is part of the constellation Canis Major, or the Greater Dog. All summer long, Sirius gets up with the sun, but on July 23, Sirius is in direct line with the sun. The Romans thought that meant extra heat on that day (it doesn't necessarily).

THE MOST POPULAR vacation destination in the summertime is the beach. The most popular vacation activity is shopping.

SCIENTISTS SAY THAT babies born in the summer tend to suffer from moodiness and mood swings more than do babies born at other times.

AMERICANS TYPICALLY EAT around 150 million hot dogs on July 4—Independence Day.

ABSOLUTELY YES, YOU can get sunburned on a cloudy day. Rule of thumb: if there's enough sun to make a shadow, there's enough sun to burn your skin.

FALL

AMERICANS TEND TO call the after-August time "fall" more than do the British. The Brits generally call the season "autumn." Ancient people sometimes referred to it as "harvest" because the harvest moon (the full moon that occurs closest to the autumn equinox) happens then.

IN THE FALL, leaves change color because they stop producing food for the tree. This cessation breaks down whatever chlorophyll is left in them, and the green color is lost. Yellows and brown colors come from the amount of chlorophyll; those gorgeous, red leaves come from the amount of sugar that's in the leaf.

JUST BEFORE THE leaf falls off the tree, a layer of cells develops that protects the branch from damage by leaving a scar at the site of the stem of the leaf. Most leaves, then, will fall and wait for you to rake them up; oak trees and a few other species will hang on to their leaves until spring, though, so keep that rake handy.

GENERALLY SPEAKING, BABIES born in the fall months are better scholars and enjoy longer lives.

RESEARCH SUGGESTS THAT squirrels are able to find nearly every single thing they stash during the fall months. Other studies also show that the hippocampus in the squirrel brain and in that of some bird species increases in the autumn months, which, theoretically speaking, makes them smarter then.

TESTOSTERONE LEVELS IN both men and women spike in the fall months. Some statistics suggest that we also have more reason to move our social media status from "single" to "in a relationship" in the fall than at any other time.

THE FALL EQUINOX usually falls between September 22 or 23, but in 1931, it fell on September 24 because the calendar didn't quite line up. But don't worry. This little blip in time won't happen again until the fall of 2303.

HANDY FACT

"Even though Earth rotates at a startling speed (1,036 miles [1,668 kilometers] per hour at the equator) and orbits the Sun even faster (at 67,000 miles [107,000 kilometers] per hour), we don't feel it because the rate of motion is a constant one, never slowing down or speeding up. We can only really feel motion when the speed changes. If you were in a moving car and couldn't see the scenery passing by, couldn't hear the wind blowing, and couldn't feel the car vibrating, you wouldn't be able to tell how fast you were going or even if you were moving at all. Objects don't fly all over the place when Earth moves because gravity keeps everything firmly in place as the planet orbits and turns."

—*The Handy Answer Book for Kids (and Parents)*

BOTANY: THE IVY IN THE LIVING ROOM AND OTHER PLANTS

Though so many people love them, sometimes to the point of having a veritable jungle in their homes, the act of having houseplants seems a bit odd when you get right down to it: you can go outside any time, so why bring the outside … *in*? And yet, the whole idea of having greenery indoors is an old one.…

Historians know that as early as 3,000 years ago, Chinese homeowners brought plants inside their homes, both as decoration and as a way to keep their plant-tending talents fresh, for when the weather was too cold for outdoor gardening. In around the year 605 B.C.E., Babylonian emperor Nebuchadnezzar built lavish hanging gardens for the woman he loved, his queen, Amytis, who was from Persia and missed the greenery of her home country. Twenty-five hundred years ago, Greek and Roman upper-class citizens caught on to the idea of having houseplants to decorate their homes, but this hobby did double duty: having the time and resources to cultivate noncrop plants was a showy way to indicate wealth and the leisure that went with it.

After the fall of the Greek and Roman empires, the use of houseplants mostly diminished in Europe; plants as decorations became more of a courtyard thing, which, again, was only for the wealthy. The occasional cook or monk might try growing herbs indoors on a windowsill or sunny spot, but, mostly, plants were kept outside.

The Victorians, romantic creatures that they were, practiced an art called floriography, or the language of flowers. If, for instance, a man wanted to tell his beloved that he admired her purity, he might send her a gardenia. If she reciprocated his admiration, she would reply with lavender. If not, she might send him a bouquet of burdock to quash his romantic ideas.

Aside from the plants brought to Europe through global exploration in the earliest part of the last millennium, the average person didn't think much about bringing the outdoors inside. There was a bit of a plant renaissance during the Renaissance when greenhouses (called "orangeries") were in vogue, and the invention of glass made true greenhouses possible, but the widespread use of houseplants for decoration didn't completely return until the eighteenth century when the French first began to use floral greenery in their salons. By this time, nurserymen had learned to cultivate plants for better strength and color, and becoming an expert at it was pretty lucrative. A salon full of plants was pricey, and again, only the upper crust could afford such a thing.

All that changed during the Victorian and Edwardian eras of the mid- to late nineteenth and early twentieth centuries. Then, cultivating, growing, and even having houseplants was a popular hobby if not nearly an obsession—in part because it had become somewhat easy to procure plants that were perceived as "exotic." It helped that people were able to heat their homes in more efficient ways, and the very nature of architecture—high ceilings, large windows, airy rooms—then practically *begged* for indoor greenery. Doctors pointed to the health benefits of bringing friendly plants indoors, and Victorians and Edwardian-era citizens on both sides of the ocean obliged.

411

Starting as early as the year 200, the Japanese and Vietnamese began focusing on the ancient art of pun-tsai. Practitioners would take a seedling from an average tree, root it in a shallow container called a "pun," and, through pruning of both leaves and roots, create a living sculpture that ultimately resembles an ancient tree in miniature—even though the plant itself might be just a few years old. We know pun-tsai as "bonsai."

Bonsai trees require many years of attention to achieve the perfect effect of a miniature world.

Oops; then, houseplants fell out of favor, perhaps because the Great Depression meant that such frivolity was off the table when jobs and food were more important. It didn't last long: with the end of World War II, landscaping and growing houseplants was once again a popular hobby and an easy way to beautify one's office or workspace and one's home, inside and out.

Today, there are literally thousands of kinds of houseplants you can find from plants that were first found on this continent to plants that weren't seen until recently when someone engineered them. Houseplants come in several basic kinds, including vines that can ring a room, cacti that need very little fuss, and showy plants that will reliably flower. You can find plants to clean the air in your home, plants that you can share with friends, even plants that thrive on downright neglect.

And if you've *really* got a brown thumb, you can always buy a *Simulacraceae*, also known in certain botanical circles as an artificial plant.

HANDY FACT

"Carnivorous plants are plants that attract, catch, and digest animal prey, absorbing the bodily juices of prey for the nutrient content. More than four hundred species of carnivorous plants exist. The species are classified according to the nature of their trapping mechanism. All carnivorous plants have traps made of modified leaves with various incentives or attractants, such as nectar or an enticing color, that can lure prey. Active traps display rapid motion in their capture of prey."

—*The Handy Science Answer Book*, 5th edition

One of the best-known carnivorous plants is the Venus fly trap (Dionaea muscipula).

TECHNOLOGY: IT'S A DATE!

Today, tomorrow, yesterday, last week, August 12. We live our lives according to what seems to be an arbitrary way of timekeeping, which is about half right.

Most of how we mark our days is governed by the sun: ancient peoples recognized early that the sun had a cycle, and many of them began by noting one-day cycles and then annual cycles. The sun moved, the stars moved, the moon waxed and waned, and eventually, they all returned to where they were at first notice; when the pattern emerged and humans started keeping records, this early astronomical knowledge allowed them to keep track of the seasons. This was especially helpful when humankind became more agrarian than nomadic (though such things helped hunter-gatherers, too).

Scientists believe that somewhere around 10,000 years ago, a sort of calendar was devised in Scotland. It's believed that this calendar was made available so that anyone could line up the solstices, the sun, and the moon to figure out approximately where they were in the year or the season.

The problem with these gigantic stone calendars that aligned with the sun was that you couldn't exactly carry one in your pouch. And they weren't entirely accurate, either—although it's true that most ancient humans had no pressing need to know that.

The Monolith of the Stone of the Sun, or Aztec stone calendar, had 260 days divided into 13 periods each with 20 days.

Apparently, about 5,600 years ago, though, a need for a regular calendar was determined and one was created, complete with the months laid out and named, and that became the base from which the Babylonian calendar was devised. The Babylonians had a seven-day week, just like ours, but—to make up the "extra" time for which we add a Leap Day every four years because of the time it takes for Earth to rotate around the sun—the Babylonians had a longer-than-usual end-of-month week every so often.

About 5,000 years ago, the Egyptians devised their own calendar, which made each year an equal 360 days, divided into 12 equal months. To make up for that "extra" time, they tacked a five-day-long month at the end of every year.

By the time the Roman Empire was fully established, keeping track of the days was quite a mess. The Romans had several calendars from which to choose, and none of them were exactly right because of that "extra" time. Toward the end of the Empire, things

411

Here's where we come to the "arbitrary" part in the beginning of this section: having seven days is a made-up thing. Yes, seven days are mentioned in the Bible as the time it took for God to create the world, but this week length may also spring from ancient astronomy and the seven known planets. In the end, though, having a seven-day week is a culturally agreed-upon notion that we use in the United States and elsewhere—although some modern cultures recognize a ten-day week, while others live by a five-day week.

became even worse because the Greeks had by then created *their* own calendar.

In 48 B.C.E., Julius Caesar finally started to make sense of it all by proposing a deeper look at the Roman calendar. For several decades, because of wars and skirmishes, timekeeping had gotten muddled, and politics and religion had gotten mixed in with calendar-making. Fixing things made it worse, at least at first: Caesar's ideas added days to what had been a woefully short calendar, making the 46 B.C.E. calendar one of 445 days. The new calendar for 45 B.C.E. started on January 1, and it looked a lot like the one we use now, complete with a Leap Day. Coincidentally, forty-five years later, the Mayas came to the same annual conclusion that Caesar had, and they devised a calendar based on eighteen shorter months.

The later use-by date for the Gregorian calendar by Great Britain caused a little bit of a problem for the fledgling America. Here's how: George Washington was born on February 11, 1731, twenty-one years prior to the British adoption of the Gregorian calendar. When the British Parliament made the switch, the new year suddenly started on January 1 rather than March 25, as the Brits' former calendar had it. The adoption of a new calendar and the resulting absorption of dates made George Washington one year younger if his mother were going by the new, adjusted calendar. What's more, the Julian calendar was at that point eleven days off from the solar year, so even more accommodations were squeezed into the Gregorian calendar then as an additional adjustment, moving Washington's birthday from the 11th to the 22nd.

Other cultures devised their own calendars through the early parts of history, though Caesar's was the most widely used. Still, it wasn't totally accurate, and by the eighth century, it was off by three days. During the Middle Ages, the calendar was off by about seven days, and by the beginning of the Renaissance, the calendar was seriously out of whack. Among other things, this caused problems for farmers and for the Church when it came to determining Easter. Ultimately, the Church fixed its calendar problem, but it wasn't until 1577 when mathematicians were asked to weigh in on one true calendar that could be implemented widely. Aloysius Lilius (1510–1576) came up with a plan that canceled Leap Day for four decades—yes, the calendar was *that* out of alignment!—and re-added it in years divisible by four (thus, the year 2100 will not have a Leap Day). This new calendar was adopted on October 15, 1582, and printed, thus also becoming the day of the first known instance of calendar printing. Because it was approved during the reign of Pope Gregory XIII, it was known as the Gregorian calendar.

And even though this new calendar was approved and adopted then by many European countries, Great Britain didn't start using the

HANDY FACT

"In the English language, some of the days of the week take their names from the celestial bodies that, according to ancient beliefs, ruled that day. So, the day ruled by the Sun became Sunday, the day ruled by the Moon became Monday, and the day ruled by the planet Saturn became Saturday. The remaining days of the week take their names from figures in Anglo-Saxon or Norse mythology. Tuesday is named for the Anglo-Saxon god of war, Tiu (which is Mars in Roman mythology). Wednesday is named for Woden, the Anglo-Saxon name for the chief Norse god Odin. Thursday gets its name from Thor, the god of thunder in Norse mythology. Friday is named either after Freya, the Norse goddess of love and fertility, or Frigg, the wife of Odin and the representative of beauty and love."

—The Handy Answer Book for Kids (and Parents)

Gregorian calendar until 1752, and Greece and Turkey didn't start using it until the 1920s.

Today, the Gregorian calendar is international, although there are still many other calendars in use, even in the United States. You can buy a fancy, meaningful calendar starting in the middle of the year, you can wait and get one free from your friendly neighborhood merchant, or you can impress everybody by actually finding one of those alternative calendars and using it. Or, if you're like a lot of people, you can just use the calendar that came with your smartphone.

CHEMISTRY: THIS MEANS WAR!

We live in dangerous times. Nuclear weapons can—and do!—fall into nefarious hands. We can try, but the truth is that it's unlikely we could ever stop the use of nuclear force or even slow down the Pandora's Box of such weaponry, but we can remember one thing: chemical and biological weapons have been around for millennia.

Scholars disagree on the particulars, but it's generally believed that poison may have been the first of many ancient chemical weapons; Indian armies were instructed on the use of covert poisons

*In Central and South America, the bush rope plant (*Strychnos toxifera*), which produces curare, was used on poisoned arrows. The poison would cause respiratory failure in either hunted animals or enemy warriors.*

in the food of their enemies, and the Chinese knew how to make poisoned smoke bombs as early as 500 B.C.E. You shouldn't be surprised to know that the ancient Greeks mention poison-tipped arrows in several different texts, and burning sand was used in 332 B.C.E. by the citizens of the port of Tyre, who threw it in the air to land on Alexander the Great's advancing army. Herodotus mentions a man-made, fermented concoction that included feces and other organic matter that scholars say was surely loaded with gangrene and tetanus bacteria; it was used as both an arrow enhancer and a sort of bomb.

Eventually, warring armies upped the ante (if you can believe that), adding limestone powder and snake venom to their bag of battle tricks. Other warfare methods followed, such as sending angry bees down tunnels, choking enemy soldiers with smoke, and employing the use of jars of live scorpions to make dandy clay bombs that could be catapulted into a castle or fortress. On the other side of the moat, allies supporting the attack feverishly worked on antidotes to all this poison, including poultices and other ways to remove any deadly chemicals or venom and salves to protect the warriors' eyes and skin. Add fire to the battle and, as antidotes to burns (myrrh for injuries; vinegar to protect buildings) were discovered or developed, the use of flames escalated with the use of naphtha and pine resin, which made burning weapons harder to remove, devastating armies and navies alike. Though certain areas knew about asbestos and used it, asbestos as a fire retardant didn't occur in a widespread manner until well into the first millennium.

You can only imagine the terror.

But it gets worse: by the Middle Ages, early chemists had thought to experiment with chemicals for use on the battlefield. Biological weapons also came into play: the Mongols and the Tartars were both infamous for catapulting disease-infested corpses over fortress walls and into compounds, which was especially nasty when the Black Death hit Europe and the Mideast in the mid-1300s. It's a likelihood that later, the Great Plague did its own damage without any help from warring factions, as disease-carrying fleas spread from attacker to defender to average citizen and killed millions over the course of four hundred years.

The Mongols and Tartars weren't alone: in 1710, the Russians took up biological warfare in Estonia when they, too, catapulted plague-plagued corpses on Swedish troops and civilians during the Great Northern War.

Still, what these secondary items and concoctions could do was merely anecdotal information to American soldiers since germ theory wasn't established until very late in the nineteenth century. Indeed, warriors and soldiers didn't know why a smelly mixture could cause gangrene or a blanket could give someone smallpox, but scientists like Edward Jenner (1749–1823), Louis Pasteur (1822–1895), and Robert Koch (1843–1910) changed all that with their discoveries in vaccines, pasteurization, and microbiology, respectively. This changed a lot in the field of biological warfare, starting with the next great war: World War I.

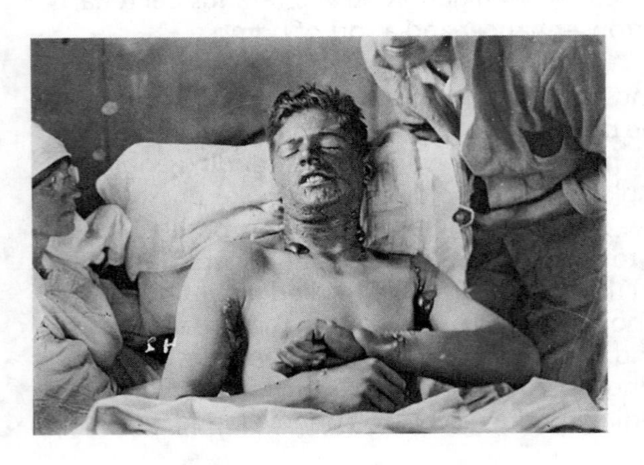

A World War I soldier lies in a hospital bed, a victim of mustard gas poisoning, causing blisters on his body. This man is lucky, though, as he did not get a lethal dose.

It's known, first of all, that Germany tried to use anthrax through the exportation of infected livestock sent to Russia and the United States. The Germans infamously also first used chlorine gas (see Science—Chemistry: Of Monsters and Men) during the war on French troops, killing hundreds by asphyxiation. Soon afterward, the Germans used phosgene, which would also kill by suffocation; though they were the first to use it, Allied forces embraced the chemical. Phosgene would result in more than eight out of every ten deaths during the war.

Finally, mustard gas killed more than 100,000 people during the war, and though its effects were long-term and devastating, it's thought that the relatively low fatality rate was because it was used outdoors, where fumes dissipated.

Though chemical and biological warfare had been a fact for thousands of years in one form or another, what happened during World War I shocked everyone. In 1925, the League of Nations adopted the Geneva Protocol, which banned the use of chemical and biological weapons but—oh-so-importantly—*did not ban the creation of or stockpiling of such weapons*. Also, importantly: if a nation was attacked by a biological or chemical weapon, they could claim the ability to retaliate in kind. It didn't take long for that to happen. Just a decade later, Italian leader Benito Mussolini used mustard gas against Haile Selassie's troops in Ethiopia.

This and other advancements in the use of chemicals made World War II notable in warfare. Before the war began, German scientists looking for a pesticide accidentally created a nerve poison they called Tabu, which ultimately became sarin. Other poisons were used against those in concentration camps and in battle against the Japanese, although chemical weapons were not used by the Nazis in Europe, perhaps because they feared that the Allies would fiercely retaliate.

HANDY FACT

"No effective antidote for human poisoning by mushrooms has been discovered. The toxins produced by mushrooms accumulate in the liver and lead to irreversible liver damage. Unfortunately, poisoning may not become apparent for several hours after ingesting a toxic mushroom. When the symptoms do present, they often resemble typical food poisoning. Liver failure becomes apparent three to six days after ingesting the poisonous mushroom. Oftentimes, a liver transplant may be the only possible treatment."

—*The Handy Science Answer Book*, 5th edition

And we all know what happened to Hiroshima and Nagasaki during World War II.

In the aftermath of the Vietnam War, and with the knowledge of the effects of the use of napalm and Agent Orange, nations went back to the table in April of 1997 and created the Chemical Weapons Convention, which once again bans the stockpiling, creation, use, or development of chemical weapons. At the time of this writing, 193 nations have signed this agreement.

As recent as 2013, Syrian civilians were victims of mustard gas from their own countrymen during the Syrian Civil War.

Stay tuned....

SPACE: FALLING FOREVER

So, let's say you're out for your daily walk with the dog when, oopsie, you accidentally trip and fall into a black hole. What happens next?

If you fell out of a space capsule without your suit, you'd be in a world of hurt: due to the effects of space vacuums, your circulation would stop within about 60 seconds. Your eyeballs and tongue would boil, and you'd be stretched to the maximum your skin would allow. Depending on where you fell, you'd either be frozen to bits or burned to a crisp within 24 hours. Not recommended. Stay in your space suit.

Ever since science has known about black holes, people have speculated about that very thing. To begin to determine the immensity of the question, though, you need to know what a black hole is: basically, it's what happens when a star is dying, and it creates a spot in space where gravity has gone crazy—specifically, gravity pulls so strongly that not even light can escape it. The black hole can be as small as the period at the end of this sentence, or it can be "supermassive" and big enough to hold millions of suns. Even more curious: since light can't escape, black holes can't be seen by the average skywatcher. They're entirely invisible without special tools available to astronomers.

Back to your little accident. First of all, black holes are only found in space, so you're relatively safe, but let's go with the story, eh?

What would happen if you fell into a black hole? Would you fall into another dimension or be stretched into a thin strand of molecules before being sucked into nothingness?

So, you've fallen near a black hole.

That pull of gravity is the first thing you'll experience: just the mere accident of getting too near a black hole makes the inevitable inescapable, as your fall propels you near the "event horizon," or a point of no return. Because you've tripped forward, your head is likely the first to hit the event horizon; therefore, it's going to be the first to feel the effect of the gravity as your noggin elongates, followed by the rest of your body at a slow pace because nothing moves faster than light (which can't escape from a black hole, remember). You'd be stretched thinner and thinner as you continued to fall until you, quite possibly, would be stretched so thin that you'd break into pieces. But wait—it's also possible that because the laws of physics are suspended, you could end up mostly intact in body or being, depending on the size of the black hole, but stuck there.

Mind blowing, isn't it? Quite literally?

At that point, if you could somehow open your eyes and forget your terror and the pain of being ripped asunder, scientists believe that you would be able to spy everything that ever fell into that particular black hole, plus all the things that will fall into it into the future forever.

Chances are, though, you're safe on your next walk.

Because, for sure, nobody's dived into a black hole and reported back on it.

HANDY FACT

Here's some more fun from *The Handy Science Answer Book*, 5th edition: "According to the Big Crunch theory, at some point in the very distant future, all matter will reverse direction and crunch back into the single point from which it began. Two other theories predict the future of the universe: the Big Bore theory and the Plateau theory. The Big Bore theory, named because it has nothing exciting to describe, claims that all matter will continue to move away from all other matter and the universe will expand forever. According to the Plateau theory, expansion of the universe will slow to the point where it will nearly cease, at which time the universe will reach a plateau and remain essentially the same."

ANIMALS: THE ONCE-A-YEAR SWARM

A flock of birds. A herd of sheep. A pack of wolves. Why do animals swarm? Let's first look at the swarm itself. Though it may look totally random, even chaotic, there's mathematics inside swarm behavior. Basically, when swarming, animals unwittingly follow three rules: they move in the same direction as others in their swarm; they remain close to others in their swarm; and they avoid collisions. Like any good list of rules, there are exceptions, but it appears that swarms come with one major goal in mind: safety for the group as a whole.

Think of a swarm of insects. If you were a predator (which, if you are intent on swatting them, you are), you couldn't possibly pick one individual out of the many during your killing frenzy. The same goes with a whale: the big guy has a hard time with the twists, dives, and turns of the group of small prey, and the chaos serves to confuse the predator … *enough*. The more creatures there are in the swarm, the chance is lessened that one individual creature could end up as dinner. Researchers know that when one animal feels danger, it alerts those near it, and a swarm results to protect the group. Or, in the case of wasps or bees, the swarm forms to protect the queen or the nest.

With locusts, the reason for swarming may be because of a tipping point in which a certain number of insects are present. With mayflies, it's because the flies have just twenty-four hours to emerge from the water, mate, and lay their eggs before they die. With some birds, it may be because of an overabundance of the insects they eat. Monarch butterflies swarm to migrate.

The fascinating thing about swarms is that the whole of the collection often moves as one unit, a phenomenon researchers call "swarm intelligence." There's still a lot of research left to do on why and how birds and insects move as they do in swarms, but it appears that leaders of the group may be in charge of knowing how movement will happen, even though individual creatures can also make their own decisions, literally on the fly.

The other cool thing that scientists know: although humans love to think we're unique among the animals, humans swarm, too.

THE ENVIRONMENT: DID YOU KNOW?

THE OLDEST TREES in the world are the bristlecone pines, located in the United States. It's estimated that some of them germinated from seeds nearly 5,000 years ago.

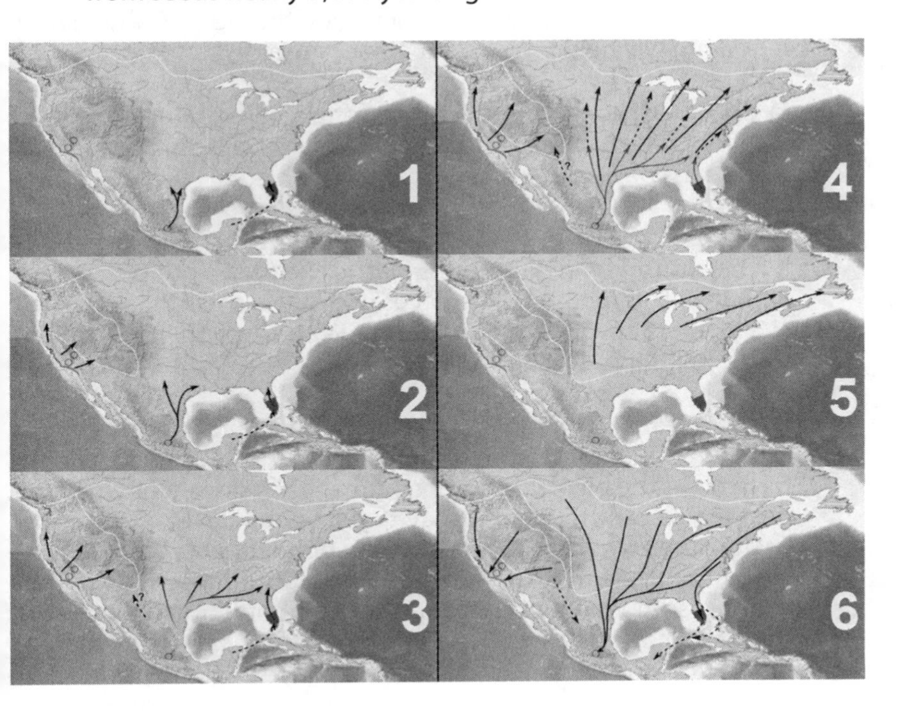

The migration of the monarch butterfly is a famous one, with the beautiful insects flying thousands of miles between Mexico, the United States, and Canada. The maps above show the route north as they fly in 1) March, 2) April, 3) end of April, 4) April to June, 5) June to August, and 6) September to November.

HANDY FACT

"Many populations of butterflies migrate, sometimes long distances, to and from areas that are only suitable for part of the year. Migration allows these species to avoid unfavorable conditions, including weather, food shortage, or overpopulation. In some lepidopetran species, all individuals migrate, while in others, only some individuals migrate. The best-known lepidopteran migration is that of the Monarch butterfly. Each fall between August and October, this insect heads south. Those who summer east of the Rocky Mountains spend the winter in Mexico, while those who summer west of the Rockies winter in California. Unlike most other migratory insects, the Monarchs who return in the spring are the same ones who left the previous winter."

—*The Handy Science Answer Book*, 5th edition

ABOUT THREE OUT of every ten pounds of trash in a landfill consists of packaging material.

PAPER CANNOT BE recycled indefinitely. After about six times through the system, the fibers are too weak to support the production, but adding new fibers (or mixing the old stuff with newer recycling) will help prolong the paper's use.

IN CASE YOU don't think every little effort helps, Hershey's Kisses are wrapped in aluminum foil, enough to cover some 40 football fields each day, and every scrap is recyclable.

THAT PLASTIC BOTTLE you threw out your car window last week? It'll be around forever because plastic like that is not biodegradable. It will ultimately fall to pieces until it's so small that you can't see it, but that will take hundreds of years. Until then, it'll be out there, polluting the soil and water.

IT MAKES SENSE that birth and death rates might even out the human population, but that's not the case. Because of declining birth rates, the birth rate among humans has slowed down in the past few decades. Scientists say that it should level off, one way or the other, when our overall human population reaches 9 to 11 billion people. And if that doesn't give you pause, this might: there is some evidence that, after the population levels off, it might take a nosedive. Eventually, scientists say, we could become extinct, just like the stegosaurus and the dodo bird.

HANDY FACT

"According to the Environmental Protection Agency, nearly 262 million tons of municipal waste was generated in 2015. This is equivalent to 4.48 pounds (2 kilograms) per person per day or approximately 1,635 pounds (742 kilograms) per person per year."

—*The Handy Science Answer Book*, 5th edition

THE AMOUNT OF water we have here on Earth is constant. Some of the water you showered with this morning might have been drunk and excreted by a dinosaur. What's in your water bottle may have come from a river that Genghis Khan traversed. You may make supper tonight with water molecules that brought Columbus to the New World.

IF YOUR FAMILY is average (and if you haven't actively stopped it), you'll get more than 840 pieces of junk mail this calendar year.

POLLUTION KILLS SOME two million people each year.

BOTANY: HOW DOES YOUR GARDEN GROW?

If there is ever a worldwide disaster that threatens your very life, the chances are that you'll be too worried about the safety of you and yours that you won't be much able to think past tomorrow. Lucky for you, a group of scientists have taken care of humanity's future for you.

Far north on the Norwegian island of Spitsbergen near the Arctic Svalbard archipelago more than 800 miles (1300 kilometers) above the Arctic Circle lies the Svalbard Global Seed Vault. In it, preserved deep in underground safety, lies the only way we might survive worldwide catastrophe.

The first person to conceive of the whole idea for a seed vault was Cary Fowler (1949–), who was then the executive director of the

Experts believe that nine out of ten of the United States' historic and heirloom fruits and vegetables are no longer around, having gone extinct (see Historymakers: Crunch, Crunch, Crunch).

Crop Trust and who began exploring the idea of having one central place to ensure that the seeds the world relies on for food would be protected as much as possible in the event of a disaster. In 1984, the Nordic Gene Bank (now NordGen) established such a vault, which it located in an abandoned coal mine outside Longyearbyen, Norway, which is the northernmost town on the planet. Just before the turn of this century, Fowler and his team assessed the state of the world's seed cache and its diversity.

In 2001, following the events on September 11 of that year, the International Treaty on Plant Genetic Resources for Food and Agriculture finalized negotiations, set rules for access and sharing, and opened for nations of the world to sign aboard. Shaken by the possibilities of world turmoil, however, Fowler and Biodiversity International (which was then called IPGRI) approached the government of Norway to construct and keep a proposed facility in the frozen mountains that would hold the world's seeds since, as it turned out, the coal mine was not suitable for seed storage due to the lack of cold (-3.5 degrees C wasn't cold enough) and occasional levels of hydrocarbon gases got too high. The Norwegian government made a commitment to such a facility (at the cost of about U.S. $9 million), building it into the permafrost near the archipelago of Svalbard, a few hundred miles from the North Pole.

On February 26, 2008, with several luminaries present, the Svalbard Seed Vault was opened. The world's nations were welcome—en-

The Global Seed Vault on the Norwegian island of Spitzbergen stores plant seeds as insurance against global disaster.

couraged—to deposit seeds and genetic seed material in the vault, with the knowledge that they could retrieve such materials back at any time. Samples of each of those seeds were also required to be a part of the Svalbard Seed Vault so that research could be done on those plants and for any future educational purposes that may be required. On the day that the Svalbard Seed Vault opened, more than 300,000 seeds were deposited, representing thousands of years of agriculture.

Were you to visit the Svalbard Global Seed Vault today, it would take awhile to get there. The nearest airport is in Longyearbyen, but you still have an 80-mile (12.9-kilometer) trek ahead of you. The vault is located literally in the side of Platåberget Mountain, appearing as if a giant hand drove a wedge into the frozen permafrost there. If you merely peeked inside the door, you'd see nothing but a vast hallway that runs back into the mountain more than 400 feet; the three seed vaults, all temperature controlled despite the already frozen ground, are at the end of the corridor. Each of the vaults is about the size of a larger ranch house, at around 32 feet wide and 88 feet long with a ceiling height nearing 20 feet.

Security is tight.

Each vault is furnished with large racks, upon which sit neatly labeled and stacked bins. The bins hold seeds, vacuum packed in special foil packaging and stored in orderly fashion to ease retrieval, should that need to happen. When it does, such as when seeds are approaching the end of their viability, they need to be replaced, especially in times of war, which was the case in 2015 after the war in Syria. Seeds from genetically modified plants are not, by Norwegian law, permitted to be stored at this vault.

Today's Svalbard Global Seed Vault holds mostly food-based seeds in a widely diverse range of species and subgroups in a vault that is kept at just below 0° F (that's −18° C). Some of the seeds that are held are for plants that are no longer cultivated but that someday may hold the right DNA to manufacture a plant that may be needed for any number of reasons. Today, somewhere just over a million unique seeds lie frozen in the Svalbard Global Seed Vault, and more than 75 percent of those seeds are for grain crops such as oats and wheat. Also represented are fruits, vegetables, herbs, cannabis, opium, and other plants we may need to survive as a species.

411

Researchers say that seeds last for as little as thirty-six months, but some are still viable tens of thousands of years after they were first collected.

HANDY FACT

"Seeds remain dormant until the optimum conditions of temperature, oxygen, and moisture are available for germination and further development. In general, the best temperatures for seed germination in most plants are 77–86° F (25–30° C). In addition to these external factors, some seeds undergo a series of enzymatic and biochemical changes prior to germination."

—*The Handy Science Answer Book*, 5th edition

There is enough room for more than two billion seeds, should the world need that kind of security. It should be noted that the vault has run into problems lately, including an inside leak that destroyed some seeds, controversy over issues of morality, and issues related to climate change. Still, as it was at its inception, the depositor of a seed holds ownership of it; sharing now is a matter of the owner's choice, but you can see what's there and who owns it by looking at the public database that's run by NordGen.

While you can't enter the Svalbard Global Seed Vault itself, a paid tour will take you close if you happen to be visiting nearby.

ANIMALS: THAT DARN WHATCHAMACALLIT

So, you've got your dogs. And you've got your cats. And you know what a rabbit is and a cow. But get a load of these strangely named creatures....

Let's start big: you already know that a tiger is a big striped cat, right? But a tiger can also be a snake, a salamander, a shrimp, a tarantula, a beetle, or any one of several kinds of fish and winged insects. They all carry the word "tiger" before their name—usually because they're striped like their feline namesake.

Those striped things probably could have just as easily had the word "zebra" in their name, except—unlike zebra spiders, zebra mice, zebra sharks, and zebrafish—"tigers" are not mostly entirely black and white and "zebra" creatures are.

Of course, none of those things have one taxonomic thing to do with tigers or zebras, but that doesn't make the etymology any different. If you see a moth, you might just call it that: a moth. But it could be a fox moth, a tiger moth, an elephant moth, a hummingbird moth, or the totally confusing bee hawk moth.

You get the picture how that happens. Looks like … must be named like.

So, now that you've got that, let's throw a wrench into the subject: Some creatures have downright hilarious names that make no sense at all (but that are still fun to know about).

The white-bellied go-away bird indeed has a white belly. It lives in Africa. Where it goes is anybody's guess (but go-away birds live in close families that generally travel together). It's actually likely that the bird got its name from its call.

Neither a chicken nor a resident of a high altitude, a mountain chicken is a frog from Dominica and Montserrat. The mountain chicken is an endangered species, most of which live in breeding facilities in captivity.

Rattus rattus and *Gorilla gorilla gorilla* are just what you think they are. Those are their scientific names, that's all.

Be glad you are not a beetle, especially the colon rectum beetle. Yes, that's what entomologist Melvin Harrison Hatch (1898–1988) named the unfortunate creature back in 1933 and, thanks to him, several other beetles in the genus carry the same first word in their Latin names. And while we're on the subject of beetles, the Agra vation and the Agra cadabra are both real beetles, as is the Pleasing Fungus beetle. More reason to be glad you're a not a beetle.

The mountain chicken (Leptodactylus fallax) is neither a chicken nor from the mountains: It's a frog that can be found on the Caribbean islands of Dominica and Montserrat. Sadly, it is critically endangered.

HANDY FACT

Here's an easy-to-figure-out name from *The Handy Science Answer Book*, 5th edition: "The giant clam, *Tridacna maxima*, is the heaviest invertebrate—it may weigh as much as 122 pounds (270 kilograms). The shells of this bivalve may be as long as 5 feet (1.5 meters)."

The pink fairy armadillo looks nothing like a fairy, but it is pink, and it's as cute as can be. Found in Argentina, this cousin of the New World armadillo loves to burrow for its dinner. It's also nocturnal, and sightings in the wild are very rare. Sightings in captivity are, too, since the pink fairy armadillo is endangered and doesn't seem to do well when captured from the wild. Screaming hairy armadillos are also armadillos—and cute little critters, at that—but they are more plentiful. You can probably guess what happens when you pick one up. (Hint: it's how they got their name.)

This one sounds like a bad skit: the maned wolf looks like a fox on stilts, but it's neither a fox nor, for that matter, a wolf. Though the maned wolf is a canid, it's the only creature of its genus, *Chrysocyon*. If you want to see one in the wild, you'd have to go to South America for that.

The Oedipina complex is a salamander with gills but no legs. I'll leave you to ponder that one awhile.

And finally, there's no need to excuse the bony-eared assfish. Found in the subtropics (but as far as British Columbia if it wants), it's a type of cusk-eel and the only one of its genus, *Acanthonus*. Don't mind its name: the "bony" part comes from the spines that the fish sports along its front tip and gills, and the last part is said to distinctly derive from a Greek word for "donkey."

CHEMISTRY: CHOOSE YER POISON

Imagine how many poisons you live with on a daily basis. As you silently count them up in your mind, you might be shocked—even more so when you remember that even the most innocuous thing, including salt, soap, and even plain old water, can be poison if used in the wrong way or overingested. So, now imagine the substances our ancestors—people who didn't have the sophisticated knowledge we possess—put on their bodies, into their mouths, and into their everyday activities.

If you were a monarch or a noble in medieval times, chances are that you had a need for a royal taster, especially if you or your loved ones were unpopular with the common man. In this case, the poison used against you might have been plain old arsenic (a longtime favorite of poisoners), or some mercury chloride might make its way into your muffins. A nearby garden or forest was often literally ripe with all kinds of tasty things that could kill in an instant. Sometimes, spoilage in the form of mold or bacteria might have been the killer of kings, but in any case, *just* in case, tasters tasted the king's food, put their lips to the king's chair, and kissed his linens and his utensils, and if the taster didn't get sick, then the monarch was considered to be safe.

If no taster was present, there were plenty of folk remedies and ways of guarding the royal stomach, including gemstones such as emerald or coral that could be passed over the food and unicorn horns that, when powdered, would surely indicate food that had been nefariously altered.

What was on the table wasn't the only way our ancestors poisoned themselves, however. What they put on themselves was equally deadly.

Starting with a woman's skin, which was absolutely supposed to be pale and flawless, medieval women liberally slathered a foundation made of white lead and vinegar called spirit of Saturn on their faces. Such a concoction may or may not have also contained arsenic and other various powders or even mercury to make a woman "beautiful." Eyes were lined with a stick that may have contained lead. Lip color was a touchy thing—many religions forbade its use—but for the boldest of women, berries mixed with animal fat worked to pink up the pucker. Cinnabar (which contained mercury) could also be used as a lip color, and it did double duty for the cheeks. For many years, red hair was the height of fashion, made possible with powdered sulfur or a dye made with, among other substances, lye and saltpeter; if baldness was becoming an issue, a shampoo containing quicklime and lead oxide was recommended. Other women swore that a sulfur-based paste would make their tresses beautiful. The problem with all this loveliness is that it didn't last: the cosmetics that women used then caused hair loss, abdominal pain, diarrhea, issues with the digestive and nervous systems, organ failure, and death.

411

Speaking of death, some crafty early seventeenth-century women were said to have added a product called Aqua Tofana to their facial powder. The secret was that Aqua Tofana was a topical poison that would kill the husband who kissed the cheek that wore the powder. The creator of Aqua Tofana, one Giulia Tofana, lived in Italy and made it her business to help women get rid of abusive, rich, or just plain annoying husbands at a time when divorce just didn't happen. Tofana was eventually caught and sentenced to death herself and, in a way, so were her customers: if the powder were lead- or arsenic-based, it generally poisoned the wearer, too, by soaking into her skin.

Also known as Manna di San Nicola, Aqua Tofana was a poison sold in the 1700s to women who wanted to kill their husbands.

But let's say you survived breakfast and your daily ablutions. Just getting through the rest of your day would have been a challenge.

Presuming that you lived in a city, almost immediately upon leaving your house, you would have been faced with an abundance of dung everywhere. It was rare for animals of any size to have been cleaned up after, and so their offal mixed with that of humans since chamber pots were often tossed out the window; even if they were emptied in a "small house" out back, you could smell it since outhouses were generally close to the medieval home and not very deep. This, of course, would have attracted flies and other vermin, leading to diseases such as dysentery, cholera, and typhoid, plus worms and other parasites. In addition, human urine was a hot commodity for use in the tanning industry since ammonia would have softened animal hides that were made into clothing and boots.

Contrary to what you may have heard, yes, medieval people bathed—or, at least, cleaned up some—at some time during the day. But there are always outliers, and that could have only added to the malaise of the area.

Heaven forbid if all this made you feel ill. Physicians of medieval times used a mixture of accidental science, goofy experiments, and folklore to fix (or try to fix) what ailed their patients. Mercury wasn't just a cosmetic; it was said to be a dandy laxative, a cure for syphilis, a fix for skin disease, and insurance for immortality. Lead was said to have been good for digestion, and it was a common product in poultices and lotions due to its supposed cooling properties. (For more weird medical cures, see Science—Botany: Weird Medical Cures from Times Gone By.)

And here's the thing: even if you decided that it was best just to stay at home and avoid your city altogether, the city might

HANDY FACT

"Some of the most poisonous mushrooms belong to the genus *Amanita*. Toxic species of this genus have been called such names as 'death angel' (*Amanita phalloides*) and 'destroying angel' (*Amanita virosa)*. Ingestion of a single cap can kill a healthy, adult human! Even ingesting a tiny bit of the amatoxin—the toxin present in species of this genus—may result in liver ailments that will last the rest of a person's life."

—*The Handy Science Answer Book*, 5th edition

come to you: early scientists called alchemists would have been working hard in their rudimentary laboratories—exploratory rooms that also emitted fumes into the air from lead, mercury, and other airborne poisons that came from their experiments.

You have to wonder sometimes how we made it as a species, eh?

MATH: DID YOU KNOW ...?

Here are some things that all math lovers should know....

ROMAN NUMERALS DO not include a figure for the number 0.

YES, THERE IS a difference between "numerals" and "numbers," but the difference is subtle and a bit complicated. A *numeral* is a symbol or written word that stands for a number, such as 3 or six or 11 or nineteen. Technically speaking, if you tapped your finger four times, that is a numeral. A *number* is the idea of how many of any item exists. For example, you see four pencils, your mind knows four items exist (the number), and you write down "four" (the numeral).

TO FURTHER MAKE a mess in your mind, a digit is what you use to write a numeral.

A statue of Aryabhata stands on the Inter-University Centre for Astronomy and Astrophysics campus in Pune, India.

USING CURRENTLY AVAILABLE coins (pennies, nickels, dimes, quarters, and half-dollars), there are 293 ways to make change for a U.S. dollar bill.

YOU'D HAVE TO count to 1,000 before you used the letter "A" (i.e., "one thousand *and* one").

THE STUDY OF math started some 8,000 years ago with the Pythagoreans.

INDIAN ASTRONOMER ARYABHATA (476–550) was the first to recognize the need for a number that signified nothing. Thus, he is widely credited with being the person to invent 0. Before you quote anybody on that, though, note that some sources beg to differ, giving the honor to an Arab, al-Khwarizmi, and to a Hindu man, Brahmagupta.

THERE ARE 86,400 seconds in a day. There are 604,800 seconds in a week. There are 31,536,000 seconds in an average (non-Leap Year) year. With these figures, you can determine how many seconds old you'll be tomorrow morning.

HANDY FACT

One reason that the number 10 is important, according to *The Handy Science Answer Book*, 5th edition, "is that the metric system is based on the number 10. The metric system emerged in the late eighteenth century out of a need to bring standardization to measurement, which had up until then been fickle, depending upon the preference of the ruler of the day, but 10 was important well before the metric system. Nicomachus of Gerasa (c. 60–c. 120), a second-century neo-Pythagorean from Judea, considered 10 a 'perfect' number, the figure of divinity present in creation with mankind's fingers and toes. Pythagoreans believed 10 to be 'the firstborn of the numbers, the mother of them all, the one that never wavers and gives the key to all things.' Also, shepherds of West Africa counted sheep in their flocks by colored shells based on 10, and 10 had evolved as a 'base' of most numbering schemes. Some scholars believe the reason 10 developed as a base number had more to do with ease: 10 is easily counted on fingers, and the rules of addition, subtraction, multiplication, and division for the number 10 are easily memorized."

THE ENVIRONMENT: IT LOOKS LIKE AN ELEPHANT

Mama always said you had your head up there. So, why not read some fun things about clouds?

THOUGH THEY ARE all differently shaped by virtue of being an assembly of water droplets, there are ten basic kinds of clouds. In ascending order of storminess, they are the wispy cirrus, cirrostratus, and cirrocumulus; the sky-covering altostratus and altocumulus; the darker stratus, stratocumulus, and nimbostratus; plus the cumulus and cumulonimbus, which are the puffy, cotton ball clouds.

CLOUDS MIGHT LOOK delicate, but looks can be deceiving: the water in a fair-weather cloud can add up to several dozen tons. If it's a storm cloud you're watching, there could be billions of tons of water hanging up there.

CIRRUS CLOUDS—THOSE sweet, little wisps in the sky—sure can appear to be innocent, but watch the skies when you spot cirrus clouds: they're often a good sign that the weather is about to change from sunny to stormy.

TORNADOES START OUT as funnel clouds. A funnel cloud is a rotating tunnel of air that churns but never reaches the ground. The second it does, though, it's considered a tornado.

PILOT WILLIAM RANKIN (1920–2009) is one of two people known to have fallen through a cumulonimbus thunderstorm cloud and lived to tell about it. Rankin was piloting an F-8 Crusader jet fighter on July 26, 1959, when the plane's engine failed, and Rankin was ejected. Forty minutes later, he landed in a forest and was rescued.

THE OTHER PERSON was Ewa Wiśnierska (1971–), a German paraglider who was accidentally caught in an updraft between conflicting thunderstorms and was pulled inside a cumulonimbus cloud. Three and a half hours later, she landed almost near where she had started her ride, having fallen more than 32,000 feet.

ACCORDING TO NASA, about two-thirds of the world is covered by clouds at any one point in time. So, if it's a nice day, count yourself fortunate and go enjoy it!

HANDY FACT

"Lenticular clouds form only over mountain peaks. They look like a stack of different layers of cloud matter. Noctilucent clouds are the highest in Earth's atmosphere. They form only between sunset and sunrise between the latitudes of 50° and 70° north and south of the equator."

—*The Handy Science Answer Book*, 5th edition

Lenticular cloud over Mt. Mayon, the Philippines.

CHEMISTRY: OF MONSTERS AND MEN

If you were lucky, when you were a kid, your parents encouraged your hobbies. Maybe what was a hobby then has become a passion now, perhaps even a way to make a little coin or an entire living. And that's good. Most of the time.

Fritz Haber was born in Breslau (now called Wrocław), Poland, in 1868 into a family of means. Three weeks after he was born, his mother died of complications of pregnancy, and little Fritz's care was relinquished to his aunts for the first six years of his life. At that point, his father remarried, and Fritz eventually had three half sisters.

Because his father was a well-to-do and well-regarded merchant, the Habers assimilated into German society, despite being Jewish. Fritz went to private schools that were a mixture of Catholic, Protestant, and

German chemist Fritz Haber earned a Nobel Prize in Chemistry in 1918 for his invention of the nitrogen fixation process that is used to help grow about two-thirds of the world's food.

Jewish students, or just the latter two. While at home, the Habers practiced the tenets of their Jewish faith but were not known as regular synagogue attendees. Fritz Haber grew up, then, feeling a strong affinity for his German heritage but not so much his Jewishness.

Haber's father had hoped the young man would join the family dye business, but Haber had other ideas and studied at various institutes of higher learning, with chemistry as his main interest. It was a subject that he had avidly studied since he was a small boy. After landing a graduate degree in organic chemistry, Haber worked for a while at his father's chemical business and the two men clashed, but not before the elder Haber gave his son opportunities to apprentice at various companies owned by friends of his father's. What Haber learned in those apprenticeships sent him back to school to study the technological processes he would need in his own laboratory. This led him to a job as a professor at the Karlsruhe Institute of Technology, where he met other chemists who encouraged him to branch out and experiment with hydrocarbons, gas combustion, and electrochemistry. His work was instrumental in furthering processes in the textile industry, and his work in other areas of chemistry allowed him to author two books.

In 1909, Haber had a breakthrough on something that had been plaguing scientists and farmers for ages: there wasn't enough organic fertilizer around, and nitrogen as fertilizer was in short supply. Haber discovered a way of using high pressure and heat to combine nitrogen and hydrogen to create ammonia, which is another natural fertilizer. Haber's colleague, Karl Bosch, helped make this process possible in an industrial setting, and this "Haber–Bosch process" was ultimately instrumental in ensuring that Germany had fertilizer and weapons during World War I.

By the time the war began, Germany had taken notice of Fritz Haber and sought his help. Haber had become an esteemed man by then; he'd converted to Christianity, married a woman who'd likewise converted, and had become a father. His reputation was well known within the German scientific community, and the Germans needed Haber's assistance. Shells used against the enemy were getting expensive. The Germans wanted Haber to help them to find a way to use poison in gas form on the battlefield.

He'd run afoul of German officials once before when his staff at the Institute for Physical Chemistry and Electrochemistry at Berlin-Dahlem was forced to resign due to Nazi race issues. That incident led to Haber spending a regretful year in Switzerland, but he wasn't gone long. Back in Germany, he was offered a job as a consultant with the German War Office, and he was given a uniform. He jumped with both feet into what the Germans espoused, signing a document that praised the kaiser for his invasions and promoting racist thought. Eventually, the kaiser, pleased with Haber's work, elevated Haber to the position of captain, and Haber set about refining some of his earlier experiments to create what is arguably one of war's most horrible weapons.

On April 22, 1915, as he stood behind the front lines in Ypres, Belgium, Haber watched as the men under his command let loose a cloud that quietly drifted across the valley and settled to the ground, where French troops had dug foxholes. It didn't take long before 150 tons of chlorine gas left 6,000 allied troops dead on the ground in what was modern history's first fully successful chemical weapons attack.

It would not be the last. This attack by Germany forced the French and British to make their own version of chemical weapons. The monster had been let out of its cage.

And yet, Haber seemed to be proud of himself. He had done his due diligence on German troops and he had killed just a few men with those experiments, so he knew that chlorine gas would work with devilish efficiency. Agreements with the Hague Convention had left chemical weapons off the table, but Germany was suffering defeats, and it vexed Haber. He defended his actions to many within the German army, telling them that dying by chlorine gas quickly was preferable to

HANDY FACT

Fritz Haber wasn't the only scientist to work with the German government during World War I. From *The Handy Science Answer Book*, 5th edition: "During World War I, the Germans needed glycerol to make nitroglycerin, which is used in the production of explosives, such as dynamite. Before the war, the Germans had imported their glycerol, but this impact was prevented by the British naval blockade during the war. German scientist Carl Neuberg (1877–1956) knew that trace levels of glycerol are produced when *Saccharomyces cerevisiae* is used during the alcoholic fermentation of sugar. He sought and developed a modified fermentation process in which the years would produce significant quantities of glycerol and less ethanol. The production of glycerol was improved by adding 3.5 percent sodium sulfite at pH 7 to the fermentation process, which blocked one chemical reaction in the metabolic pathway. Neuberg's procedure was implemented with the conversion of German beer breweries to glycerol plants. The plants produced one thousand tons of glycerol per month. After the war ended, the production of glycerol was not in demand, so it was suspended."

suffering from a wound on the battlefield. In the aftermath of the first attack in Ypres, Haber felt strongly that he'd done well for his country.

The first indication of problems ahead came one week after that morning in Ypres when Haber's wife, who was said to have been horrified by what he'd done, killed herself with a gun. Two years later, he had another wife and had started another family; he continued to work with Germany in the years after World War I; indeed, at war's end, he brought home a Nobel Prize in Chemistry for his work with hydrogen and nitrogen back in 1909. Between World War I and World War II, Haber invented an insecticide that he called Zyklon, which morphed into Zyklon-B, a chemical used in the gas chambers at Nazi concentration camps.

By that time, however, Haber and his immediate family were on the run. Though he proclaimed loyalty to Germany and to Christianity, there was no erasing that Haber was born a Jew and, therefore, would always be Jewish in Nazi eyes. He fled Germany and, for the rest of his life, he wandered from European country to country before dying of a heart attack in 1934 in Switzerland, the very nation he'd disliked not long before. His extended family was put to death in a concentration camp that likely used a modification of the chemical he'd discovered.

PHYSICS & MATH: JUST A SEC

If you can remember December 31, 2016, you might have noticed that you suddenly had more free time on your hands—one extra single second, to be exact, because that was the time when the Coordinated Universal Time needed to be readjusted to accommodate a slower spin of Earth's rotation and irregularities in the universe to keep pace with solar time. This kind of thing has to be done occasionally but not on a scheduled basis—just when it's needed, or timekeeping would eventually get horribly off-kilter.

So, what's a second among friends, anyhow? You can't do much with it, right?

Or wrong....

IN ONE SECOND, a bumblebee will beat its wings more than 200 times.

A bee beats its wings 200 times a second!

 ***IN ONE SECOND,*** more than 10,000 cans of Coca-Cola are consumed worldwide. In that same second, three Barbie dolls are sold.

 ***FOUR TO SIX*** new lives begin each second of the day on Earth. Not quite two people leave this world in that same time.

 ***IN ONE SECOND,*** more than 41,000 posts are made on Facebook and some three million Google searches are completed. Nearly three million emails are sent.

 ***OUR PLANET IS*** struck 100 times per second by lightning.

 ***ON AVERAGE, MORE*** than 200 pounds of food is thrown out every second of every day in America.

 ***JUST ABOUT ONE*** wedding happens per second somewhere in the world. It takes thirty-six times that to have just one divorce, on average.

 ***A VEHICLE DRIVING*** sixty miles per hour will travel eighty-eight feet in one second's time. And if that's not incredible enough, a bullet will travel nearly 3,000 feet in that same time.

 ***A CESIUM ATOM*** will vibrate 9,192, 631,770 times, and be glad you didn't have to count them.

 ***IN AN AVERAGE*** normal second in the United States, not quite a million people are in the air on a plane. More than 200 people are driving in cars, drunk.

HANDY FACT

"The continental United States (meaning the 48 states on the North American continent, which excludes Hawaii and Alaska) is divided into four time zones. From east to west, they are: Eastern, Central, Mountain, and Pacific. Each of these time zones is one hour apart, with times being successively earlier as you move west. So, if it's 3 P.M. eastern time, it's 2 P.M. central time, 1 P.M. mountain time, and noon Pacific time.

"The Alaska time zone is one hour behind Pacific time, so when it's noon in California, it's 11:00 A.M. in Alaska. Hawaii's time zone for part of the year is one hour behind Alaska. Hawaii does not participate in daylight saving time, however, so during that period (from April to October) when most of the U.S. states have set their clocks forward one hour, Hawaii stays at standard time and is two hours behind Alaska."

—The Handy Answer Book for Kids (and Parents)

BIOLOGY: HA HA HA HA! *SNORT*

It's been claimed that the average person laughs seventeen times a day. Have you reached your capacity for the day, or do you need to read on and see why not?

Long before we can speak clearly, we laugh. Babies—including those who are born blind or hearing impaired—start laughing at just a few weeks old; so, for that matter, do baby chimps. Scientists say, in fact, that babies and chimps share laugh patterns until human babies mature enough and grow savvier. Chimps, by the way, aren't alone: other animals, too, have senses of humor and are able to laugh—or, at least, to approximate something akin to our chortles (look up "rat giggles" online if you've never seen the video. You won't be sorry). Anecdotally, some animals are even able to understand a joke or jokingly tease.

If you've ever stood and quietly listened to a crowd—say, at a food court or auditorium—you'd know that there are ten discernible

Why do we laugh? Well, for one thing, it helps to form social bonds. It also has health benefits, such as decreasing stress, elevating mood, and increasing oxygen levels.

types of laughter: on one end, there's etiquette laughter, which oils our social interactions. There's contagious laughter, nervous laughter, belly laughs and their silent cousins, and laughter that relieves extreme stress (think: the classic *Mary Tyler Moore Show* "Chuckles Bites the Dust" episode). There's "pigeon laughter," which is a laughing snort produced when the lips are sealed. And then, there are two types that count but maybe shouldn't: canned laughter and cruel laughing.

What's interesting is that you might use one or a combination of any of these in one conversation, and nobody had to teach you how to do it. The other interesting thing: no matter where you come from and who is doing the ha-ha-ing, most people can tell a fake laugh from the real thing—which leads to questions about canned laughter or "laugh tracks"; specifically, why are they there? The answer is that laugh tracks were first used in the 1940s for radio shows to subtly tell the at-home listener that something humorous just happened; since laughter is contagious, the listener chuckled and, suddenly, the unseen show was funny. The next natural step was to use it on TV and, over time, the technique got better until you barely notice that it's there now—that is, if it is, because not all television programs use canned laughter these days. This constant updating of technology also means that the old saw about the laugh track you hear today having been recorded decades ago is pure mythology. At any rate, the reason canned laughter persists is that it still works.

But back to the *real* laughs. Why do we do it?

In a conversation, the person speaking laughs more than those listening. In social situations, women laugh far more than do men.

The big answer is that we laugh because we are social creatures. Good laughter is the glue that holds us together as a species: we like to laugh, especially when we can do it with our family and friends, but it's also a universal language that everyone speaks. You can laugh with someone from Bali or Britain or Bangladesh or Bora Bora, and you didn't even have to spend hours learning to do it.

But that's nowhere near a complete enough solution to the question of why, though. Laughter indicates status: casual observers can tell the dominance of individuals in a group by the sound of the laughter from others in the group. Surprisingly, even a submissive laugh from a dominant person indicated dominance to the observers in the study.

Laughter spurs our appetite for food (note to dinner-and-a-show fans): in one study, researchers learned that subjects who watched a funny video had more leptin and ghrelin in their bodies than those who didn't view comedy.

If you want to burn calories, laugh—just know that it's going to take awhile. One hour of good, solid belly laughs (which, you have to admit, would be hard to achieve) burns about a hundred calories. It also works your abs and your lungs, and it raises your heart rate, which are all good. Laughter also lowers blood pressure, and people who laugh statistically have fewer heart attacks. So, go ahead and laugh, but don't give up that gym membership.

There are three basic vocal sounds in laughter: the "ha," the "ho," and the "hee." Most people use one or the other the majority of the time, but the occasional switch happens in a string of laugh sounds (as in "ha-ha-ho-ho-ho") but never alternating in the same string (as in "ha-ho-ha-hee"). Once you get to know someone, you can usually identify them by this pattern of laughter. Also, once they reach adolescence, kids' laughter tends to mimic that of a parent or caregiver, suggesting that laughter might have a basis in genetics. Yes, actual studies have been done on this.

Scientists know that laughter comes from several different parts of the brain. They also say that there are three basic reasons why we might find something funny, which explains why we laugh (you knew we'd get to that, didn't you?). Incongruity theory says that we laugh because logic and familiarity don't match in a surprising and delightful way that our brains didn't anticipate. Superiority theory is when we

HANDY FACT

From *The Handy Answer Book for Kids (and Parents)*, on which action uses more muscles, smiles, or frowns: "Yes, what your parents tell you is true: It takes far more muscles to frown than it does to smile. Smiling only uses 17 muscles, while frowning requires 43."

laugh at someone else's misfortune or foolishness, whether real or invented. Relief theory says that laughter can break tension and ease nervousness or nervous energy.

So, why don't we laugh at the same things?

There are several reasons: we're more apt to laugh at anything if we're with others who find humor in a situation. Age has a lot to do with humor: babies and children don't have the load of experiences that adults have, so their incongruity theory is more pronounced; teens in a so-called "awkward phase" are more apt to embrace the superiority theory. Mood has something to do with a sense of humor, but the biggest predictors appear to be one's belief system, environment, and their stressors. We won't laugh at something we find offensive, but something that's stressful holds more snickers.

The best advice if you want to laugh more is to know what tickles your funny bone, then try to expand on that with different mediums. Also, hang out with people who make you happy or who have a similar sense of humor.

GENERAL SCIENCE: DID YOU KNOW ...?

IF YOU COULD somehow get every single person in the world together, you could stand everyone inside the borders of Rhode Island, and it wouldn't even be a tight fit.

FOR THE FIRST time in history, in 2012, the scales tipped, and more than half the world's population was under the age of thirty.

IT'S A MINOR fallacy that twins skip a generation, but it is true that the chances of having fraternal twins is genetic. In either case, the incidences of twinning have been rising steadily in the past few generations.

PINK AND BLUE weren't always assigned to, respectively, girls and boys. About a century ago, pink was perceived as a strong color and blue as delicate. The perception switched shortly after World War II.

STATISTICALLY SPEAKING, THE Dutch are overall the tallest people in the world, averaging nearly five feet, eight inches tall. The people from Timor, a small island in southeast Asia, are statistically the shortest at an average of just over five feet tall.

IT'S ESTIMATED THAT there are five million trillion trillion bacteria on Earth, give or take, if you'd like to count them up.

SEVERAL MILLION PEOPLE today are walking around with DNA indicating that they were descendants of Genghis Khan.

THE DUNNING–KRUGER effect is a real cognitive bias first explained by psychologist David Dunning and Justin Kruger, one of his graduate students. The theory says that each of us believes we are above average in certain abilities in which we are actually inadequate, and we think that certain skills we have are better than they truly are. One of their studies found that 80 percent of those surveyed believed themselves to be above-average drivers, which, given the law of averages, is not possible.

HANDY FACT

The Handy Science Answer Book, 5th edition, says, "Many snails move at a speed of less than 3 inches (8 centimeters) per minute. This means that if a snail did not stop to rest or eat, it could travel 16 feet (4.8 meters) per hour."

PHYSICS: RUBE GOLDBERG

As anyone who's endured an arduous trip or has had to carry something awkward and heavy for a long distance can attest, going from Point A to Point B can be a challenge that makes you think there's gotta be a better way. As one man showed, that better way can be hilarious.

If nothing else, one could say that Rube Goldberg was prolific: in his lifetime, he created more than 50,000 cartoons, featuring thousands of make-believe contraptions. He was so good at his work that it must've seemed effortless; remarkably, he was able to draw (and keep straight!) several cartoon series at a time.

Reuben Garrett Lucius Goldberg entered the world on July 4, 1883, in San Francisco, the third of seven children (one of four survivors) born to Jewish parents. Goldberg's father, a police and fire commissioner, encouraged his son to draw, eventually giving young Rube formal lessons at age eleven. At the same time, he encouraged Rube to consider a career in the field of engineering.

In 1904, Rube Goldberg graduated from the University of California, Berkeley, with a degree in engineering and started a job in the City of San Francisco's water and sewer departments. Apparently, though, his hobby called to him, and he made it a career when he quit working with the city and began working as a sports cartoonist for two different local newspapers. In 1907, he moved to New York City and landed a job there with a syndicated newspaper. By 1915, Goldberg was known as one of the nation's most popular cartoonists, making much more per year than some Hollywood starlets (which would be in the vicinity of a million bucks in today's money). In 1916, he became an animated-film writer and animator. He also got married that year.

By the beginning of World War II, Goldberg was undeniably famous, mostly because of his most popular cartoon character, Professor Lucifer Gorgonzola Butts, a genius inventor who never, ever could seem to make things easy. A simple task done in Butts's lab—say, for instance, flipping a switch—would require two or three dozen steps; an equal number of everyday, seemingly unrelated items; and, quite often, some degree of mathematics or science. Goldberg's creation (and his creation's creations) were ridiculously, brilliantly hilarious, and they became so iconic that competitions sprang up—and are still held today—to invent the next "Goldberg" contraption.

Self-Operating Napkin

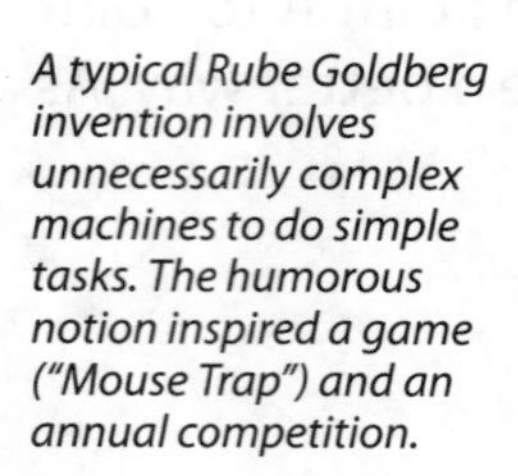

A typical Rube Goldberg invention involves unnecessarily complex machines to do simple tasks. The humorous notion inspired a game ("Mouse Trap") and an annual competition.

In 1948, Goldberg won the Pulitzer Prize for Editorial Cartooning; alas, his political cartoons also brought no small amount of hate mail—enough that Goldberg supposedly asked his two sons to change their surnames in order to protect them.

In the postwar years and until his retirement, Professor Butts continued to have free rein in the laboratory while Goldberg became enamored of modern technology, particularly of those things that were meant to make work easier. It's been said that Goldberg believed it to be human nature that we'd pick a more difficult method to finish a simple task than we would to do the chore directly.

In addition to his various awards and accolades, Goldberg was the founder of the National Cartoonists Society, he was the man for whom the Reuben Award (for cartoonists) is named, he was the inspiration for the 1960s game "Mouse Trap," and his work was honored by an exhibit at the National Museum of American History in Washington, D.C., in 1970.

Rube Goldberg died in December 1970.

HANDY FACT

"The most important invention of the Electrical Age (1891–1934) was the automobile, which enjoyed widespread use by 1910. Other consumer items that were introduced and changed the everyday life of the average person during these years were safety razors, thermos bottles, electric blankets, cellophane, rayon, X-rays, and the zipper."

—*The Handy Science Answer Book*, 5th edition

BIOLOGY: THE FINAL FRONTIER

So. The worst has happened, and ... so sorry, but you're dead. You could have a normal, run-of-the-mill funeral, complete with a fancy coffin and a nice, new outfit to wear for eternity, but think about it: those things are for common people. No, you want an afterward as unique as you were in life, right? So, here are some unusual things you (or your long-suffering heirs) can do with your carcass:

IF YOU ENJOY nature now, you might enjoy it for a few years longer if you donate your body to some sort of cadaver laboratory, like the Body Farm in Tennessee or the facility that the Texas State University runs. Your mortal remains might end up for a few months or weeks in a car trunk or sitting in a tree or on the ground or two feet under. What happens to you next is not pretty, except for the researchers who'll use information on body degradation, insect infestation, and weather to solve crimes and answer questions for someone else's loved ones, giving them closure. Think of it as a literally rotten way to do something really good for someone else.

WHILE WE'RE ON the "donate" subject, you can become a museum exhibit if you suffered from an unusual malady in real life. Some medical facilities and colleges still accept bodies for anatomy and physiology classes (dissection), but you may have to look a little harder for them. Also, ask your doctor for donor card information: your skin, eyes, bone marrow, heart, lungs, liver, kidneys, and other body parts may give someone else a chance at a normal life. It's another nice thing you can do for someone else's loved ones.

IF YOU LOVED bling in real life, you can become bling. You could have your ashes mixed and encased in a locket, key chain, or other kind of jewelry, so your loved ones can hang around with you. You could choose to have some cremains encased on the inside of a stuffed animal or other folk art or sculpture. Or, a tiny amount of your remains can be added to ink for a tattoo or a piece of art. Likewise, your cremains can be mixed and pressed into a record album—maybe a recording of your voice?

A good broker could get upward of $10,000 for your carcass. He will likely divide it into separate parts with separate prices. It's not legal for your heirs to sell your bod themselves.

 ***AS OF LATE,*** it's not uncommon for pets to be freeze-dried. Now you can get the process done for yourself, but it's not cheap. Along this same line, you can give your body up for scientific education by having it plasticized.

 ***ASK YOUR FAMILY*** to literally encase your feet (and the rest of your ashes) in cement and toss you into the ocean to become part of a reef for coral to grow on. It's environmentally friendly, and it's good for the fishes, too.

 ***THEY SAY THAT*** in real life, we are all made of stars. So, become a star once again by having your cremains shot into space via satellite. Or if that sounds close to your idea of a good funeral, you can be shot into the air as part of a fireworks show. In both cases, tiny bits of you would mingle with the earth, eventually. Until then: twinkle, twinkle.

 ***FOR THE GARDENER,*** there are many options. You can be buried in a compostable coffin, or you can *become* compost in a park all by yourself (your remains will be cleaned and sanitized, don't worry). Your remains can also be planted with a tree seed or seedling by burial in a vessel that fertilizes the plant over time.

 ***YOU CAN BECOME*** dust faster than you ever thought by going through promession, which chemically freeze-dries you. Then the

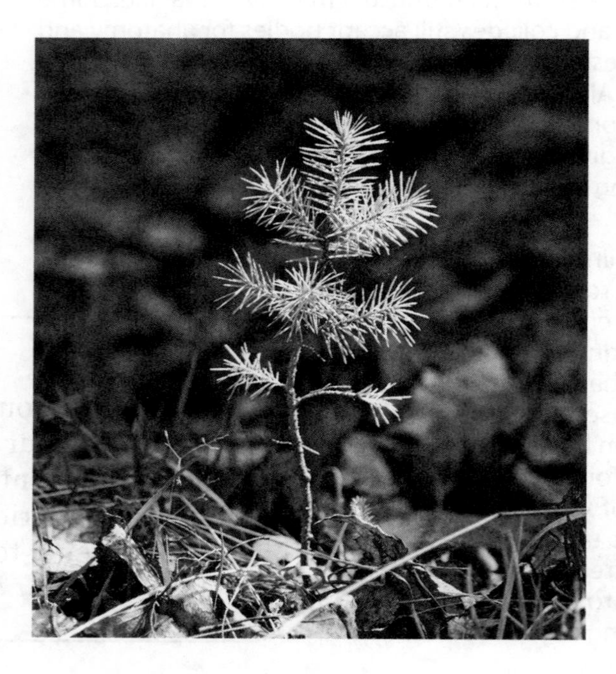

One option that appeals to many people is to have your ashes used to fertilize a tree seed. Essentially, your body becomes a tree, a living legacy to your life.

HANDY FACT

"Cells die for a variety of reasons, many of which are not deliberate. For example, cells can starve to death, asphyxiate, or die from trauma. Cells that sustain some sort of damage, such as DNA alteration or viral infection, frequently undergo programmed cell death. This process eliminates cells with a potentially lethal mutation or limits the spread of the virus. Programmed cell death can also be a normal part of embryonic development. Frogs undergo cell death that results in the elimination of tissues, allowing a tadpole to morph into an adult frog."

—*The Handy Science Answer Book*, 5th edition

moisture and metals inside your body are removed, and what's left is packed in a biodegradable container that requires a small plot for burial. The pros are that you'll take up much less land for interment.

No matter what you choose, experts say that you should have your instructions written down, preferably by a lawyer, and be sure to tell your doctor and your loved ones. In the meantime ... take care of yourself out there.

TECHNOLOGY: TV OR NOT TV?

So, you spent a sleepy Saturday afternoon recently tooling through some old TV-show favorites. It was a blast, watching those shows when you were a kid, in part because they were so fantastical. But now you're wondering: will you ever get to live like the Jetsons?

If you've got dreams, read on....

IS IT POSSIBLE that we'll ever teleport, like Captain Kirk did on *Star Trek*? Not anytime soon. The laws of physics say that teleportation is impossible, although in 1993, a group of scientists sent photons to a satellite some 300 miles away through "quantum entanglement." Well, *kind of*: what they did relies on a phenomenon in which a pair of photons, no matter how far apart, share the same state. Change one particle, and the other, miles away, changes, too. By no remote means does this indicate that it might be possible for a full-size, thicker-than-a-photon human to safely be teleported. It just means that science is still working on it, and they likely will be for a while.

WILL HUMANS EVER time travel, as on *The Twilight Zone*? Theoretically, yes: Einstein's Theory of Relativity aside, it has to do with perception of passing time and travel against the rotation of Earth. When we travel fast, our perception of time slows; in reality, scientists have shown that a clock in a plane will lag a fraction of a second behind a clock on the ground. As for traveling to a whole different century forward or backward, no, that's not possible, and it likely won't ever be.

WILL WE EVER be like Superman, able to leap tall buildings in a single bound? In a word: no. The longest vertical leap ever made by a human was accomplished by Canada's Evan Ungar in 2016, with a jump of just over sixty-three inches. Considering that a "tall building" of two stories (not so tall, right?) is between eighteen and twenty feet high, there's a way to go before a human could *leap* a tall building. However, all that other Superman stuff: Lifting cars? Check. Twisting metal? Check. Speedier than a locomotive? Nope. Able to stop bullets? Not a bit.

CAN I REALLY voyage to the bottom of the sea? Yes and no. At the time of this writing, the farthest a human has dived in an underwater vehicle has been just a little over 35,000 feet (nearly seven miles). The Challenger Deep, located at the very end of the Mariana Trench, is just over 36,000 feet deep, so you might be able to go to the bottom in that single location with a diving device, but nobody has done that yet. As for diving without the device, the pressure of the water at that depth would make it impossible now (and probably in the future) to free-dive that deep and live to tell about it.

WILL WE EVER have a robot pal like Will Robinson had on *Lost in Space*? It's closer than you think, actually. Japan already employs robotic teachers. Some hotels use robotics to take care of guest requests in a more expedient manner. Robots have been used in industrial settings already for years, and they're starting to show up in hospitals, nursing homes, and in situations where security is needed. There are bugs to work out, of course—it's said that robots are not great with small children or animals, both of which can be unpredictable in their activity, and they're not so great for stairs or uneven terrain—but robots in everyday life? That's coming soon; in fact, if you've got a robotic vacuum cleaner in your home, you're a little ahead of the curve.

COULD I HAVE a pet dinosaur, like on *The Flintstones*? Technically speaking, yes! Chickens descended from a common ancestor of dinosaurs and are considered to be "living dinosaurs." As for having your very own Dino running around the house, looking for a mastodon bone—although scientists are working with DNA to clone the woolly mammoth—it's highly unlikely.

CAN I BE frozen and thawed out in a couple centuries, like Buck Rogers? Not completely yet. Although, here's the thing: we already know that frozen DNA can survive for many centuries (see dinosaurs, above). Science has already revived frozen seeds that yielded plants after tens of thousands of years in ice. Human sperm and ova can be frozen in dry ice (liquid carbon dioxide) for years and still be viable. But a whole, complete you? Not yet, but cryonics experts have hope.

CAN I SWIM in a pool full of money, like Scrooge McDuck? Probably not. While no one will say how much physical money exists in the world, it's likely somewhere north of $5 trillion. Assuming that much of it is coin, that would make a hard landing in a pool rather than a splash. Even if half of it is comfortable paper—which, yes, you could *probably* "swim" in—you'd have to collect all that moola in one place, and that's likely not going to be possible.

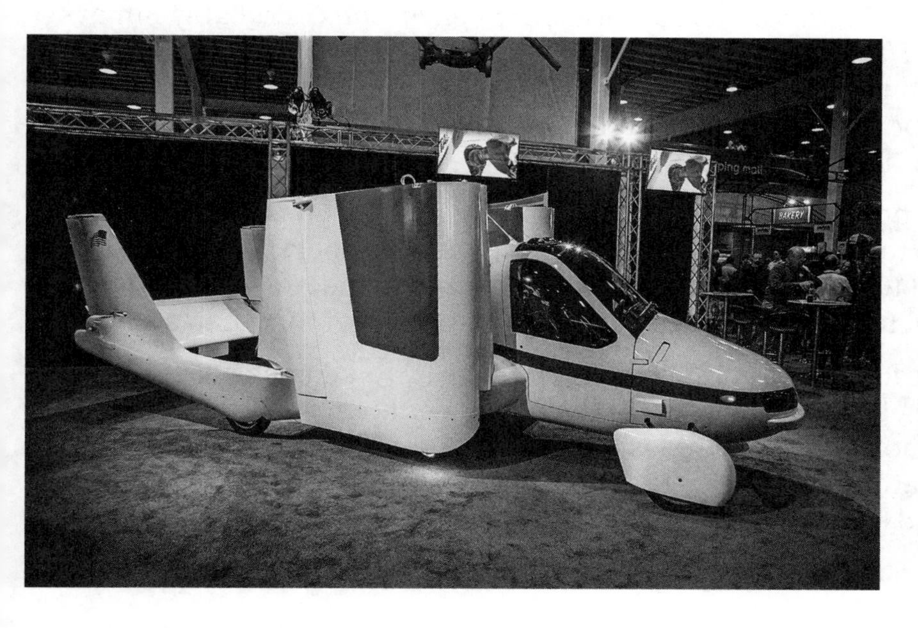

This concept car on display at a 2012 New York City car expo is a hybrid of a car and an airplane. The idea of flying cars has been bandied about for years, but as yet there are no practical versions of the vehicle from the cartoon The Jetsons.

WILL WE HAVE flying cars, like George Jetson had? Yes and no. Engineers are busy perfecting driverless cars at this time, but at least one flying car has been made into a prototype, and it's very, very pricey. On the positive side, small, light, personal aircraft are available now, as are hoverboards—although, at the time of this writing, both of them have their downsides and dangers, and neither are anywhere near to a Jetson-like vehicle. Still, we are closer to achieving something like a consumer-based, inexpensive, flying car than we were back when *The Jetsons* aired in 1962.

PHYSICS: COLOR MY WORLD

So, we already know all about the color black (see Science—Physics: The Color Black), but if you're still feeling a little blue, how about some more colorful facts?

PINK IS A color-not-a-color. You can buy all kinds of pink things and you can chew pink bubblegum all you want, but there is no pink on the rainbow, and pink does not have a wavelength.

HANDY FACT

Could you harness the wind, like Samurai of *Super Friends*? Probably not exactly like it, says *The Handy Science Answer Book*, 5th edition: "A cyclone has rotating winds from 10 to 60 miles (16–97 kilometers) per hour, can be up to 1,000 miles (1,600 kilometers) in diameter, travels about 25 miles (40 kilometers) per hour, and lasts from one to several weeks.

"A hurricane (or typhoon, as it is called in the Pacific Ocean area) has winds that vary from 75 to 200 miles (120–320 kilometers) per hour, moves between 10 and 20 miles (16–32 kilometers) per hour, can have a diameter up to 600 miles (960 kilometers), and can exist from several days to more than a week.

"A tornado can reach a rotating speed of 300 miles (400 kilometers) per hour, travels between 25 to 40 miles (40–64 kilometers) per hour, and generally lasts only minutes, although some have lasted for five to six hours. Its diameter can range from 300 yards (274 meters) to 1 mile (1.6 kilometers) and its average path length is 5 miles (8 kilometers) with a maximum of 300 miles (483 kilometers)."

IN ANCIENT TIMES, Homer said that rainbows consisted of one color: purple. Isaac Newton (1642–1727) was the one who came up with seven colors on a rainbow: red, orange, yellow, green, blue, indigo, and violet.

THAT'S ROY G. BIV, for you acronym lovers who want to remember the rainbow colors in order.

THERE ARE COLORS in existence that most humans cannot see. The cones in our eyes just aren't made for it, although there are some humans who can see further along the color spectrum than most of us are able to see—possibly as many as nine million more colors. Those tetrachromats are almost always female and often have a colorblind father. Tetrachromacy is a genetically inherited condition.

PERCEPTION OF COLOR is an individual thing. Internet memes might have taught you that, with the "what color do you see?" games that go around now and then. It might be hard to believe, but since some cultures lack words for certain colors, they don't "see" the color they have no corresponding word for.

IF YOU WANT someone to remember something, write it down with a red pen on blue paper. Black and white items and words are forgotten quicker than things that are presented in color.

AROUND THE WORLD, blue is the color most identified as a "favorite color."

WOMEN ARE BETTER than men at telling two very similar colors apart.

BEWARE BEFORE YOU paint that room: red and yellow will make you hungry; red is also said to be a color to excite. Pink and light blue are both soothing. Lilac and pastel colors are said to be mood boosting.

EIGENGRAU IS THE name of the gray color you see right after you turn off the lights and before your eyes adjust to dim light or lack of light.

HANDY FACT

From *The Handy Science Answer Book*: "Rare reports of all-red or all-white rainbows have indeed occurred."

GENERAL SCIENCE: IS IT REAL?

It seems exactly right to end *The Big Book of Facts* with a scientific fact that leans a little on history and a whole bunch of questions—the biggest of which is: Are we alone in the universe?

The answer to that might lie on the globe's 37th parallel.

THE BACKSTORY

If you were to trace your finger roughly from the border between Virginia and North Carolina all the way across the country to about San Jose, California, your finger would have followed the 37th parallel north, which is the point of latitude that's thirty-seven degrees above the planet's equator. Knowing where a specific latitude or longitude on the planet lies is necessary for mapmakers, navigators, pilots, geocaching fans, and others who might need to find an exact point on Earth.

In 2006, siblings and ufologists Chuck Zukowski and Debbie Ziegelmeyer were poring over records they'd amassed over years of research and documentation when they saw something they'd never noticed before: a disturbingly large number of the livestock mutilations, UFO sightings, and other odd paranormal phenomena they'd investigated happened along the 37th parallel (with a leeway of seventy miles south and north).

THE EVIDENCE

For nearly two centuries, everyday Americans have reported weird phenomena along the 37th parallel: so-called "spook lights" sighted on the east side of the nation. Weird, floating orbs that expanded and contracted. Floating, blue or red, glowing balls in the sky. Objects such as hovering discs that defy common knowledge—some that were audacious enough to follow planes. Hundreds or perhaps thousands of mutilation deaths of livestock, particularly cattle, in which the animals are cut and organs removed with surgical precision and little to no blood loss. Native American legends tell of some of these phenomena; other accounts have been written and reported, going back to the early 1800s.

Believers also point out that a lopsided number of devastating earthquakes have happened within the area straddling the 37th parallel, including the 1906 San Francisco earthquake and, overseas, the 2011 Fukushima, Japan, quake.

For what it's worth, a lot of iconically curious landmarks also lie along the 37th parallel, including Fort Knox, Mammoth Cave in Kentucky, the "Four Corners" where four states meet, the Grand Canyon, and Area 51. A number of underground bunkers are in the area. There have been reports of chupacabra (a creature known to cryptozoologists) and a southern version of Sasquatch. Areas sacred to some Native American tribes are also located along the 37th parallel.

THE VERDICT

So, is the 37th parallel really a "paranormal highway"?

To recap:

A lot of important and iconic buildings and natural land formations are found within seventy miles, either way, of the 37th parallel … but a lot of important and iconic buildings and natural land formations are *not*. It seems to be more coincidental than otherwise that man-made constructions are where they are; ditto for what nature has given us.

Attempts have been made to explain away the livestock mutilations over the years, but the sheer numbers, the frequency, the stealth of the perpetrators, the location of the acts, the length of time that this has been happening, and the methods with which mutilations are done don't seem to match with those explanations. Furthermore, blaming the animals' owners makes no sense: it defies logic to think that hundreds of farmers and ranchers over thousands of miles would band together to do what has been done.

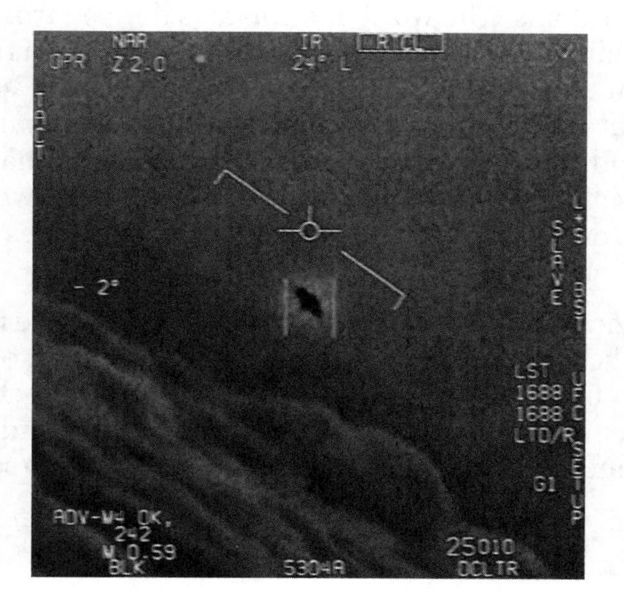

Three videos have recently been released by the U.S. Navy (video still shown above) showing UFOs spotted by fighter pilots. In 2020, the Pentagon made a public statement that, indeed, these are incredible flying objects that no one has been able to explain.

Strictly speaking and naysayers aside, UFOs exist until they don't—meaning that as long as they're *unidentified*, they are *U*FOs. It's true that unidentified flying objects have been seen in areas all over the world, but the frequency slightly above and below the 37th parallel overwhelms the evidence reported elsewhere. That's a lot of UFO reports, many of which can't be denied.

Other countries that straddle the 37th parallel have reported unexplained phenomena, but not as much as has come out of the United States. Cultural pressure to keep quiet about something so controversial may explain any reticence there might be overseas.

The U.S. government released supposed UFO tapes in April of 2020 hoping to quell the spread of misinformation that happens online. There's not much to see on the video; the audio has more to tell, making viewers wonder if there are more tapes moldering in some vault, but the fact is that we'll probably never know. Officials won't discuss classified matters, but it seems they eliminated a once secret operation to investigate UFOs in July of 2020.

IN THE END

This is one of those things that just doesn't have an answer. And it might never have one. It's maddening, yes, but you'll have to make up your own mind on this one.

HANDY FACT

"The north–south lines on a map run from the North Pole to the South Pole and are called meridians. The word 'meridian' is derived from the Latin word *meridianus*, meaning 'noon.' When it is noon at one place on the line, it is noon at any other point on the imaginary line as well. The lines, called longitudes, are used to measure how far east or west a particular place might be from zero degrees of longitude. The prime meridian is the meridian that passes through the Royal Observatory at Greenwich, England. In 1884, it was adopted internationally as zero degrees of longitude. The longitudinal lines are 69 miles (111 kilometers) apart at the equator.

"The east–west lines on a map are called parallels and, unlike meridians, are all parallel to each other. They measure latitude, or how far north or south a particular place might be from the equator. One hundred and eighty lines circle Earth, one for each degree of latitude. The degrees of both latitude and longitude are divided into 60 minutes, which are then further divided into 60 seconds each."

—*The Handy Science Answer Book*, 5th edition

FURTHER READING

BOOKS AND PERIODICALS

Bird, William L., Jr. *Paint by Number: The How-To Craze That Swept the Nation*. New York, NY: Princeton Architectural Press, 2001.

Bondeson, Jan. *Amazing Dogs: A Cabinet of Canine Curiosities.* Ithaca, NY: Cornell University Press, 2013.

Bondar, Carin. *Wild Moms: Motherhood in the Animal Kingdom.* New York, NY: Pegasus Books, 2018.

Broudy, Oliver. *The Sensitives: The Rise of Environmental Illness and the Search for America's Last Pure Place.* New York, NY: Simon & Schuster, 2020.

Chandler, Adam. *Drive-Thru Dreams: A Journey through the Heart of America's Fast-Food Kingdom.* New York, NY: Flatiron Books, 2019.

Cusick, Dawn. *Cool Animal Names.* Watertown, MA: Imagine Books, 2011.

Dilenschneider, Robert L. *Decisions: Practical Advice from 23 Men and Women Who Shaped the World.* New York, NY: Kensington Publishing, 2020.

Donald, Graeme. *The Long and the Short of It: How We Came to Measure Our World.* London: Michael O'Mara Books, 2016.

Eckstein, Bob. *The History of the Snowman: From the Ice Age to the Flea Market.* New York, NY: Simon Spotlight Entertainment, 2007.

Fagan, Abigail. "The Evolution of Hygiene," *Psychology Today*, Nov./Dec. 2019: 19.

Firpi, Jessica, editor. *Ripley's Believe It or Not! Presents Sideshow and Other Carnival Curiosities.* Orlando, FL: Ripley Entertainment Inc., 2020.

Fishman, Charles. *One Giant Leap*. New York, NY: Simon and Schuster, 2019.

Giblin, James Cross. *From Hand to Mouth; or, How We Invented Knives, Forks, Spoons, and Chopsticks & the Table Manners to Go with Them.* New York, NY: Thomas Y. Crowell, 1987.

Herman, Eleanor. *The Royal Art of Poison*. New York, NY: St. Martin's Press, 2018.

Langton, Jerry. *Rat: How the World's Most Notorious Rodent Clawed Its Way to the Top*. New York, NY: St. Martin's Press, 2006.

Levy, Joel. *Poison: An Illustrated History*. Guilford, CT: Globe Pequot Press, 2011.

Mask, Deirdre. *The Address Book: What Street Addresses Reveal about Identity, Race, Wealth, and Power*. New York, NY: St. Martin's Press, 2020.

Moberg, Julia. *Presidential Pets: The Weird, Wacky, Little, Big, Scary, Strange Animals That Have Lived in the White House*. Watertown, MA: Charlesbridge Publishing, 2012.

Morrison, Ellen Earnhardt. *Guardian of the Forest: A History of the Smokey Bear Program*. Alexandria, VA: Morielle Press, 1976, 1989.

Murrie, Matthew, and Steve Murrie. *The Screaming Hairy Armadillo and 76 Other Animals with Weird, Wild Names*. New York, NY: Workman Publishing, 2020.

O'Brien, Michael. "The US Cavalry Camel Corps." History, Dec./Jan. 2020: 40–42.

Peterson, Coyote. *The King of the Sting*. New York, NY: Little, Brown and Company, 2018.

Reid, Jack. *Roadside Americans: The Rise and Fall of Hitchhiking in a Changing Nation*. Chapel Hill, NC: University of North Carolina Press, 2020.

Rowan, Roy, and Brooke Janis. *First Dogs: American Presidents and Their Best Friends*. Chapel Hill, NC: Algonquin Books of Chapel Hill, 1997.

Smith, Andrew F. *The Tomato in America: Early History, Culture, and Cookery*. Columbia, SC: University of South Carolina Press, 1994.

Stein, Mark. *The Presidential Fringe: Questing and Jesting for the Oval Office*. Lincoln, NE: Potomac Books, 2020.

Stevenson, Katherine Cole, and H. Warn Jandl. *Houses By Mail: A Guide to Houses from Sears, Roebuck and Company*. New York, NY: John Wiley & Sons, 1986.

Stewart, Amy. *Wicked Bugs: The Meanest, Deadliest, Grossest Bugs on Earth*. New York, NY: Workman Publishing, 2017.

Sullivan, Robert. *Rats: Observations on the History & Habitat of the City's Most Unwanted Inhabitants*. New York, NY: Bloomsbury, 2004.

Tackach, James. *Lincoln and the Natural Environment*. Carbondale, IL: Southern Illinois University Press, 2019.

White, April. "Origins: On Hold: How a Telephone Monopoly and a Fear of Wiretapping Hung Up the Answering Machine for Decades." *Smithsonian*, Dec. 2019: p. 30.

Wiltshire, Patricia. *The Nature of Life and Death*. United States: G. P. Putnam's Sons, 2019.

Woolf, John. *The Wonders: The Extraordinary Performers Who Transformed the Victorian Age.* New York, NY: Pegasus Books, 2019.

WEBSITE ARTICLES

"1816: The Year Without a Summer." *New England Historical Society*, 6 Dec. 2020, www.newenglandhistoricalsociety.com/1816-year-without-a-summer/.

"1940s Stockings: Hosiery, Nylons, and Socks History." Vintage Dancer, 21 Aug. 2014, vintagedancer.com/1940s/1940s-stockings-history/.

Adams, Zoe. "25 Interesting Facts about Peas." *The Fact Site*, 17 Mar. 2021, www.thefactsite.com/interesting-pea-facts/.

"Albert Staehle, 74, Creafor (*sic*) of Smokey the Bear, Dies." *New York Times*, 5 April 1974, www.nytimes.com/1974/04/06/archives/albert-staehle-74-creator-of-smokey-the-bear-dies.html.

"Alleta Sullivan: A Mom Like No Other." The Sextant, Naval History and Heritage Command, 9 May 2014. https://usnhistory.navylive.dodlive.mil/2014/05/09/alleta-sullivan-a-navy-mom-like-no-other/.

Andersen, Charlotte Hilton. "What It Was Like Giving Birth In Every Decade Since the 1900s." *Redbook*, Redbook, 16 Oct. 2017, www.redbookmag.com/body/pregnancy-fertility/g3551/what-it-was-like-giving-birth-in-every-decade/.

Andersen, Charlotte Hilton. "What It Was Like Giving Birth In Every Decade Since the 1900s." *Redbook*, Redbook, 16 Oct. 2017, www.redbookmag.com/body/pregnancy-fertility/g3551/what-it-was-like-giving-birth-in-every-decade/.

Arbuckle, Alex. "The First Person to Go over Niagara Falls in a Barrel Was a 63-Year-Old Teacher." *Mashable*, Mashable, 30 Oct. 2016, mashable.com/2016/10/30/niagara-falls-barrel/.

Armstrong, Katherine. "Are You a Tetrachromat?" *Lenstore.co.uk*, 18 Jan. 2019. www.lenstore.co.uk/eyecare/are-you-a-tetrachromat.

Arney, Kat. "Methaemoglobin and Diaphorase 1." *Chemistry World*, Chemistry World, 27 Jan. 2020, www.chemistryworld.com/podcasts/methaemoglobin-and-diaphorase-1/3008547.article.

Bedewi, Jessica. "How Exactly Did Australia Become England's Penal Colony?" *Ranker*, 2 Jan. 2020. www.ranker.com/list/how-did-australia-become-british-penal-colony/jessica-bedewi.

Bellis, Mary. "Who Invented Crayola Crayons?" *ThoughtCo*, www.thoughtco.com/crayola-crayon-history-1991483.

Bellis, Mary. "How Was the Lawn Mower Invented?" *ThoughtCo*, 1 Mar. 2019, www.thoughtco.com/first-lawn-mower-1991636.

Bell-Murray, Sandy. "The History of the Greeting Card." *Northern Cards*, northerncards.com/blogs/nc/the-history-of-the-greeting-card.

Blakemore, Erin. "Meet Stagecoach Mary, the Daring Black Pioneer Who Protected Wild West Stagecoaches." *History.com*, A&E Television Networks, 14 Sept. 2017, www.history.com/news/meet-stagecoach-mary-the-daring-black-pioneer-who-protected-wild-west-stagecoaches.

Blauvelt, Christian. "Strange Tales of the Vanished Oscars." *BBC Culture*, BBC, 6 Mar. 2018, www.bbc.com/culture/story/20180305-strange-tales-of-the-vanished-oscars.

Blitz, Matt. "Who—or What—Was the First Sports Mascot and How Did the Practice Start?" *Today I Found Out*, 30 Sept. 2016, www.todayifoundout.com/index.php/2016/09/first-sports-mascot/.

Blumenthal, Ralph, and Leslie Kean. "No Longer in the Shadows, Pentagon's UFO Unit Will Make Some Findings Public." *Baltimoresun.com*, Baltimore Sun, 23 July 2020, www.baltimoresun.com/news/nation-world/ct-nw-nyt-pentagon-ufo-unit-20200723-b3akzzy44zdgxc3bmhgko6nkgm-story.html.

Boboltz, Sara. "A Brief Yet Complex Color History of Crayola Crayons." *HuffPost*, HuffPost, 7 Dec. 2017, www.huffpost.com/entry/crayola-crayon-color-history_n_7345924.

"Bohemian Grove: The Secret Society Summer Camp." MessyNessyChic.com, 25 May 2016, www.messynessychic.com/2016/05/25/bohemian-grove-the-secret-society-summer-camp/.

Bradford, Alina. "History of Fireworks." *LiveScience*, Purch, 30 Aug. 2018, www.livescience.com/63468-fireworks-history.html.

Bradford, Alina. "Kangaroo Facts." *LiveScience*, Purch, 2 Mar. 2016, www.livescience.com/27400-kangaroos.html.

Breiding, Dirk H. "Arms and Armor—Common Misconceptions and Frequently Asked Questions." Oct. 2004, *Metmuseum.org*, www.metmuseum.org/toah/hd/aams/hd_aams.htm.

Bridget. "Slaves to Fashion: A Brief History and Analysis of Women's Fashion in America." *Bellatory*, Bellatory, 17 Oct. 2014, bellatory.com/fashion-industry/A-Brief-History-of-American-Womens-Fashion.

Britton, Bianca. "Frank Hayes: The Jockey Who Won a Race despite Being Dead." *CNN*, Cable News Network, 10 Dec. 2018, www.cnn.com/2018/12/10/sport/frank-hayes-sweet-kiss-belmont-park-intl-spt/index.html.

Brockell, Gillian. "Jackie Kennedy's Fairy-Tale Wedding Was a Nightmare for Her African American Dress Designer." *The Washington Post*, WP Company, 1 Sept. 2019, www.washingtonpost.com/history/2019/08/28/jackie-kennedys-fairy-tale-wedding-was-nightmare-her-african-american-dress-designer/.

Brooks, Rebecca Beatrice. "How the Civil War Broke Chang and Eng Bunker." *Civil War Saga*, 28 July 2018, civilwarsaga.com/how-the-civil-war-broke-chang-and-eng-bunker/.

Buchanan, Matt. "Object of Interest: The Flash Drive." *The New Yorker*, The New Yorker, 6 July 2017, www.newyorker.com/tech/annals-of-technology/object-of-interest-the-flash-drive.

Bump, Philip. "The Breathtaking Scale of Santa Claus's Task on Christmas Eve." *The Washington Post*, 20 Dec. 2019, www.washingtonpost.com/lifestyle/2019/12/20/breathtaking-scale-santa-clauss-task-christmas-eve/.

Caballero, Gene. "The History of Lawns: The Complete History of Lawns (Illustrated)." GreenPal, 8 Oct. 2019, www.yourgreenpal.com/blog/every-wonder-about-the-history-of-lawns.

Cabral, Carrie. "Who Are the Blue People of Kentucky? Why are They Blue?" PrepScholar.com, 2 Nov. 2019, blog.prepscholar.com/blue-people-of-kentucky.

Cartwright, Mark. "The Armour of an English Medieval Knight." *Ancient History Encyclopedia*, 13 June 2018, www.ancient.eu/article/1244/the-armour-of-an-english-medieval-knight/.

Chaikin, Andrew. "Neil Armstrong's Spacesuit Was Made by a Bra Manufacturer." Smithsonian.com, Nov. 2013, www.smithsonianmag.com/history/neil-armstrongs-spacesuit-was-made-by-a-bra-manufacturer-3652414/.

Chalakoski, Martin, "Bobby Leach, First Man to Survive Niagara Falls Barrel Plunge, Died after Slipping on an Orange Peel." Vintage News, 19 Dec. 2017, www.thevintagenews.com/2017/12/19/bobby-leach-niagara-falls.

Champion, Leo. "The History of USB Flash Drives." *Techwalla*, www.techwalla.com/articles/the-history-of-usb-flash-drives.

Chan, Amy. "Camels Go West: Forgotten Frontier Story." *HistoryNet*, HistoryNet, 20 Sept. 2018, www.historynet.com/camels-go-west-for gotten-frontier-story.htm.

Cherry, Kendra. "The Color Psychology of Black." Verywellmind.com, 20 September 2019, www.verywellmind.com/the-color-psychology-of-black-2795814.

Chua, Jasmine Malik. "The Girdle-Inspired History of the Very First Spacesuits." Racked.com, 5 September 2018, www.racked.com/2018/9/5/17771270/spacesuit-girdles-playtex-seamstresses-nasa.

CityFursuits. "What Does it Feel Like to Be Wearing a Mascot Costume?" Reddit, 2016, www.reddit.com/r/questions/comments/33w7bn/what_does_it_feel_like_to_be_wearing_a_mascot/.

Cleaning Institute. "2009 National Clean Hands Report Card Survey Findings." www.cleaninginstitute.org/newsroom/surveys/92109-sum mary.

Cohen, Jennie. "Fireworks Vibrant History." History.com, 3 July 2019, www.history.com/news/fireworks-vibrant-history.

Colleary, Eric. "Apples in America: A Very Brief History." American Table, 31 Oct. 2011, www.americantable.org/2011/10/apples-in-america.

Collins, Mike, 2017. "The Theremin Explained." Doctor Mix, 25 July 2017, www.doctormix.com/blog/the-theremin-explained.

Cook, Rosie. "Chemistry at Play." Science History Institute, 6 April 2010, www.sciencehistory.org/distillations/chemistry-at-play.

Covino, Elizabeth. "State Funerals: Planning the Last Wishes of a President." JWU Online, 3 Dec. 2018, online.jwu.edu/blog/planning-state-funerals.

Crew, Bec. "Where Was the Last Place on Earth Discovered by Humans?" ScienceAlert.com, 28 Oct. 2016, www.sciencealert.com/where-was-the-last-place-on-earth-discovered-by-humans.

Curtis, Valerie A. "A Natural History of Hygiene." Pulsus Group, Jan. 2007, www.ncbi.nlm.nih.gov/pmc/articles/PMC2542893/.

Cutlip, Kimbra. "How 75 Years Ago Nylon Stockings Changed the World." Smithsonianmag.com, 11 May 2015, www.smithsonianmag .com/ smithsonian-institution/how-75-years-ago-nylon-stockings-changed-world-180955219.

Dalzell, Rebecca. "The Spirited History of the American Bar." Smithsonian.com, 2 Aug. 2011, www.smithsonianmag.com/history/the-spirited-history-of-the-american-bar-42912195/.

"Dan's Definitive History of Houseplants." Medium.com, 7 July 2017, medium.com/welltended/dans-definitive-history-of-houseplants-fca56a07b0cf.

Dash, Mike, 2013. "For 40 Years, This Russian Family Was Cut Off from All Human Contact, Unaware of World War II." Smithsonian.com, 28 Jan. 2013, www.smithsonianmag.com/history/for-40-years-this-russian-family-was-cut-off-from-all-human-contact-unaware-of-world-war-ii-7354256/.

Davis, Nancy, and Amelia Grabowski. "Sewing for Joy: Ann Lowe." National Museum of American History, 12 Mar. 2018, americanhistory.si.edu/blog/lowe.

D'Costa, Krystal. "The American Obsession with Lawns." *Scientific American*, 3 May 2017. blogs.scientificamerican.com/anthropology-in-practice/the-american-obsession-with-lawns.

Debczak, Michele. "The Tumultuous History of Tinsel." mentalfloss.com, 5 Dec. 2019, www.mentalfloss.com/article/609741/christmas-tree-tinsel-history.

Distasio, Steph, and Colton Kruse. "Frank Hayes: The Dead Man Who Won a Horse Race." *Ripley's Believe It or Not!*, 29 May 2019, www.ripleys.com/weird-news/frank-hayes/.

Drillinger, Meagan. "Forbidden Destinations You Can Never Visit." FarandWide.com, 12 April 2018, www.farandwide.com/s/forbidden-destinations-398b80e7688c4c77.

Duffy, Jim. "A Frederica Story: From Ladies Undergarments to the Man on the Moon." Secrets on the Eastern Short, 17 April 2018, www.secretsoftheeasternshore.com/frederica-moonsuit/.

Dunbar-Ortiz, Roxanne, and Dina Gilio-Whitaker. "What's behind the Myth of Native American Alcoholism?" *Pacific Standard*, 14 June 2017, psmag.com/news/whats-behind-the-myth-of-native-american-alcoholism.

EarthSky Voices in Space. "Some Black Holes Erase Your Past." EarthSky.org, 25 Feb. 2018. earthsky.org/space/some-black-holes-erase-your-past.

Eckels, Carla. "The Curious Case of a Missing Academy Award." NPR.org, 22 Feb. 2009, www.npr.org/templates/story/story.php?storyId=100937570.

Edmonds, Molly, and Joseph Miller. "10 Different Types of Laughter." HowStuffWorks.com. Accessed 5 Nov. 2020, science.howstuffworks .com/life/inside-the-mind/emotions/5-types-of-laughter.htm.

Edwards, Aaron. "7 Ways Medieval Knight Armor Was More Dangerous Than Just Wearing Nothing." Ranker.com, 30 Mar. 2020, www.ranker .com/list/the-dangers-of-wearing-a-medieval-knight-suit-of-armor/ aaron-edwards.

"Everything You Need to Know about July 1 Moving Day in Montreal." *Montrealgazette*, Montreal Gazette, 26 June 2019, montrealgazette .com/news/local-news/everything-you-need-to-know-about-july-1- moving-day-in-montreal.

Ewing, Jake. "The Graphite Pencil." BBC: A History of the World, www .bbc.co.uk/ahistoryoftheworld/objects/AKHAcJoNQQ-DaYEz1dvZ9Q.

Everts, Sarah. "A Brief History of Chemical War." Science History Institute, 11 May 2015, www.sciencehistory.org/distillations/a-brief-his tory-of-chemical-war.

Faulkner, Claire A. "A Presidential Funeral." The White House Historical Foundation, www.whitehousehistory.org/a-presidental-funeral. Accessed Nov. 2, 2020.

Ferrell, Ciara. "#61 Turnspit Dogs." foodnonfiction.com, 28 Jan. 2017, www.foodnonfiction.com/episodes/item/15-61-turnspit-dogs.

Finch, Bill. "The True Story of Kudzu, The Vine That Never Truly Ate the South." *Smithsonian*, September 2015., www.smithsonianmag .com/science-nature/true-story-kudzu-vine-ate-south-180956325.

Flaccus, Gillian. "10 Pioneer-Era Apple Types Thought Extinct Found in US West." Associated Press, 15 April 2020, abcnews.go.com/Lifestyle/ wireStory/10-pioneer-era-apple-types-thought-extinct-found-70156 282.

Franz, Justin. "For 40 Years, Crashing Trains Was One of America's Favorite Pastimes." Atlas Obscura, 1 July 2019, www.atlasobscura.com/ articles/staged-train-wrecks.

Frater, Jamie. "Top 10 Places You Don't Want to Visit." ListVerse.com, 16 June 2014, listverse.com/2010/03/22/top-10-places-you-dont-want-to-visit.

"From Divorces to Beheadings, Here Are the Grim Fates of Henry VIII's Wives." *All That's Interesting*, 5 Feb. 2021, allthatsinteresting.com/henry-viiis-wives.

Frontier Gap. "10 Disastrous Consequences of Humans Importing Invasive Species." The Dodo, 20 Jan. 2015, www.thedodo.com/invasive-species-wreaking-havo-941016023.html.

Fulton, Wil, 2016. "From Chopsticks to Sporks: A Brief History of Eating Utensils." thrillist.com, 8 Jan. 2016, www.thrillist.com/eat/nation/history-of-chopsticks-forks-spoons-and-sporks-who-invented-the-spoon-who-invented-the-fork.

"Fun Facts about Your Tongue and Taste Buds." Physician's Review Network (PRN), onhealth.com, 21 June 2016, www.onhealth.com/content/1/tongue_facts.

Gaia Staff. "America's Paranormal Highway: The 37th Parallel." Gaia.com, 14 June 2018, www.gaia.com/article/37th-parallel-americas-paranormal-highway.

Garvey, Kathy Keatley. "The First Native American Honey Bee." University of California Agriculture & Resources, 28 July 2009, ucanr.edu/blogs/blogcore/postdetail.cfm?postnum=1544.

Gault, Matthew. "How to Survive a Nuclear Bomb." Vice.com, 9 Jan. 2020, www.vice.com/en_us/article/qjd8bq/how-to-survive-a-nuclear-bomb.

Gecko Hospitality. "Trends of the Hospitality Industry: Hotels: 100 Years of Trends in the Hospitality Industry: Hotel Edition." Gecko Hospitality, 4 Jan. 2017, www.geckohospitality.com/2017/01/04/trends-hospitality-industry-hotels/.

Gefter, Amanda. "The Strange Fate of a Person Falling into a Black Hole." BBC.com, 25 May 2015, www.bbc.com/earth/story/20150525-a-black-hole-would-clone-you.

Gershon, Livia. "A Brief History of U.S. Drinking." JStorDaily, 12 Aug. 2016, daily.jstor.org/a-brief-history-of-drinking-alcohol/.

Gibbens, Sarah. "A Brief History of How Plastic Straws Took Over the World." National Geographic Australia, 9 July 2018, www.nationalgeographic .com.au/nature/a-brief-history-of-how-plastic-straws-took-over-the-world.aspx.

Gibson, Brittany. "Yes, There's Actually a Difference between the Terms "Dinner" and "Supper"." Reader'sDigest.com https://www.rd.com/culture/difference-between-dinner-and-supper/.

Glamorously Vintage. "Kitchen Colors—Colors Through the Years 1950, 1960 and 1970." *Glamorously Vintage*, 15 Jan. 2017, glamorouslyvintage.com/kitchen-colors-colors-through-the-years-1950-1960-and-1970/.

Gorman, Alice. "60 Years in Orbit for 'Grapefruit Satellite'—the Oldest Human Object in Space." *The Conversation*, 8 June 2020, theconver sation.com/60-years-in-orbit-for-grapefruit-satellite-the-oldest-human-object-in-space-93640.

Gornstein, Leslie. "Crazy Things to Do with Your Body after You Die (pictures)." C/Net.com, 7 July 2015, www.cnet.com/pictures/crazy-things-you-can-do-with-your-body-after-you-die/.

Greenfieldboyce, Nell. "Swarming Up a Storm: Why Animals School and Flock." NPR.org, 17 Aug. 2012, www.npr.org/2012/08/17/158931 963/swarming-up-a-storm-why-animals-school-and-flock.

Grimes, Gerlinda. "How a Theremin Works." HowStuffWorks.com, 25 July 2011, electronics.howstuffworks.com/gadgets/audio-music/ther emin.htm.

Guarino, Gen. "Speaking of Science: Home Is Where the Tongue Is." *Washington Post*, 25 Mar. 2020, khgkgkgkj.blogspot.com/2020/03/ speaking-of-science.html.

Hartman, Mitchell. "Here's How Much Money There is in the World—and Why You've Never Heard the Exact Number." Marketplace.org, 17 Nov. 2017, www.businessinsider.com/heres-how-much-money-there-is-in-the-world-2017-10.

Hayes, Adam. "Skirt Length (Hemline) Theory." Investopedia, 30 Sept. 2019, www.investopedia.com/terms/s/skirtlengththeory.asp.

Haze, Jackie "10 Strangest New Year's Eve Traditions." ListVerse.com, 30 Dec. 2019, listverse.com/2019/12/30/10-strangest-new-year39s-eve-traditions.

Hendrix, Steve. "Mail That Baby: A Brief History of Kids Sent Through the U.S. Postal Service'" *Washington Post*, 24 May 2017, www.washing tonpost.com/news/retropolis/wp/2017/05/24/mail-that-baby-a-brief-history-of-kids-sent-through-the-u-s-postal-service/.

Hill, Selena. "Black History Month: The Son of a Slave Who Ran for President, George Edwin Taylor." BlackEnterprise.com, 3 Feb. 2020, www .blackenterprise.com/black-history-month-son-slave-ran-president/.

Hiskey, Daven. "What Was It Actually Like to Be a Knight in Medieval Times?" TodayIFoundOut.com, 11 Mar. 2020, www.todayifoundout .com/ index.php/2020/03/what-was-it-actually-like-to-be-a-knight-in-medieval-times/.

"History." *GCA*, 14 Feb. 2018, www.greetingcard.org/industry-resources/ history/. Accessed 1 Mar. 2020.

"A History of Valentine's Day Celebrations—from Fertility Festivals to the First Cards." *HistoryExtra*, 10 Feb. 2021, www.historyextra.com/period/modern/a-brief-history-of-valentines-day-cards/.

Howell, Helen Murphy. "A History of Humanity's Disgusting Hygiene." Owlcation, 27 Mar. 2013, owlcation.com/humanities/A-History-of-Dirty-Habits.

Hutton, Paul Andrew. "Camels Go West: Forgotten Frontier Story," HistoryNet, Dec. 2007, www.historynet.com/camels-go-west-forgotten-frontier-story.htm.

"Inside the Global Seed Vault, Where the History And Future of Agriculture Is Stored." *NPR*, NPR, 24 July 2017, www.npr.org/2017/07/24/539005688/inside-the-global-seed-vault-where-the-history-and-future-of-agriculture-is-stor.

"Is Time Travel Possible?" NASA.gov, 30 April 2020. https://spaceplace.nasa.gov/time-travel/en/.

James, Susan Donaldson. "Fugates of Kentucky: Skin Bluer Than Lake Louise." ABC News, 21 Feb. 2012, abcnews.go.com/Health/blue-skinned-people-kentucky-reveal-todays-genetic-lesson/story?id=15759819.

Jenkins, Beverly L. "The Inspiring True Story of Franklin, Peanuts' 1st Black Character." Inspiremore.com, 2 Aug. 2018, www.inspiremore.com/charles-schulz-franklin/.

Johnson, Elizabeth Ofosuah. "Meet the Little-Known African-American Journalist Who Ran for President in 1904." Face2FaceAfrica.com, 13 Feb. 2019, face2faceafrica.com/article/meet-the-little-known-african-american-journalist-who-ran-for-president-in-1904.

Johnson, Eric. "Black Color Meaning: 10 Fascinating Facts about Our Darkest Color—Typeform Blog." *Go to Typeform.com*, www.typeform.com/blog/ask-awesomely/black-facts.

Johnson, Scott. "The Truth Behind the Infamous 2000 Oscar Heist." Hollywood Reporter.com, 21 Feb. 2017, www.hollywoodreporter.com/news/truth-behind-infamous-2000-oscar-heist-977085.

Johnson, Thomas J. *A History of Biological Warfare from 300 B.C.E. to the Present*, c.aarc.org/resources/biological/history.asp. Accessed September 13, 2020.

Jones, Josh. "When Sears Sold Houses, Board by Board." Open Culture, 29 Oct. 2018, www.openculture.com/2018/10/sears-sold-75000-mail-order-homes-1908-1939-transformed-life-america.html.

Jones, Tegan. "The History of Knives, Forks and Spoons." Gizmodo.com, 3 Oct. 2013, gizmodo.com/the-history-of-knives-forks-and-spoons-1440558371.

Kelly, Debra. "10 Remarkable Facts about Animals Put in Human Perspective." ListVerse.com, 8 Dec. 2019, listverse.com/2019/12/08/10-remarkable-facts-about-animals-put-in-human-perspective/.

Kennedy, Bryan. "The Fearless Character of One Eyed Charley." Outhistory.org, outhistory.org/exhibits/show/tgi-bios/charley-parkhurst.

King, Gilbert, 2012. "Fritz Haber's Experiments in Live and Death." *Smithsonian*, 5 June 2012, www.smithsonianmag.com/history/fritz-habers-experiments-in-life-and-death-114161301/.

Kinney, Pat, and Meta Hemenway-Forbes. "Brother's Deaths 'Took a Toll' on Surviving Sister." *Waterloo-Cedar Falls Courier*, 10 Nov. 2017, wcfcourier.com/sullivanbrothers/brothers-deaths-took-a-toll-on-surviving-sister/article_6c99042a-e23c-5540-a187-7c2c225564f2.html.

Knight, Daniel. "Personal Computer History: 1975–1984." Low End Mac, 26 April 2014, lowendmac.com/2014/personal-computer-history-the-first-25-years/.

Knight, Eliza. "History of Hygiene: Bathing, Teeth Cleaning, Toileting & Deodorizing." historyundressed.com, 14 July 2008, www.historyundressed.com/2008/07/history-of-hygiene-bathing-teeth.html.

Knight, Eliza. "History of Socks and Stockings." HistoryUndressed.com, 1 Dec. 2008, www.historyundressed.com/2008/12/history-of-socks-and-stockings.html.

Lacaze, Katherine. "History and Hops: How Women Played a Key Role in Exploration of the West." SeasideSignal.com, 11 May 2019, www.seasidesignal.com/news/local-news/history-and-hops-how-women-played-a-key-role-in/article_cacc68c8-727f-11e9-a76f-9b1bc04be402.html.

Laskow, Sarah. "A Machine that Made Stockings Helped Kick off the Industrial Revolution." Atlas Obscura, 19 September 2017, www.atlasobscura.com/articles/machine-silk-stockings-industrial-revolution-queen-elizabeth.

Lee, Amber. "How Sports Teams Got Their Names." Bleacher Report, 7 Aug. 2015, bleacherreport.com/articles/2538917-how-sports-teams-got-their-names.

Lefler, Leah. "Blue People of Kentucky: Why the Fugate Family Had Blue Skin." Owlcation, 11 Aug. 2018, owlcation.com/humanities/Blue-People-in-Kentucky-A-True-Story-of-a-Family-with-Blue-Skin.

Levy-Bonvin, Jacques. "Hotels." Hospitality.net, 15 Dec. 2003, www.hos pitalitynet.org/opinion/4017990.html.

Lewis, Danny. "A Brief History of Children Sent through the Mail." Smithsonian.com, 21 Dec. 2016, www.smithsonianmag.com/smart-news/brief-history-children-sent-through-mail-180959372/.

Lewis, Jamie. "Peeling Back the Bark." Forest History Society, 14 Dec. 2016, fhsarchives.wordpress.com/2016/12/14/president-bans-christ mas-tree-from-white-house-cites-environmental-concerns/.

Lisa, Andrew. "10 Reasons Not to Play the Lottery." Cheapism.com, 3 September 2019, blog.cheapism.com/reasons-not-play-lottery/.

Longley, Robert. "When It Was Legal to Mail a Baby." ThoughtCo.com, 27 Nov. 2017, www.thoughtco.com/when-it-was-legal-mail-babies-3321266.

"The Lottery: Is It Ever Worth Playing?" Investopedia.com, 27 Jan. 2020. https://www.investopedia.com/managing-wealth/worth-playing-lot tery/.

Lindeman, Tracey. "In Montreal, 70,000 Households Move on the Same Day." Bloomberg City Lab, 28 June 2019, www.citylab.com/life/2019/06/montreal-moving-day-history-housing-apartments-rent-holiday/591865/.

Little, Becky. "Nothing Says 'I Hate You' Like a Vinegar Valentine." Smithsonianmag.com, 10 Feb. 2017, www.smithsonianmag.com/history/nothing-says-i-hate-you-vinegar-valentine-180962109/.

Little, Becky. "When People Used the Postal Service to Mail Their Children." History.com, 9 April 2019, www.history.com/news/mailing-chil dren-post-office.

Malone, Noreen. "What Happens If You Fall into a Black Hole?" Slate.com, 9 September 2008, slate.com/news-and-politics/2008/09/what-happens-if-you-fall-into-a-black-hole.html.

Manca, Davide. "Shavarsh Karapetyan, Maybe Not Everybody Knows That . . ." Finswimmer.com, 30 Mar. 2019, www.finswimmer.com/sha varsh-karapetyan-maybe-not-everybody-knows-that/.

Martin, John. "Meet the Last Lykov." Vice.com, 1 April 2013, www.vice .com/en_us/article/dp4mzj/meet-the-last-lykov-000001-v20n4.

Martin, Laura Jane. "The Secret (and Ancient) Lives of Houseplants." YourWildLife.org, 21 April 2013, yourwildlife.org/2013/04/the-secret-and-ancient-lives-of-houseplants/.

McCarthy, Erin. "The Reason Elections are Held on Tuesdays." mental floss.com, 4 Nov. 2019, www.mentalfloss.com/article/12901/why-are-elections-held-tuesdays.

McGraw, Eliza. "The Kentucky Derby's First Female Jockey Ignored Insults and Boycott Threats. She Just Wanted to Ride." *Washington Post*, 4 May 2019, www.washingtonpost.com/history/2019/05/04/kentucky-derbys-first-female-jockey-ignored-insults-boycott-threats-she-just-wanted-ride/.

Mcleod, Saul. "Nature vs. Nurture in Psychology." *Nature Nurture in Psychology | Simply Psychology*, www.simplypsychology.org/naturevsnur ture.html.

Mersereau, Dannis. "15 Facts about The Year Without a Summer." Mentalfloss.com, 15 Jan. 2016, www.mentalfloss.com/article/73585/15-facts-about-year-without-summer.

Minutaglio, Rose. "Ann Lowe Is the Little-Known Black Couturier Who Designed Jackie Kennedy's Iconic Wedding Dress." Elle.com, 12 Sept ember 2019, www.elle.com/fashion/a29019843/jackie-kennedy-wed ding-dress-designer-ann-lowe/.

Mirzoyan, Gayane. "Twenty-Five Seconds Per Life." *Aurora Humanitarian Initiative*, auroraprize.com/en/aurora/article/heroes/9235/twenty five-seconds-per-life. Accessed July 20, 2020.

Mitchell, Nancy. "What Ever Happened to Colored Toilet Paper?" ApartmentTherapy.com, 24 Mar. 2018, www.apartmenttherapy.com/col ored-toilet-paper-history-255476.

Montanaro, Domenico. "Why Do We Vote on Tuesdays?" NPR.org, 1 Nov. 2016, www.npr.org/2016/11/01/500208500/why-do-we-vote-on-tuesdays.

Moore, Christine. "Fun Facts about Horse Racing at Your Next Trivia Night." America's Best Racing, 4 Jan. 2017, www.americasbestracing .net/lifestyle/2017-fun-facts-about-horse-racing-your-next-trivia-night.

Morgensen, Jackie Flynn. "Whoopi Goldberg's Oscar Was Once Found in an Airport Trash Can—And Other Insane Oscar Stories." MotherJones.com, Mar./Apr. 2018, www.motherjones.com/media/2018/03/whoopi-goldbergs-oscar-was-once-found-in-an-airport-trash-can-and-frances-mcdormand/.

Morrison, David. "Invasion of the Bunny Rabbits: What Happens When Non-Native Species Conquer Their New Territories." Urbo.com, 24 July 2018, www.urbo.com/content/what-happens-when-non-native-species-conquer-their-new-territories/.

Murphy, Pauline. "Flogging a Dead Jockey: The Bizarre Death of Frank Hayes." Headstuff.org, 17 July 2018, www.headstuff.org/culture/history/1900-present/flogging-dead-jockey-frank-hayes/.

Nalewicki, Jennifer. "Half of the Inhabitants of This Australian Opal Capital Live Underground." Smithsonianmag.com, 3 Mar. 2016, www.smithsonianmag.com/travel/unearthing-coober-pedy-australias-hidden-city-180958162/.

Natanson, Hannah. "Lincoln's Forgotten Legacy as America's First 'Green President.'" *Washington Post*, 16 Feb. 2020, www.washingtonpost.com/history/2020/02/16/lincoln-green-president-environmentalist/.

Nicholson, Sibel. "The Moon Is Covered with 400,000 Pounds of Human Trash." Interesting Engineering, 1 Feb. 2018, interestingengineering.com/the-moon-is-covered-with-400000-pounds-of-human-trash.

Nix, Elizabeth. "9 Things You Might Not Know about 'Peanuts.'" History.com, 22 Aug. 2018, www.history.com/news/9-things-you-might-not-know-about-peanuts.

Nosowitz, Dan. "What Exactly Are Africanized Bees, and How Scary Are They?" Modern Farmer, 16 June 2016, modernfarmer.com/2016/06/africanized-bees/.

O'Brien, Jane. "The Time When Americans Drank All Day Long." BBC News, 9 Mar. 2015, www.bbc.com/news/magazine-31741615.

O'Connor, Brendan. "Inside the Surprisingly Dark World of Rube Goldberg Machines." *The Verge*, 22 Apr. 2015, www.theverge.com/2015/4/22/8381963/rube-goldberg-machine-contest-history-ideas.

O'Dell, Nick. "When the First Clock Was Invented, How Was It Known What the Actual Time Was? How Was the Clock Set?" Quora.com, 1 September 2019, www.quora.com/When-the-first-clock-was-invented-how-was-it-known-what-the-actual-time-was-How-was-the-clock-set.

O'Donnell, Noreen. "Jockeys Urge Higher Weight Restrictions to Live Healthier." NBCWashington.com, 18 June 2018, www.nbcwashington.com/news/national-international/jockeys-urge-higher-weight-rules-to-live-healthier/2015567/.

Oliver, Mark. "25 Dangerous Animals That Would Mess Up Any Human." *All That's Interesting*, All That's Interesting, 13 Apr. 2018, allthatsinteresting.com/dangerous-animals.

O'Neill, Tim. "How the Middle Ages Really Were." Huffington Post, 6 Dec. 2017, www.huffpost.com/entry/how-the-middle-ages-reall_b_5767240.

Ouellette, Jennifer. "The Explosive Physics of Pooping Penguins: They Can Shoot Poo Over Four Feet." ARS Technica, 4 July 2020, arstechnica.com/science/2020/07/poopy-projectiles-penguins-can-fling-their-feces-over-four-feet-study-finds/.

Palmer, G. D. "Colors of Appliances in the 1970s." HomeSteady, 21 July 2017, homesteady.com/info-7804189-70s-refrigerator-colors.html.

Pellowski, Michael J., and Bryan Wendell. "14 Funny and Unusual College and Pro Sports Mascots." *Scout Life Magazine*, 21 Jan. 2020, boyslife.org/features/23227/unusual-college-sports-mascots.

"Peter the Anteater." UC Irvine Sports, 26 May 2017, ucirvinesports.com/sports/2017/5/26/anteater-index-html.aspx.

Poppick, Susie. "How to Survive a Nuclear Bomb: 3 Steps to Save You in Case of a Missile or Attack by North Korea." Mic.com, 12 Jan. 2018, www.mic.com/articles/187306/how-to-survive-a-nuclear-bomb-or-north-korea-war-3-steps-to-save-you-from-the-nuclear-blast-and-fallout-if-north-korea-attacks.

Pullin, Kate. "Greeting Card Industry Facts and Figures." TheSpruceCrafts.com, 27 Jan. 2019, www.thesprucecrafts.com/greeting-card-industry-facts-and-figures-2905385.

Ramendranath, Kajal, Chetana Ramesh Ratnaparkhi, Bapuji Shrawan Gedam, and Kushal Ashok Tayade. "An Unusual Case of Retained Abdominal Pregnancy for 36 years in a Postmenopausal Woman." U.S. National Library of Medicine, National Institutes of Health, September-Dec. 2015, www.ncbi.nlm.nih.gov/pmc/articles/PMC4606584/.

Rampton, John. "The History of the Calendar." Calendar.com, Jan. 2020, www.calendar.com/history-of-the-calendar/.

Raymond, Chris. "U.S. Presidential Funeral Traditions." Funeral Help Center, 1 Dec. 2018, www.funeralhelpcenter.com/us-presidential-funeral-traditions/.

"The Real Story Behind the Presidential Turkey Pardon." National Constitution Center, 27 Nov. 2019, constitutioncenter.org/blog/the-real-story-behind-the-presidential-turkey-pardon.

Reay, David. "The German Dr. Evil Who Invented Chemical Warfare." Handelsblatt Today, 20 Apr. 2018, www.handelsblatt.com/today/politics/fritz-haber-the-german-dr-evil-who-invented-chemical-warfare/23581908.html.

Reevell, Patrick. "Siberian Man Claims to Have Created the World's Smallest Book." ABCNews.com, 1 Mar. 2016, abcnews.go.com/International/siberian-man-claims-created-worlds-smallest-book/story?id=37306465.

"Remembering Potoooooooo, the Racehorse with the Best Name Ever." *Atlas Obscura*, Atlas Obscura, 17 Apr. 2019, www.atlasobscura.com/places/potoooooooo.

Renault, Marion. "Reading the Past in Old, Urine-Caked Rat's Nests." *New York Times*, 20 Feb. 2020, www.nytimes.com/2020/02/20/science/packrat-nests-dna.html.

Rhodes, Jesse. "Unwrapping the History of the Doggie Bag." *Smithsonian*, 25 Jan. 2011, www.smithsonianmag.com/arts-culture/unwrapping-the-history-of-the-doggie-bag-28056680/.

Rice, Doyle. "200 Years Ago, We Endured a 'Year without a Summer.'" *USA Today*, 9 June 2016, www.usatoday.com/story/weather/2016/05/26/year-without-a-summer-1816-mount-tambora/84855694.

Roberts, Amy. "A By-the-Numbers Look at Valentine's Day." CNN.com, 14 Feb. 2019, www.cnn.com/2018/02/14/us/valentines-by-the-numbers-trnd/index.html.

Rocheleau, Jackie. "50 Famous Firsts from Science History." *The Stacker,* 20 Feb. 2020, thestacker.com/stories/3959/50-famous-firsts-science-history.

Romero, Frances. "A Brief History of U.S. Mail Delivery." *Time*, 15 Mar. 2010, content.time.com/time/magazine/article/0,9171,1969717,00.html.

Roser, Max, Hannah Ritchie, and Bernadeta Dadonaite. "Child & Infant Mortality." Ourworldindata.org, Nov. 2019, ourworldindata.org/child-mortality.

Rossen, Jake. "14 Colorful Facts about Crayola." MentalFloss.com, 31 Mar. 2017, www.mentalfloss.com/article/65779/13-colorful-facts-about-crayola.

Rutigliano, Olivia. "On the Trail of Hollywood's Stolen Oscars." CrimeReads.com, 5 Feb. 2020, crimereads.com/stolen-oscars/.

Sawe, Benjamin Elisha. "Dry Counties of the United States." World Atlas, 25 Apr. 2017, www.worldatlas.com/articles/dry-counties-of-the-united-states.html.

Schmucker, Kristine. "From Innovation to Laughable Derelict: 8-Track Tapes." Harvey County Historical Museum, 25 Sept. 2015, hchm.org/8-track-tapes/.

Scott, Maiken. "How Did Birth Move from the Home to the Hospital, and Back Again?" WHYY.org, 13 Dec. 2013, whyy.org/segments/how-did-birth-move-from-the-home-to-the-hospital-and-back-again/.

Semeniuk, Ivan, "DNA Deepens Mystery of Newfoundland's Lost Beothuk People." *Globe and Mail,* 12 Oct. 2017, www.theglobeandmail.com/news/national/dna-deepens-mystery-of-newfoundlands-lost-beothuk-people/article36560469/.

Shapira, Ian. "She Was the Last American to Collect a Civil War Pension—$73.13 a Month. She Just Died." *Washington Post*, 4 June 2020, www.washingtonpost.com/history/2020/06/04/she-was-last-american-collect-civil-war-pension-7313-month-she-just-died/.

"She Became History's Youngest Mother—When She Was Just 5 Years Old." *All That's Interesting*, All That's Interesting, 18 Feb. 2021, allthatsinteresting.com/lina-medina.

Sisson, Patrick. "How Sears Kit Homes Changed Housing." Curbed.com, 16 Oct. 2018, www.curbed.com/2018/10/16/17984616/sears-catalog-home-kit-mail-order-prefab-housing.

Smallwood, Karl. "Why the Dodo Went Extinct." Todayifoundout.com, 28 June 2013, www.todayifoundout.com/index.php/2013/06/why-the-dodo-went-extinct/.

Smith, Ernie. "A Brief History of the Modern-Day Straw, the World's Most Wasteful Commodity." Atlas Obscura, 7 July 2017, www.atlasobscura.com/articles/straws-history.

Smith, Heather R. "What Is a Black Hole?" NASA.gov, 21 Aug. 2018, www.nasa.gov/audience/forstudents/k-4/stories/nasa-knows/what-is-a-black-hole-k4.html.

Smith, K. Annabelle. "Why the Tomato Was Feared in Europe for More Than 200 Years." *Smithsonian*, 18 June 2013, www.smithsonianmag.com/arts-culture/why-the-tomato-was-feared-in-europe-for-more-than-200-years-863735/.

Song, Juliet. "Surprisingly Advanced Ways the Ancient Chinese Bathed and Did Laundry." *Epoch Times,* 21 Mar. 2016, www.theepochtimes.com/surprisingly-advanced-ways-the-ancient-chinese-bathed-and-did-laundry_1993963.html.

Stephenson, Philip A. "The Science of Building the Perfect Snowman." Quartz, 13 Dec. 2013, qz.com/153318 the-science-behind-the-art-of-building-a-snowman/.

"The Story of a Gilded Age Anti-Noise Campaign." Ephemeral New York, 6 Aug. 2018. https://ephemeralnewyork.wordpress.com/tag/julia-rice-society-for-the-suppression-of-unnecessary-noise/.

Swancer, Brent. "Alien Hunting on America's Bizarre UFO Highway." MysteriousUniverse.org, 8 May 2018, mysteriousuniverse.org/2018/05/alien-hunting-on-americas-bizarre-ufo-highway/.

T., Eric, 2012. "What Is the Origin of the Doggie Bag?" Culinary Lore, 13 Nov. 2012, culinarylore.com/food-history:origin-of-the-doggie-bag/.

Taggart, Emma. "The Fascinating History of 'Paint-by-Numbers' Kits." MyModernMet, 8 July 2019, mymodernmet.com/paint-by-numbers-history-dan-robbins/.

Tanque Verde Ranch. "The Disguised Life of Charley Parkhurst." TanqueVerdeRanch.com, 26 May 2010, www.tanqueverderanch.com/the-disguised-life-of-charley-parkhurst/.

Taylor, Christie. "From 'Nettles' to 'Volcano,' a Pain Scale for Insect Stings." *Science Friday*, 28 Nov. 2018, www.sciencefriday.com/segments/from-nettles-to-volcano-a-pain-scale-for-insect-stings.

Taylor, Oliver. "10 Facts We All Get Wrong about Colors." Listverse.com, 7 Jan. 2020, listverse.com/2020/01/07/10-facts-we-all-get-wrong-about-colors/.

Tharoor, Ishaan. "Why Chemical Warfare Is Ancient History." *Time*, 13 Feb. 2009, content.time.com/time/world/article/0,8599,1879350,00.html.

"This Family In Siberia Was So Isolated, They Didn't Even Know About WWII." *Bored Panda*, 1 Jan. 1968, www.boredpanda.com/reclusive-family-siberia-taiga-agafia-lykov.

Thomas, Pauline Weston. "Rational Dress Reform, Victorian Bloomers and Cycling Costumes." *Fashion*, 5 Sept. 2018, www.fashion-era.com/rational_dress.htm.

Thompson, Helen, 2015. "Do You Want to Build a Snowman? Physics Can Help." Smithsonian.com, 27 Jan. 2015. https://www.smithsonianmag.com/science-nature/do-you-want-build-snowman-physics-180954024/

Thorpe, J. R. "The History of Men & Skirts." Bustle.com, 22 May 2017, www.bustle.com/p/the-history-of-men-skirts-58088.

"Thought to Be Extinct, Beothuk DNA Is Present in Living Families, Genetics Researcher Finds | CBC News." *CBCnews*, CBC/Radio Canada, 8 May 2020, www.cbc.ca/news/canada/newfoundland-labrador/beothuk-dna-steven-carr-1.5559913.

Trent, Sydney. "Slavery Cost Him His Family. That's When Henry 'Box' Brown Mailed Himself to Freedom." *Washington Post*, 28 Dec. 2019, www.washingtonpost.com/history/2019/12/28/slavery-cost-him-his-family-thats-when-henry-box-brown-mailed-himself-freedom/.

"True Story of a Real Life Super Hero: Shavarsh Karapetyan." People of AR, 8 Feb. 2014, www.peopleofar.com/2014/02/08/true-story-of-a-real-life-superhero-shavarsh-karapetyan/.

Trueman, C. N. "History of Hygiene Timeline." History Learning Site, 1 Sept. 2015, www.historylearningsite.co.uk/a-history-of-medicine/history-hygiene-timeline/.

Uncle Steve. "Chile Pepper History." Uncle Steve's Hot Stuff, 17 Dec. 2008, ushotstuff.com/history.htm#:~:text=Chile%20peppers%20are%20native%20to,pepper%20native%20to%20South%20Asia.

Uncle Steve. "Just How Hot are My Chiles?" Uncle Steve's Hot Stuff, 26 Feb. 2001, ushotstuff.com/Heat.Scale.htm.

Vickery, Amanda. "A Brief History of Human Filth." History Extra, 23 July 2019, www.historyextra.com/period/georgian/history-human-dirt-how-people-keep-clean-bath/.

Vidyasagar, Aparna. "Facts About the Fungus Among Us." *LiveScience*, Purch, 5 Feb. 2016, www.livescience.com/53618-fungus.html.

Walker, Tim. "New Year 2014: Smash a Plate, Drink a Wish, Eat a Grape—How Did the Rest of the World See in 2014?" *The Independent*, 13 Dec. 2013, www.independent.co.uk/news/world/politics/new-year-2014-smash-a-plate-drink-a-wish-eat-a-grape-how-did-the-rest-of-the-world-see-in-2014-9032137.html.

Wallace, Greg. "Ranking the Top College Football Rivalry Game Trophies." Bleacherreport.com, 20 Nov. 2016, bleacherreport.com/articles/2677287-ranking-top-college-football-rivalry-game-trophies.

Wei-Haas, Maya. "Space Junk Is a Huge Problem—and It's Only Getting Bigger." National Geographic USA, 25 Apr. 2019, www.nationalgeographic.com/science/space/reference/space-junk/.

Weiler, Mallory. "All of the Odd Contents of Abraham Lincoln's Pockets the Night He Perished." Ranker.com, 19 May 2020, www.ranker.com/list/what-was-in-lincolns-pockets/mallory-weiler.

Weiner, Sophie. "Russian Espionage and Electromagnetic Fields: The Story of the Theremin." *Red Bull Music Academy Daily*, 27 Oct. 2017, daily.redbullmusicacademy.com/2017/10/theremin-instrumental-instruments.

Wells, Chris. "The Origin of Our Calendar and Why George Washington Has Two Birthdays." "Beyond Bones," Houston Museum of Natural

Science, 31 Dec. 2018, blog.hmns.org/2018/12/the-origin-of-our-calendar-and-why-george-washington-has-two-birthdays/.

"Who Invented the USB Flash Drive?". Premium USB by USDigitalMedia, 11 Aug. 2011, www.premiumusb.com/blog/who-invented-the-usb-flash-drive.

Witze, Alexandra. "The Quest to Conquer Earth's Space Junk Problem." Nature.com, 5 Sept. 2018, www.nature.com/articles/d41586-018-06170-1.

Wolchover, Natalie. "What Would Happen if You Fell into a Black Hole?" LiveScience.com, 13 Apr. 2012, www.livescience.com/19683-happen-fall-black-hole.html.

"Women's History Matters: The Legend of Mary Fields." WHM, 18 Apr. 2014, montanawomenshistory.org/the-life-and-legend-of-mary-fields/.

Wong, Kevin. "Franklin Broke *Peanuts'* Color Barrier in the Least Interesting Way Possible." Kotaku.com, 31 July 2018, kotaku.com/franklin-broke-peanuts-color-barrier-in-the-least-inter-1793843085.

Wootson, Cleve R., Jr. "The Five Sullivan Brothers, Serving Together, Were Killed in World War II. Their Ship Was Just Found." *Washington Post*, 20 Mar. 2018, www.washingtonpost.com/news/retropolis/wp/2018/03/20/five-sullivan-brothers-serving-together-were-killed-during-world-war-ii-their-ship-was-just-found/.

Word Wizard, "The Origin Of 'Foot' for Measurement (+ Why 12 Inches?)." *Spark Files*, sparkfiles.net/foot-whats-special-12-inches.

Yeager, Ashley. "Human Fleas and Lice Spread the Black Death." *The Scientist*, 16 Jan. 2018, www.the-scientist.com/the-nutshell/human-fleas-and-lice-spread-black-death-30409.

Yong, Ed. "The Incredible Thing We Do During Conversations." TheAtlantic.com, 4 Jan. 2016, www.theatlantic.com/science/archive/2016/01/the-incredible-thing-we-do-during-conversations/422439/.

Yuko, Elizabeth. "Here Are All the Ways You Can (Legally) Dispose of a Dead Body." Lifehacker.com, 3 July 2019, lifehacker.com/here-are-all-the-ways-you-can-dispose-of-a-dead-body-1836055910.

Zarrelli, Natalie. "The Best Kitchen Gadget of the 1600s Was a Small, Short-legged Dog." Atlas Obscura, 11 Jan. 2017, www.atlasobscura.com/articles/the-best-kitchen-gadget-of-the-1600s-was-a-small-shortlegged-dog.

Zarrelli, Natalie. "The Rude, Cruel, and Insulting 'Vinegar Valentines' of the Victorian Era." Atlas Obscura, 11 Feb. 2019, www.atlasobscura.com/articles/vinegar-valentines-victorian.

Zielinski, Sarah. "The Rise and Fall and Rise of the Chemistry Set." *Smithsonian*, 10 Oct. 2012, www.smithsonianmag.com/science-nature/the-rise-and-fall-and-rise-of-the-chemistry-set-70359831/.

Zimmer, Carl. "From Ants to People, an Instinct to Swarm." *New York Times*, 13 Nov. 2007, www.nytimes.com/2007/11/13/science/13traff.html.

INDEX

NOTE: (ILL.) INDICATES PHOTOS AND ILLUSTRATIONS

C

D

E

F

G

H

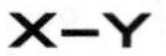

X–Y

Z